T0201968

THE POWER OF IMPERFECTIONS

THE POWER OF IMPERFECTIONS

A Key to Technology, Love, Life and Survival

Peter Townsend

OXFORD
UNIVERSITY PRESS

OXFORD
UNIVERSITY PRESS

Great Clarendon Street, Oxford, OX2 6DP,
United Kingdom

Oxford University Press is a department of the University of Oxford.
It furthers the University's objective of excellence in research, scholarship,
and education by publishing worldwide. Oxford is a registered trade mark of
Oxford University Press in the UK and in certain other countries.

© Peter Townsend 2022

The moral rights of the author have been asserted

Impression: 1

All rights reserved. No part of this publication may be reproduced, stored in
a retrieval system, or transmitted, in any form or by any means, without the
prior permission in writing of Oxford University Press, or as expressly permitted
by law, by licence or under terms agreed with the appropriate reprographics
rights organization. Enquiries concerning reproduction outside the scope of the
above should be sent to the Rights Department, Oxford University Press, at the
address above.

You must not circulate this work in any other form
and you must impose this same condition on any acquirer.

Published in the United States of America by Oxford University Press
198 Madison Avenue, New York, NY 10016, United States of America

British Library Cataloguing in Publication Data

Data available

Library of Congress Control Number: 2021942515
ISBN 978–0–19–285747–7

DOI: 10.1093/oso/9780192857477.001.0001

Printed and bound by
CPI Group (UK) Ltd, Croydon, CR0 4YY

Links to third party websites are provided by Oxford in good faith and
for information only. Oxford disclaims any responsibility for the materials
contained in any third party website referenced in this work.

Preface and Audience

Technology is the basis of the modern world and we need an intuitive feel of how it functions. My aim is to offer this to a wide audience and simultaneously introduce the excitement of the underlying science for teenagers, undergraduates and people of all ages and backgrounds. The topics are wide ranging, simplistic for access, but with enough detail to sense deeper issues and possibilities. Totally surprising is every example is crucially dependent on imperfections in the materials we use. In cookery, herbs and spices enhance our foods, in science it is imperfection that gives performance. My examples deliberately use many that will be unfamiliar, but they include wood, glass, metals, semiconductors and catalysis to genetics and viruses. Asides offer a background to the science of imperfections, and advice on how to make a successful career and what information one should trust. The final sections consider more complex issues of love, life and the survival of humanity. Once again imperfections are the key, and if we will recognize them and change our behaviour then we may survive, even with a rapidly expanding world population. Failure to do this implies the collapse of civilization as we know it and a backward step of many millennia.

Acknowledgements

Writing needs continuous feedback and I therefore thank many friends who have taken the time to read, discuss and make helpful comments. These include input from Professor Norman Billingham (in discussion of catalysis) and helpful criticisms across the entire book from Angela Goodall, Jean Hotchkiss, Ian Purvis and many others.

Contents

1

Before we begin

Perfection is a fiction, but for most of us, a book title that praises and sees virtue in imperfections and defects is completely contrary to our normal conditioning. However, I have a very firm and secure conviction in my choice of examples that will justify praising imperfection. In reality it is remarkably easy, since most of our familiar technology from metals, glass, electronics and computers, to cars and building materials exist, and all have desirable properties, only because of their imperfections. There are many examples that are familiar, understandable and readily explicable. My aim is that no prior science background is required and I will provide any that is needed. For the life sciences, imperfections are equally relevant but the details can be more complex, although clearly just as important. The biological variations caused by intentional or accidental imperfections not merely define how we have evolved and why no two creatures are identical, but also can determine how we lead our lives. Such imperfections are rarely seen as positively as for the technological examples, and their function and consequences are, at best, only partially understood.

I have a number of strong passions in my life, ranging from experimental physics to music, fencing and many friendships. It is, therefore, a pleasure to combine and include some of these apparently diverse segments in a single volume with a clear unifying thread of extolling the many benefits that appear as the result of imperfections in technological materials. Imperfections are equally key features·in how we process information from the signals of sight, sound or smell etc. Music is relevant as an example of valuable imperfections, so in Chapter 15 I discuss musical oddities which result from lots of information but a limited brain capacity.

Whilst being totally fascinated by technology and hugely impressed by how we have managed to unravel and understand so much about it, it is sad that much of the effort has been directed to produce better armaments, weapons of war, or for suppression, often with the

The Power of Imperfections. Peter Townsend, Oxford University Press.
© Peter Townsend (2022). DOI: 10.1093/oso/9780192857477.003.0001

destruction of both the planet and fellow humans. If I can inspire actions to counteract this apparently ancient human trait then I will be very happy, and surprised. Such idealism is often held by young people, even though it can evaporate as people age and want power and wealth. Human behaviour is certainly never perfect and the ways in which we treat one another by waging wars, religious persecutions, greed etc., are totally selfish acts, which seem to be entrenched in our psyche. For me these are highly undesirable imperfections and I will not (indeed cannot) attempt to justify them. Nevertheless, in the concluding chapters I will at least mention them as, once identified, it is possible to comment on how humans might improve. This is probably highly idealistic as it is invariably contrary to anything we have achieved in the past. Unfortunately, fundamental change in human behaviour is essential as we are already overpopulated and destroying the greatness of our planet. Unless we can quickly modify our behaviour, and not just stabilize the population but actually reduce it, we face the possibility of rapidly becoming an extinct species. To make such changes, the first step is to identify our human imperfections before deciding how to resolve them.

Whilst I have no doubt that imperfections are essential in the wider areas of life, friendships and love, any discussion is much more challenging as there will be highly personal input. I once started a science book with a comment that 'Crystals are like people and it is their imperfections that make them interesting.' This is true, but unlike scientific research, there is no way we can do repeat experiments or have large statistics when picking friends, a love life or family. Therefore, the exact nature of particular desirable defects and imperfections in people will be hard to define and quantify as there are too many sources of individuality and attractiveness which vary with time. There is also the minor problem that if I am too specific, my friends may recognize themselves and not have the same view. What is certain is that we are unique and develop in many different ways through the course of life, and attributes that are a fault in one person may be a bonus in another. For example, selfishness and insensitivity to others may be a virtue for a company director or top politician, but a disaster in more intimate relationships. This is not a modern realization as identical observations have been made by people as diverse as Niccolo Machiavelli and Abraham Lincoln.

In reproductive terms, imperfections make us unique and enable development between generations which we term 'evolution'. These are essential factors for our ability to cope with climatic and biological variations as well as new diseases. Reproductive 'perfection' would imply we are all identical and unchanging. This is clearly highly undesirable and, indeed, many plants and species have become extinct precisely because they could not adapt to changing environments. Imperfection in humans has been the subject of comments from many people with quite unrelated backgrounds. Marilyn Monroe said that 'Imperfection is beauty' — a succinct and valid comment. Winston Churchill made the less direct observation: 'The maxim — Nothing prevails but perfection — may be spelled PARALYSIS.'

In my research attempting to improve technological materials, I have no problem in recognizing that defects, imperfections and impurities in materials are absolutely essential for their performance. Indeed, the skill of the scientist and engineer is not to produce 'perfect' materials but to control and exploit their inherent and unavoidable flaws to our advantage. Thermodynamics unequivocally states that achieving any 'perfect' state is impossible. Emotive words, such as 'imperfections' or 'defects', unfortunately conjure up a poor image with both theoretical or ivory-tower scientists and the general public. We should have invented a more positive vocabulary. Nevertheless, I hope I will change your perspective.

Many articles, books and observations have commented that imperfections in people are desirable. One example is that *nearly* symmetric faces are seen as attractive in all cultures. Nevertheless, real faces are never 100% symmetric. If we cheat by making a symmetric image of one, it produces a very unsettling impression. Computer graphics can generate excellent and expressive images for cartoon animation of people and avatars for films and TV, but if they are both realistic and symmetric, they are disconcerting. The industry therefore deliberately degrades them. In other aspects of life, the gurus of style and fashion also proclaim that too much perfection is undesirable. Rustic construction, handmade objects etc. benefit from this, and there are major markets for jeans that appear to have been previously used, e.g. with rips in the legs (and higher price tags). An unkind view may be that our liking for imperfection is a factor in the popularity of the work of many artists.

By contrast, with intentionally added imperfections it is simple to be very exact in showing detailed benefits, and the scale of imperfection that will give the very best performance. With inanimate crystals and materials, the results are repeatable, quantitative and experimentally testable. These 'defects' have become essential throughout all our products, from glass to steel to semiconductors. Indeed, one cannot overstate that understanding the science of imperfections and impurities in materials is the basis of **all** modern technology. For non-scientists it is easy to make analogies with more familiar situations. For example, many of the ideas and patterns are apparent in geological formations and/or in minerals, and I will exploit this.

The initial chapters focus on technologies which are long established, as sketched in Figure 1.1 My approach will be to take a semi-chronological journey through tools and materials, starting with the developments using wood and Stone Age tools, and then progressing to some advances with metals and glass.

In our early history, our science was addressing materials such as wood or stone and a very key practical use of fire for cookery, which

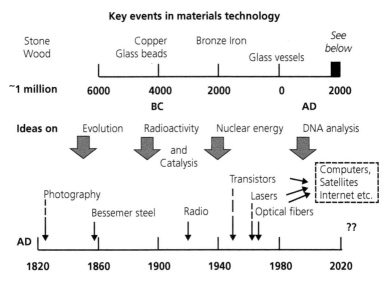

Figure 1.1 Historic examples of materials technology and more recent developments of ideas. The common theme is that all these key examples are controlled and enabled by the presence of imperfections.

had the side effects of modifying materials to reveal metals such as copper and form glass beads. This kick-started our move into production of metals such as bronze, iron and other metallic alloys. Simultaneously, furnaces led to production of various glasses and vessels.

This simple 'cookery' could be incredibly complex and so I will pick an example of Samurai sword production as a way to emphasize that we can make significant progress without understanding the details of the science. Cookery is an ideal familiar technology, as the results are highly variable, where minute traces of spices and controlled heating can offer greatly improved products. This is an informative analogy as we tackle essential and complicated experiments of cookery on a daily basis, and do so even though the underlying chemistry is far beyond our knowledge and comprehension.

Figure 1.1 emphasizes that the role of science has brought major changes to our lives within living memory. Therefore, I have chapters discussing major new technologies such as semiconductors, lasers, optical fibres and the possible current and future developments of twenty-first-century electronics and photonics. These will be presented in rather more detail (as fortunately the science is easy to understand). My prejudices and personal experience, as well as my own research, will often be obvious. In part this makes writing simpler, but it also allows me to deliberately include less familiar examples than are typical in mass market articles and TV programmes.

Science is certainly not perfect, so being doubtful, and challenging information from many sources is wise. Chapter 5 offers words of caution on how much one should trust published data and statistics. Even reliable data can be presented in many ways depending on the audience one is addressing, and often tables, graphs and numbers will be misconstrued by people (including scientists) who work in unrelated areas. Unfortunately, many politicians have minimal understanding of science, data and statistics. This is an imperfection that needs to be seriously addressed. Perhaps an exception was the British Prime Minister George Canning who, in 1827, said, 'I can prove anything by statistics, except the truth.'

Variations and differences are certainly desirable in music. Musical taste and styles differ between cultures and generations, but in all cases, from genres of music to performers and instruments, they offer individuality which we enjoy. Equally we accept inherent imperfections in the design of instruments, musical scales, recording and sound transmission and personal differences in hearing and experience. These 'imperfections' are actually the key to giving us individuality and pleasure. Similarly, there is beauty in minerals and gem stones caused by the imperfections that define both colours and crystal shapes.

For electronics, the speed of technological change has increased incredibly rapidly within living memory, most obviously with the advent of commercial uses of semiconductor materials. It is hard to imagine how we would now survive without them. Nevertheless, the only real difference from the gem stones is that, because of the commercial potential of semiconductors, we have invested effort and gained a deeper understanding of the role of the imperfections in producing the electronic properties that we want.

In the space of a mere sixty years, new glass products have leapt into the forefront of technology to bring optical fibre communication and stimulated faster electronics and compact laser light sources. Glass fibre production is a key element of modern living, and its construction is entirely based on control of imperfections, as well as having spawned a revolution in communication with billions of photographs being transmitted each year.

To most people all these changes seem like progress, but I will be contentious because I sense many serious problems could appear in the relatively near future. One such example that will be presented is the fact that in our transition from data storage on paper, or photographs, to the dynamic range of electronic media, we are in severe danger of having only very transient storage. The new systems may well fail to retain historical records, photographs, knowledge and even legal documents. Progress also comes at a cost and our transmission of billions of electronic correspondence, photos, videos and downloads etc. requires electricity. One consequence is that it is the third most prolific source of carbon dioxide (i.e. far greater than air transport).

Within mainstream materials science and physics, there is a rising tide of new products and ideas based on purely optical processing, and this field is called 'photonics'. It will probably dominate the early part of the twenty-first-century technology, so I will offer some simple clues and hints on future directions.

In addition to topics which have received lots of positive publicity, there are many other areas which are still on the brink of making an impact on our lives, even after a quarter of a century. The publicity and hype needed to generate their funding has not yet delivered the goods on the scale that had been predicted. Examples include nanoparticles, room temperature superconductors, graphene and fusion energy etc., but controlling and understanding their inherent imperfections is difficult. The reasons require more detail than seems appropriate here, but do not be surprised if they eventually break through into valuable products.

Although I am a physicist, I recognize parallel advantages of imperfections in fields of chemistry and modern biology. For the chemistry and biology, I have focussed only on a few limited examples, such as catalysis and the value of imperfections in cell reproduction. Both are topics of major importance. The latter controls both evolution and new approaches to medical interventions. They also generate emotional responses and difficulties of separating science from our lives, cultures and prejudices.

The bonus of recognition of personal and cultural imperfections is to turn them to our advantage and, as with technologies, exploit the faults to our own purposes for our career development and social relationships. With this in mind, one chapter includes some thoughts on realities of scientific progress, recognition and fame which might be useful for scientists, especially at the start of a career.

Overall, my examples deliberately do not require prior scientific knowledge. Indeed, by recognizing the features of inanimate materials, we may see parallels in social behaviour and benefit from it throughout our own careers and lives. Spotting our own imperfections is the first step even for such emotive topics as race, religion and how we interact with one another, both globally and at the individual level. It is essential for the continuation of our species.

Not least is that the social concept of 'perfection' is flawed, unattainable and deeply destructive. It is a mythical image, although improvements are feasible in every aspect of life, such as health, understanding, greater distribution of wealth or politics. These are real, possible and desirable. Claims of some ultimate perfection are merely an incentive to encourage us to seek improvements in the current situations. Unfortunately, this can often involve propaganda defined by political or religious beliefs that are intended to control our activities and inhibit logical thinking. Understanding and responding to them is more challenging than dealing with imperfections in materials. Once recognized, we might just be able to create a better world. This is an extreme glimmer of hope, but it is absolutely certain that, at the personal level, it is feasible to make very positive changes to bring greater pleasure, contentment and love to our lives.

Rather than separate descriptive thoughts from the science, I have deliberately incorporated both in the text. I am sure I am not alone in that when the two are well separated, with science contained in boxes, I tend to skip the hard bits. I hope that there are no 'hard bits' in my descriptions and analogies.

2

The case for technological imperfections

Part I Civilization through imperfection

Civilization has evolved slowly and steadily as mankind has shaped
the natural world. Despite many shortcomings, this is deemed to be
progress, but it has only been possible because we developed tools and
technological skills. Humans are not alone in using them, and simple
items such as sticks are used by both birds and monkeys to help extract
food of termites or honey. Birds also use stone anvils to break shells
of snails; sea otters swim on their backs with a stone balanced on their
front when they crack mussels. More surprising is that many animals
can make several sequential steps to find food, and some may even be
willing to have delayed gratification by dropping bread into water in order to catch fish. However, humans moved beyond this stage. We can
consider the start of civilization as being the point when we began to use
simple technology. Items for hunting (and warfare) would have been
on the priority list. In part this was essential, as hominids are not well
equipped physically to be hunters since we lack the teeth and claws of
other major predators.

Less obvious to most of us is that the only reason we were able to
enter the Stone Age was because the stones and flints we used were imperfect. So, we could break and shape them. Their key property is that
they readily split and fracture because they have planes of weakness.
Exploiting these natural imperfections moved us from feeble hominids
to being the top predator and dominant species in terms of shaping
the world to our own ends. Rather than admit that we are destroying
the planet and wage wars, we euphemistically see all the changes as the
growth of civilization.

The weaknesses of flint were absolutely essential in making shaped
tools and arrow heads as, without defects, flint would be a thousand

The Power of Imperfections. Peter Townsend, Oxford University Press.
© Peter Townsend (2022). DOI: 10.1093/oso/9780192857477.003.0002

times stronger. Tougher, defect-free flints would have been impossible to break, or chip them into the form of Stone Age tools. This is a momentous fact, and imperfections are indeed the key to our success.

The science of flint growth and fracture

The use of natural stones by humans appeared some 2.5 million years ago, and well-shaped flint tools, such as arrowheads, appeared maybe 500,000 years ago. As flint is the basis of civilization, it would be nice to be able to describe the process of their growth and controlled fracture. Unfortunately, we are still struggling to do this and, despite its significance in our long history, we are not totally clear how it was formed. Flints occur in both chalk and limestone areas. Among the competing theories is a suggestion that flints derive from seabed sediments made from death and decomposition of marine creatures and sponges. Shells of sea creatures made thick layers of chalk (which is primarily calcium carbonate). However, periodic inclusion of sponges adds in more localized regions of material rich in silicon dioxide and other related silicates. Crystals of silicon dioxide are very familiar to us in the form of quartz, as this is the basis of sandy beaches. Temperature and sea conditions meant that although the chalk was deposited continuously, the sponge population only peaked at certain time intervals. The result is that the sponges and silicates settled in layers in the chalk. This non-uniform deposition was crucial. Flint formation then took place as water dissolved material from the chalk and carried the chemicals down to the level where there had been sponges. Flint growth started to seed around the less soluble quartz grains, which grew larger and nucleated further growth and large-scale flint stones. This means flints are younger than the chalk in which they occur. In the local chalk cliffs here in Sussex in the UK, there are often flint stone layers separated typically by 3 to 10 feet (1 to 3 metres) of chalk. The end result can be large isolated flints or continuous sheets of flinty material.

The formation model says we needed both large quantities of chalk and localized layers of sponge debris. This is an interesting idea as it explains why the flints are 90 to 96% silicate but include traces of calcium, iron and aluminium etc. These sponge 'imperfections' in the chalk are a critical part of flint formation. If the chalk and sponge inclusions had been uniformly distributed, we might never have entered the Stone Age. Figure 2.1 shows how they can appear in a local Sussex large cliff face which is some 150 feet (50m) high.

Figure 2.1 A view of a Sussex chalk cliff with flint formation either in layers or sheets of material

If the flint composition were pure and simple (i.e. perfect) then the material might have grown as single crystals, in which there was a neatly ordered array of each type of atom. Instead, the structure is more like a glass with a semi-random array of the silica building blocks. The difference is similar to an ordered parade of soldiers (crystals) and a random crowd (glass). Although the average spacing between the people is similar, it will vary in the random crowd. Growth layers offered Stone Age craftsmen occasional weaker regions caused by deposition of different types of chemical. If we strike the flint with a hard rock then the shock waves cause a fracture along the lines of weakness where there are changes in composition.

A truly perfect and crystalline flint would be extremely difficult to break and might only form a powder, rather than the nicely shaped shards of sharp material which are so well suited to making cutting tools, axes and arrowheads. Two positive features are that fault lines allow fracture, and the flints are made of strong fragments with sharp edges, which are a consequence of the strong silicon to oxygen bonds.

Figure 2.2a shows how the original flints can be knapped by striking with another stone to make a crude scraper. The interior of the flint is obviously non-uniform and of varying composition as the layers have steadily grown. Figure 2.2b emphasizes that for a skilled Stone Age craftsman it is possible to shape the flint into a high quality sharp arrowhead for hunting.

(a)   (b)

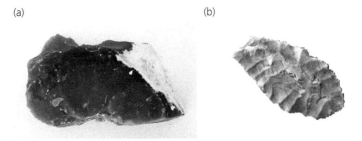

Figure 2.2 (a) The photograph on the left shows a partially knapped Sussex flint intended for use as a scraper. The work is only partially complete as there is still chalk attached to one surface; (b) by contrast the image on the right shows a highly crafted flint arrowhead.

Equally beneficial was that flints not only made cutting tools, axes, arrowheads and knives, but later on, once metals had been produced, the flints could be struck on metal to give sparks for lighting fires. Their production meant we could kill and eat bigger animals and could explore other ways to improve the current lifestyle. The improvements included the first steps to glass manufacture by melting of sands which produced glass beads, and the first steps into metallurgy as the fires separated metal from mineral ores. Flint did not immediately go out of fashion as it was used to produce sparks for fires and in flintlock weapons to ignite gunpowder. Here in Sussex it is still used as a building material because it is strong and tough. Such buildings are admired and are highly rated in many regions, as in the churches in Suffolk. The downside is that flints hidden in the chalk are a real nuisance when digging the garden as they can break a steel fork prong. They are equally a problem where playing fields have been developed on the chalk, as the flints come to the surface and can cause injuries.

Before we discount Stone Age technology as history it is worth noting that, even in the twentieth century, new tribes were discovered in the Pacific who are living very successfully, but totally reliant on stone tools.

One can cite a familiar parallel example of slate. The material originated as layers of mud, that over the last few hundred million years has been heated and compressed to form the rock we term slate. Nevertheless, the layer deposition of the original mud means it is strongly bonded in the plane of the layer, but is only weakly joined to adjacent layers. This allows it to be very easily split into thin sheets, and so it has been used for roofing tiles and other products. To misquote Professor Higgins 'The strain in a grain falls mainly in a plane'.

Technology through imperfections: have we left the Stone Age?

Early flint knappers had no idea why stones fractured so nicely, but it did not matter. They produced sharp tools. Even in our scientifically more advanced society we still only have tentative growth. We have kept this pragmatic Stone Age mentality, even for the modern technologies, from steel-making to photography to electronics. In every case, the materials and their uses always developed in advance of understanding the detailed science. With photography, the fine details of the relevant imperfections in the crystalline inclusions are still emerging, after nearly 200 years of use, and the demise of it as a popular medium. The pragmatic approach is effective. With more detailed scientific knowledge we usually gain better control of the processes and develop superior performance. Often a technology will initially benefit from natural imperfections, as with semiconductors. When the potential markets are vast (as for semiconductors) there is the financial incentive for a huge investment in research, and then performance and understanding quickly develop.

Progress before detailed understanding may seem to be an unexpected realization, but humans are survivors, so if something can be made to work then we accept it and can live without questioning how it is happening. Many people drive cars without the faintest idea how the engine operates. Most of us use computers and software which show highly individual behaviour, but again we have no idea why some days they give us trouble. The only difference is that in earlier generations, we would have said the daily differences, including the weather and our health, were caused by the gods. Particular problems, or good fortune, would have been blamed on whichever gods were fashionable at the time. At least for computers the odd quirks are now blamed on humans who wrote poor software or computer viruses.

Advances in both materials and biology have come by recognizing the properties we want to use, and identifying the differences between nominally identical materials caused by imperfections. The examples of semiconductor electronics, computer technology, or high- speed communication with optical fibre cables are unequivocal. They are all totally based on control, and a reasonable understanding, of imperfections in semiconductor and glass materials. 'Imperfections' is a misnomer if we like the consequences, but that is the way we use language.

Caveats

I have implied that imperfections are our technological friends. This is certainly not universally true. The presence of localized defects in materials, splits, cracks, inclusions etc., are all part of the panoply of imperfections which, in ideal circumstances, we can exploit. In many cases the balance between 'useful' and 'problematic' defects is on a very fine dividing line. The situation is like playing with fire. When careful it provides warmth and cooking, but a minor digression can scorch us, ruin the food or burn the house down.

We also need to recognize that funding for technological advances has frequently come from our desire for better weapons to kill or subjugate our neighbours, or steal their land and wealth. Even in agriculture we rarely have cared about better ploughs or quality and diversity of the land, but happily destroy natural forests and other resources to produce more food. We fail to realize that, for many prosperous nations, some two thirds of the food we grow is never eaten, yet half the world population is undernourished and lacks clean water. Rather than use our wealth and intelligence to resolve this imbalance, the majority remain poor and/or starving. Instead, we lavish vast sums of money on technological gimmicks (such as mobile phones or sending astronauts to the moon) and developing weapons that could destroy the world. Such imperfections in behaviour are unacceptable. By contrast, understanding imperfections does indeed lead to very advanced and excellent technologies.

In this book I am attempting to show that, by understanding defects and their properties, we can make great technological advances, but one caveat is that many technologies could undermine our survival. In *The Dark Side of Technology*, I highlighted a wide range of technology-driven problem/disaster scenarios. Not least is the fact that there are natural or terrorist events that could destroy our electrical grids, and/or kill long range communication systems. These would lead to a total collapse of advanced societies. Additionally, exploitation of technologies, as for cyber-crime, criminal surveillance and new weapons, can be extremely destructive. Since agriculture is now heavily dependent on technology, there are further negative implications. The topic is not a happy one, but it emphasizes that we are consciously unwilling to face up to these self-generated problems. We need a change of viewpoint that includes a comprehensive overview of our actions, not just short-term gains and profits.

Part II The development of metals

We moved forward from the Stone Age by developing tools that used metals. Metallurgy has improved steadily over the millennia as the result of trials, serendipity, new types of ore and methods of heating or separating the metals in them. A theoretical physicist might disparage this as being purely empirical progress. It is, but in reality, with such complex problems this is the only effective route. Stone Age fires melted some minerals and released traces of metals. Very rarely, people may also have found metal meteorites. Recognition of these new materials meant there were experiments to separate and benefit from them. Initially the metals would have been copper, gold, iron, tin etc., but it would soon have been apparent that alloying various metals together, and the ways in which they were heated and cooled, had major influences on their properties. This set us on the path to modern metallurgy. We have not changed the purely pragmatic route of trial and error, used for bronzes and steels, but by the twenty-first century we understand enough of the science that we can make computer predictions of which materials might combine. We also now know that the tiny flakes of materials that comprise everyday metals can be enhanced if we modify their defect structures, or in extreme cases, grow entire items out of a single crystal. This is a fantastic challenge, but used in the aerospace industry.

Casting entire objects in metal is known to have been possible from at least 6,000 years ago (there is an example of a copper frog from Mesopotamia of that vintage). Similarly, casting in bronze was standard by two millennia ago. The extreme working conditions of aero-engine turbine blades pose demands far greater than normal materials can undertake. The burning gas environment can be as high as 1,700 degrees

Centigrade and have turbine shafts spinning at 12,000 rpm. Conditions are so extreme that the blades must contain cooling channels (*cooling* here means about 1,150°C). Single crystal blades made from nickel/aluminium alloys can survive, whereas normal polycrystalline materials cannot. This is an extreme evolution of technology from Stone Age fires, and it is only possible because we understand enough about the defect structures and imperfections in the metals.

Bronze making for beginners

Copper from chunks of ore found on the surface (i.e. before any mining) would probably be the first to reveal their metal content when heated in a fire. Copper is an ideal starting material as, once separated from the original minerals, the metal is quite soft and easy to shape, so it was likely to have been the first metal to have been worked. In terms of hardness it is softer than flints for arrowheads, but it could be shaped more accurately and hammered into forms, such as plates or bowls, which were impossible with flint.

Copper is a soft metal but addition of some tin makes a harder alloy called bronze. The first examples of bronze tools appeared about 6,000 years ago and to me this was a very surprising development. Copper and tin ores do not normally co-exist except for rare sites in Iran or Thailand. Therefore, in hindsight, it is not too surprising that some of the oldest bronzes come from sites in Iran. This fortuitous combination of metals probably resulted in a conscious effort to try to harden the copper by including other mineral ores in the melting stage. Using separate copper and tin ores would have allowed experimentation with the ratio of the two metals to form the best alloy. There was also a benefit in terms of the temperature of the fire, or furnace, needed to melt the metal. Pure copper melts at 1,085°C (almost 2,000 degrees in Fahrenheit) but the softer metal tin has a much lower melting point of just 232°C (one of the reasons it is used in solder). Crucially the bronze alloys of copper/tin melts more than 100 degrees lower than pure copper. A lower melting point was a real bonus for early furnace technology. Higher concentrations of tin give a roughly proportional drop in melting point, but too much tin does not give a strong alloy, so the 10% tin level is normally the limit. Development of bronze led to a very lucrative tin trade, which involved shipping tin ores great distances (including from Britain, from areas such as Cornwall), to the bronze

Figure 2.3   An example of an early bronze age sword from China

manufacturing sites in the ancient world, as well as making bronze locally with Cornish copper. For international trade, the tin was shipped
in preference to the copper ores as bronze is roughly 90% copper and
only 10% tin.

   The early bronze artefacts are not only superior to the stone imple-
ments in terms of controlled shapes, but they also indicate that progress
towards 'civilization' already had some serious faults. Figure 2.3 is an
image of a Chinese bronze, but instead of it being designed to improve
the quality of hunting, the technology has already become a basis for
weapons to use against other humans. Sword designs differed between
regions but, unfortunately, similar trends have occurred worldwide
with each such technological advance.

Preferences of arrangements of atoms in alloys

All mixtures of different metals melted together are termed alloys.
Their properties vary with composition and heat treatments that modify their defect structures. For a copper/tin alloy, the amount of tin
dissolved into the copper strongly influences the atomic arrangement
of the different types of atom in the metal. The alternative ways in
which they can pack together are called phases. We might not realize
that within a solid there are different ways of arranging atoms, but we
will recognize the more obvious and familiar phase changes of water.
For water the material goes between phases of ice, liquid and steam.
Finding the detailed structure of a particular phase needs techniques
such as X-ray crystallography, and this has only been possible within

the last 120 years. There are many ways to arrange a mixture of atoms in a solid. A hypothetical example could be as follows: adding 10% of a metal A into a host B could give an arrangement where a row of B atoms has every tenth B type atom neatly replaced by an A atom. Since I want to emphasize analogies between imperfections in materials and people, we could suggest that isolated A atoms might feel lonely, so would rather go into the row as pairs of A atoms and the row would continue with 18 Bs before the next A pair. This is a familiar experience of how, say, two men would sit in a room of 18 women. Rather than be randomly placed, they are likely to sit near one another.

The two options of 9B then an A then 9B etc., or 18B and a pair of A, would be two examples of phases of an alloy made from A and B type atoms. Clearly one could think of huge numbers of variations, all of which would result in slightly different properties. For the particular case of 10% tin dissolved in copper, the actual situation is that there are three clearly different arrangements of the atomic mixture. There is a stable arrangement at room temperature which can survive up to ~350°C, a second phase then develops up to ~800°C and, finally, a third arrangement of the atoms occurs up to the melting point.

Modern detailed analyses for tin in copper, over the entire composition range from 0 to 100% tin, is complicated. There are some twenty six stable and different identified phases between room temperature and the melting point. In addition, there are unstable phases (termed metastable) which either change with time or can be triggered to switch their structure! Understanding such complexity is extremely difficult for a modern metallurgist, and was clearly impossible and irrelevant for a Bronze Age metal smith.

During slow heating or cooling of an alloy of several metals, the system will steadily switch between the phases, but if the material is cooled extremely rapidly it is possible to retain a high temperature phase frozen into the material. In cookery the equivalent metastable situation is a baked Alaska dessert, which uses fast baking of a meringue exterior with ice cream still frozen in the interior. For many materials, trapped phases can offer surprising results as they can have different mechanical properties. There is an apocryphal story that, for economy, Napoleon used a tin alloy button, instead of the normal brass buttons, on the clothing of the lower ranks in his army when he attacked Russia. The Russian winters are particularly cold and the claim is that the tin alloy underwent a phase transition and the buttons crumbled. There indeed is a tin crystallization temperature at just -4°C so, in a Russian winter, the officers' clothing buttoned with brass would have remained closed, but a tin alloy might well have turned into powder and the French privates would have literally been exposed to the cold weather.

Modern bronzes, as used in coins and springs, include a lower percentage of tin. Bronzes are relevant for our current fashionable technological materials. The bronze mixture of copper and zinc, which is really another example of a bronze, but with the 10% of tin in the copper replaced with zinc, forms a new alloy with a different colour, and we call it brass.

Arsenic and copper also form a bronze but this seemed less fashionable, although arsenic bronze vessels were found with the Dead Sea Scrolls. Arsenic bronze was not popular, either because it was harder to make, or just that working with the arsenic alloys produced so much toxic vapour that the bronze smiths died at a very early age. This meant that there were no family businesses in this field.

In many ways, bronze is far superior to iron as it has a similar hardness and density to iron, but it has one very real advantage. Exposure to the atmosphere produces an oxide (plus other chemicals) layer that acts as a protective coating. By contrast, the rust on the surface of iron continues to develop and penetrate, and the object becomes weaker. To some extent we accept this in our language, as surface coatings on bronze are termed 'patina', which is seen as a desirable and decorative feature, whereas 'rust' on iron is only thought of as a negative factor and a problem. It is unclear whether the aesthetic view or the practicality defined our difference in language in these cases.

Some historians believe the Bronze Age was displaced by the Iron Age, not because of the metallurgy, but because of the rising price of tin about 3,000 years ago. Once iron work became the front runner then the iron alloy development commenced. This led to a range of irons and steel which were cheaper than bronze and, eventually, superior in properties.

A touch of steel

We have a vanity in that we tend to think of the Iron Age as something very primitive and empirical, whereas a modern steel industry is based on a deep scientific knowledge. This is not totally true. The reality is that there has been 4,000 years of steady progress with experimentation of new alloy compositions, different heat treatments for melting and hardening and some very sophisticated equipment to study the composition and lattice arrangement of the elements in the crystallites of the steels. Such continuous progress with materials of immense practical, commercial and military applications has meant the industry has generated new developments of iron casting, production of high tensile steels and stainless steel, plus many far more exotic metal alloys with very high strength and light weight, as used in aerospace and military applications. Metallurgy is very difficult. It is, however, a highly respected science with textbooks describing the structure and compositions of metallic grains, theoretical models of alloy formation, and a wealth of data that lead to production of new alloys. Nevertheless, the results come from knowledge and developments that are squarely based on empirical additions of impurities and heat treatments, albeit fairly controlled and carefully executed. Even detailed analysis of what we make does not imply we fully understand what is happening. I have an old friend who, when a boy, worked in a steel works and he remembers adding green wood to adjust the carbon balance of steel when the furnace colour visually altered. Similarly, temperature measurements were by skill and intuition, not precise instrumentation. Instrumentation and on-line analyses have improved, but modern steel production still relies on experience and control of impurities, rather than a total understanding of the processes. With so many variables, this situation is never going to totally change.

The cutting edge of Samurai weapons

If I am realistic then progress, despite only partial understanding, is inherent in all technologies. The difficulty of working with unknown, but important, factors is nicely demonstrated via the historic mystique and religious ritual associated with sword making, particularly for the high quality steels used by the Japanese Samurai. Whilst swords had existed in many forms, their shape and size were determined by the particular actions for which they were used as well as the available materials. The Romans used a short stabbing sword, because they could make them. When fighting armoured opponents, a heavy mass sword with a cutting edge was needed, whereas for street fighting a pointed short sword or rapier were popular. The Samurai swords were curved (as a result of the technology) but this is, in fact, an ideal situation and superior to the action from a straight sword. If the centre of the curve matches the centre of percussion as the blow is struck, a greater pressure is generated and the cut is deeper, and this persists as the swing of the curved blade progresses.

One assumes this tried and tested shape is the ideal shape for both a cavalry sword and for foot soldiers. Curved weapons from sabres to the Gurkha kukri, scimitar (Middle East), talwar (India) etc., have all benefitted from a curved design. It is therefore unusual that the 1908 British cavalry sword (the last sword to be standard issue) was designed by a committee as a straight thrusting weapon. In a cavalry charge it had great penetration, but typically could unhorse, or break the wrist of the cavalryman. This was an imperfection of the desk bound army committee rather than a problem with the steel.

The Samurai set a very challenging design of a sword as it was light enough for rapid action, and had both cutting and point action. This produced a major conflict in design, as the steels originally available were either able to have a sharp edge, or be strong and not snap under impact. The solution was a material made with a composite structure of a central strong core and a different hard steel which could be sharpened as an outer layer.

This was a difficult construction involving many types of metal working skills and maybe as much as six months in production per sword blade. The price was high, but realistically sword failure was somewhat serious during a fight, so price was not the defining factor. Contrary to many film versions, Samurai fights were brief. They often involved just a single blow following smoothly as one action when the Samurai drew his curved sword from the scabbard. A major advantage of a curved sword is that the shape aids the speed of bringing the blade into action, and victory may precede before the opponent has drawn his weapon. Indeed, the martial art of Iaido (or Iai) focusses on just the rapid action and a single cut.

To emphasize the complexity of the problem, I will outline the key steps in manufacture. We need not feel inadequate if some of the science looks difficult, instead we should be amazed that people have succeeded despite the difficulties and lack of understanding of the science. Knowledge helps, but only as a guide. All scientists need funding and so we claim we have a high level of understanding, but need just a little more research money to fully appreciate what is happening and then predict where to go next. This is the game played between scientists, funding bodies and governments, but as this Samurai sword example shows, only the most naïve really believe they will fully understand and control the total process. Rather, the funding comes because there is the hope of some advance and, particularly in the case of weapons, everyone feels this is in the national or personal interest, so research must be supported. The Samurai sword example is actually a classic model of progress with difficult complex problems that involve more factors than we will ever understand.

The first step in sword making is to identify the best starting materials. It is fairly obvious that iron that contains lots of rubbish is of poor quality. The best approach is probably to try to make the purest iron that is possible, and then start to add in the necessary impurities in a reasonably controlled fashion. (This is identical to the plan for semiconductors or optical fibres). For steel, the purest source of iron that was available was quite often from iron sands, rather than mined ore which has more contaminants. Finding very large pieces of nearly raw iron is occasionally possible in the form of iron meteorites. The fact that the iron arrived from space (i.e. the heavens) would be ranked extremely highly as everyone making and using it would instantly assume the weapon had gained some added mystical power. The results from such steel derived from a meteorite would, of course, support this as anyone wielding the sword would gain confidence from believing this, and if the opponent knew it as well, he would be more nervous and a certain loser. Confidence wins, and confirms the mystic properties. (Equally true in modern fencing.) Fortunately, we are blessed with mental powers over our actions that can influence outcomes. The links between confidence and ability are complex and, even in medicine, cancer patients who believe they will get better have a measurably better survival rate than those who feel they will not.

Heat treatments and improvements of steel

Iron has a significantly higher melting point (1,538°C) than copper (1,085°C), so a furnace that approached this temperature was needed, and it was produced using charcoal. However, the strength and properties of iron and steel vary considerably with the carbon content that results from contact with charcoal. Difficulty in melting iron is, in fact, a bonus as, contrary to our first instincts, melting the iron degrades the steel. Great care was taken to only let it soften (not melt). The science behind this is that when it is not quite molten there is only a limited inward diffusion of carbon from the charcoal. Limited carbon content gives a better grade of steel. At the same time the iron oxides break up and there is an outward diffusion and loss of oxygen from the source material (again a desirable outcome). Overall, this implies better steel. The chemistry of carbon uptake into iron is complicated as it varies with the presence of other impurities and forms a wide range of grain sizes and compositions. Even more confusing is that the percentage of carbon within the grains will differ from the amounts of carbon in the inter-grain regions. Hence control is a very tricky process as it depends on the starting materials, the temperature, background gases in the furnace and careful monitoring over several days of softening. A high carbon level, of say 1.5%, produces very hard steel which can be sharpened, as for a razor. Lower carbon steel, say ~0.5%, produces steel which does not suit sharpening but is ideal for shock absorption during impact.

Introducing the correct carbon content at the high furnace temperature is only part of the solution, as the mixture still needs to cool down to room temperature and the speed of cooling the furnace is critical. Many changes need to take place and modern metallurgists will show complex diagrams to demonstrate just how many alternative compositions and structural phases could develop during the cooling. This is a major difficulty even in the computer control of furnaces of the twenty-first century, but all that the early Japanese foundry workers had was the benefit of experience. After furnace treatments, the fragments of steel were sorted into hard and soft types by the way they could be hammered. This was an incredibly tedious task which needs heating and hammering to drive out residual furnace slag from the ore. At the same time the metal was repeatedly folded to encourage certain types of crystal growth. All this effort could take many weeks' work just for a single sword blade.

As I have already said, sword production poses a metallurgical problem as the weapons need to be both tough to survive blows, but sharp enough to cut. In terms of the carbon content these are incompatible requirements. A brilliant and innovative step was to form a channel in the hard steel and exactly fill it with the tougher soft steel. Coating this package in clay protected it for the next heating stage. Even this stage required skill as the sharp edge was only thinly coated in clay. This had two consequences; the first was that when the edge was eventually polished and sharpened it developed a wavy pattern resulting from all the folding steps. The second feature was much less obvious (even to a modern physicist) but it meant the stress of the final rapid cooling step caused the sword blade to curve. Success gave an ideal end result for usage, but often the bending caused a disaster in that the blade shattered. There is probably a word used for such an event by a Japanese sword smith, but I am sure it is not printable.

We need to recognize that these techniques were developed 1,000 years ago, so the technology was a fantastic achievement. We might also assume that only the Japanese sword smiths had the necessary skill, but in the Norse Beowulf sagas the swords are mentioned with a snake-like pattern along the edge (i.e. as for the Japanese katana). Beowulf was even earlier, from the sixth century, so perhaps some of the same sword making skill that existed then was subsequently lost. Modern techniques may give more consistency for such a task, but we still do not understand the fine details of what is happening. The entire procedure is totally dominated by the presence of impurities, a range of iron-carbon crystallite materials, their size distribution, interface materials and many steps of thermal processing. Mastering such imperfections was indeed a technological miracle, but it seems an unfortunate example of a developing civilization since the aim was to produce better instruments for killing. Perhaps I should have presented this as an advance in metallurgy which could have produced better ploughshares. The defect of preferring killing to agriculture resides in society, rather than any inherent destructiveness of the technology.

Figure 2.4a shows an example of a Samurai katana with a typical ornate handle. It demonstrates the typical blade curvature away from the sharp edge. In this photograph the wavy pattern along the edge where the clay thickness was thinned during the formation is scarcely visible. Therefore, for Figure 2.4b I have photographed it with a green coloured filter which emphasizes the edge pattern.

(a)

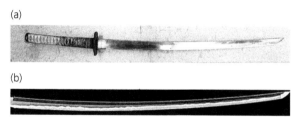

(b)

Figure 2.4 (a) An example of a Japanese sword blade with naturally developed curvature; (b) the lower colour image offers an obvious demonstration of the wavy pattern along the length of the blade edge.

Part III Iron and steel since the nineteenth century

Iron manufacture has long been a mass production material, but the better grade examples of steel were often limited to crucible size quantities as needed for cutlery or small knives. Nevertheless, reports from eleventh century Asia, plus examples from China and Japan in the seventeenth century, indicated that better steels could be made if the materials were fairly pure. In the mid-nineteenth century, Bessemer realized there was an immense market for steel canons and designed a melting chamber with a clay type liner in which the iron could be melted. Impurity villains in the iron ores were uncontrolled quantities of elements such as carbon, manganese and silicon. The trick employed in the Bessemer furnace (not necessarily his idea) was to blast air through the melt and this removed the impurities in the form of oxide gases. Having obtained a purer melt, he then added controlled amounts of manganese and carbon to generate a high grade steel. Many other impurities formed a floating slag, which was easily separated from the steel. Indeed, the slag was valuable as it was used as a fertiliser. The Bessemer steel process transformed steel by providing high production rates at low cost, and enabled the use of steel in everything from canons to railway lines and girders.

Alloys and steel processing

The history of sword production is fascinating as it shows how far one can progress by experimentation and empiricism. Modern metallurgy includes far more detailed knowledge about the crystalline structures of metals and alloys. It has also expanded to include metals with far higher melting temperatures, as well as making combinations of materials which were not previously feasible. Hence stainless steel or light weight aerospace alloys etc., are all possible. As is familiar to any blacksmith, if the red hot material is plunged into a cold liquid (called 'quenching') it freezes the surface faster than the interior, and this sets up a permanent stress which makes for a tougher material casing. Modern methods can even produce this effect in small localized areas by using pulses of high power intense laser light to heat a small area of metal, if necessary without heating nearby regions. The laser melting or quenching expands the range of materials that can be used together and offsets many of the twentieth century rules of thermodynamics and physics, which set limits to the metal combinations that could form alloys. In these examples the surfaces' stresses and defects are generally beneficial.

Batteries designed for electric cars require welded copper with high precision definition. Furnaces and flames are not suitable as the metal can run or bubble. The very recent solution is to use blue laser beams, as copper absorbs the blue light and gently melts and flows to make very clean welds.

There is now a fairly good understanding of the thermodynamics which define why some metals can be alloyed together, and what is the possible alloy concentration range, as well as how different alloy compositions switch into different structural arrangements of the atoms (i.e. changes in phase and crystallography). Such rules of alloy formation are inherently limiting features, and so we cannot always meet competing demands. For example, razor blades need to be rust resistant but stainless steel does not hold a sharp edge. Alternatively, hip replacement parts may need to be from one alloy type, but the ideal alloy may be attacked by the chemical defences of the body.

Problems of steel composition and temperature

We normally think of steel as an extremely tough material. Most of us who are not metallurgists are unaware that, at lower temperatures, steels can readily show a brittle fracture where cracks spread and the material splits and fails. This can happen where cracks start at sites where two steel plates have been welded together. Metallurgy textbooks (and Web images) often include photographs of a Second World

War ship, the SS Schenectady, which was of a type called a Liberty ship that was built very rapidly (within less than a week), at a time of ship shortages. The steel, manufacturing quality and welding of these were sometimes inadequate and, in cold water, cracks in the steel could develop and spread, so the steel could fracture and split. This happened to the Schenectady and a number of other Liberty ships which literally split into two during cold conditions. Adding different trace impurities into the steel and some redesigns were needed to stop the problem. Avoiding similar design faults are still key problems, for example in containment pressure vessels for nuclear reactors where, not only must the steel survive at low temperatures, but it must also operate at high temperature and pressure, and survive very significant amounts of nuclear radiation damage to the steel atomic structure during reactor operation. Unfortunately, the problem of embrittlement increases with exposure to radiation damage. Empirical experimentation with a wide range of steel compositions and processing is, therefore, central to these most modern developments of steel technology. Understanding the properties induced by defects, impurities and imperfections is clearly a minefield, as nominally similar defects can variously be positive or negative depending on the application!

Implanting impurities into a very thin surface layer

In the last fifty years, there has been steady progress to add impurity atoms into the surface of steels which can solve problems of conflicting needs for the applications of the steel, and is not inhibited by the thermodynamics and rules of alloy formation. One method used is termed 'ion beam implantation'. The concept is quite simple as all that is required is to make a gas of the atoms that we would like to have in the surface. The gas source for a metal may just be from boiling the metal. The gas could be anything from another metal, such as gold, to a chemically inert gas, such as argon. The neutral gas atoms are then ionised (i.e. an electron is removed from the atom to leave a positively charged ion). Such an ionisation can be made with a beam of electrons, or an electrical discharge, exactly as in a fluorescent lamp. The next step is to extract some of these positive ions from the discharge region and accelerate them with a large voltage of, say, 200,000 volts (which is simple to do). They are then moving with very high speed and impact on the surface that is to be treated. With such high impact energy, the ions bury themselves into the surface of the target. (Think of throwing small stones into a bank of hard snow, or firing shot gun pellets into a pan of treacle, or making pebble dash building coatings by throwing small pea-size pebbles into wet mortar). The result is a bulk material with a modified and different set of surface properties. In the metal examples the bulk gives stability etc. as required for the object, but it has a surface layer with different properties defined by the implants. The mixture is made so crudely, with high energy ions crashing into the surface, that the mixing is not limited by the normal thermodynamics, which apply to steady changes.

Early razor blade examples (around 1970) used platinum implantation. The metal was added in such small quantities that it did not add to the cost of the blade, but it allowed the production of a sharp edge which did not rust. Hip joint replacements have used several alternatives, including additions of titanium into stainless steel. Drills which have been implanted with nitrogen last at least twice as long as the normal steel. This is economic, despite the cost of the implantation, as they produce many more holes, and the cost of labour to change a worn drill in a production line is far more than the price of the drill.

Toughening of steel washers, by ion beam implantation, is used in the aviation industry on washers used in the stabilizers of helicopter rotors. It may seem to be an unlikely and expensive way of improving washers, but in an application where the helicopter is flying at low level over a desert, there are very serious problems. Not only are windows sand blasted, but exposure to the sand destroys the washers in the stabilizing rotors rather quickly (i.e. a few hours). Washer failure means a helicopter crash. Doubling the working life of the washer then seems a rather helpful and cost-effective use of the surface imperfections resulting from implanted impurities, as a military helicopter can not only fly to a target, but also return.

Parallels between metallurgy and cookery

The science of metallurgy is very daunting as the language of the metallurgist has much jargon: numerous crystal phases with different names, all of which can form in different grain sizes and change with the heating cycles. The scale of the processing can be vast, and for non-experts, such as myself, it is hard to see how to cope with the enormous number of variables over which we have so little real control.

To gain some self-confidence I need a more prosaic analogy. Maybe this is cookery. According to the numerous TV chefs, cookery is a difficult science which needs dedication and training, but also involves a range of materials and heat treatments with stages that we cannot quantify. If we consider quality cake making then we try to start with the best available products. They are natural materials, so will be quite variable. They need to be mixed, heated in an oven with careful control of the temperature gradients and timing, as well as changes in temperature to sort out specific features that control the texture or final surface appearance. Failure at any stage can result in a non-uniform cake with fruit that has sunk to the bottom, or a surface that has burnt. Certainly, setting an identical cooking task to six chefs will result in six different cakes, all of which may be acceptable. Here you may think that we diverge from the metallurgist as the chefs will add traces of salt, spices and flavouring in ways which have a real influence on the final cake, even though the additives may represent only a ten thousandth of the weight of the cake. These traces of components are essential ingredients or the result of different sources for the recipe. We taste the differences and enjoy them. We and the chef will delight in their individuality. No way will the chef describe the spices and trace components as impurities and imperfections.

I previously mentioned meringues (made from whipped egg whites and sugar), but good chefs use traces of an acidic ingredient such as lemon, vinegar or cream of tartar. We now understand how this changes the chemistry of the process and results in a more uniform foamy meringue. Chefs achieved this decades before the chemistry was understood.

A metallurgist will choose a very different set of emotive words for these trace materials. They will call small compositional variations imperfections and impurities, and will be annoyed when the changes were unintentional or uncontrollable. The problems therefore look similar to that of a chef, but the difference is in our attitude and the language we use. We are happy with cookery and maybe would find metallurgy easier if we used more positive terms.

3

Cookery and technological spices

Bonus effects of impurities

Whilst still focussing on pre-history, the use of fire for cookery and metal working would have resulted in the first examples of glass formation. Initially it might have been accidental droplets of glass made from heated sand. Even these accidents made products which were interesting and used as ornamental beads. Improved furnaces and experience would have allowed the production of many decorative products, as well as practical vessels for food etc. Glass gained popularity in household vessels such as plates, jugs and vases as well as coloured jewellery and stained glass windows. The understanding of glass production is also the basis of glazes on pottery and china. The colour of glass is largely defined by the presence of traces of impurities, but these 'imperfections' can equally influence many other properties, such as strength, or resistance to chemical attack, and physical or thermal shock (as in ovenware). Structural imperfections of the glass network can, as with metals, be deliberately introduced in the surface by rapid cooling. Such treatments introduce thermal stress that is highly beneficial in making glass less likely to shatter when it is dropped. Examples abound, from the introduction of lightweight milk bottles to tableware and car windscreens. (Museum examples of milk bottles are very heavy compared with modern ones.) Modern usage and refinements (to be discussed in a later chapter) include a vast diversity of glass type products, from spectacles, telescopes, microscopes, decorative glassware, and double glazing to optical fibres. The current production of optical fibres for telecommunication sets the highest production standards for any type of glass product, in the sense that our control of impurities is absolutely critical and the transparency of optical fibre glass is roughly one million times better than window glass.

In the original manufacture of glass, the secrets of the trade were to know which compounds needed to be added to allow glass to melt or

The Power of Imperfections. Peter Townsend, Oxford University Press.
© Peter Townsend (2022). DOI: 10.1093/oso/9780192857477.003.0003

soften it at a low temperature, or to produce a particular colour. The melting point of natural sand, which is mostly grains of quartz, is at too high a temperature for fires and furnaces that existed 2,000 years ago. Therefore essential 'impurities' were added to allow glass to melt or soften at a lower temperature. Impurities are clearly a misnomer as they can make up to more than 50% of the glass composition in materials such as car windscreens or windows. Historically, glass items from the late Bronze Age have been found and material we would recognize as similar to modern glass emerged from the area around Syria, Mesopotamia and Egypt and, by about 650 BCE, there were manuals on glass manufacture. Chinese examples of comparable age have also been discovered. In terms of design and colour variations, many glass objects from the last two millennia have been successful, and reinvented many times.

Rubbish and imperfections existing in materials

In a later chapter I will discuss modern glass production including the world-changing innovation of float glass, which makes available to modern architecture an assortment of glass products ranging from enormous sheets of household double glazing to cladding on entire skyscrapers. Equally relevant for the last fifty years has been the use of optical fibres for communications, since without it we would not have the internet. Both examples will involve very precise control of impurities and arrangements of the glass networks ranging from additive 'impurities' (i.e. dopants) in large quantities to control the refractive index and the speed of light in the fibres, to almost total exclusion of impurities which absorb light at the wavelengths we are using. In terms of levels of the two types of impurity atoms, they can range from several hundred parts per million of the glass down to as low as parts per billion of the undesirable impurities. This is a scale of numbers where we have little intuitive feeling. Shorthand for parts per billion is 1 in 10^9 (the -9 implies 9 zeroes, so $10^{-9} = 0.000,000,001$).

Both glass makers and semiconductor industries can be very misleading as they proudly mention the extreme levels of control they have achieved in terms of key dopants (e.g. often in parts per billion) but rarely mention the higher *percentage* of other additives or background rubbish which is optically or electrically unimportant, or has been used to lower the melt temperature of the silica host glass, or raise the refractive index.

Semiconductor technology

Semiconductors are midway between insulators and conductors and will be discussed later as their electrical properties, and all modern electronics, rely totally on precise control of impurity dopants. These imperfections have dominated and totally changed our lives during the last eighty years. The industry wisely did not describe these impurities as imperfections, but for marketing purposes it uses the more positive word of 'dopants'. This sent out the message that they were deliberately putting them there, and understood how they were located and behaving. Once again there is some self-delusion, and intentional hype, associated with adding dopants.

Imperfections are not difficult to visualize and simple variants are called a 'vacancy' where an ion is missing from a lattice site, and an 'interstitial' where one is between lattice sites. Impurities can sit on lattice sites, or as interstitials, or cluster together. More extreme examples are inclusions of new compounds. Simple textbooks somewhat incorrectly describe these as 'point defects'. It is true that the core of the defect may be near a single site in the lattice, but in reality, this is an oversimplification as a single defect site causes relaxation, or compression, and makes changes and distortions over a much larger volume of material. In a cookery analogy, a bad egg produces a smell over a far larger volume than the original site in the egg box. An alternative view of defects in a lattice might be the mesh of a pair of tights. One flaw in the mesh distorts a much larger region than the original snag or knitting error. The topic will be expanded as needed in later examples.

Nevertheless, imperfections are the key to all the metal, glass, semiconductor and modern photonic industries. Intrinsic imperfections are hard to combat, but external rubbish (the impurities) can largely be removed or limited in a material. From the examples of semiconductors and optical fibres, we see that this only occurs when there is sufficient commercial incentive to clean up the processing methods to reduce undesirable impurities in the system, even down to parts per billion (1 in 10^9). Such control is an extreme and expensive challenge for the choice and preparation of the chemicals used in manufacture and, although feasible for many materials, is totally uneconomic in virtually any other materials example currently considered.

What is our normal scale of purity?

I want to emphasize that parts per billion are far away from anything we normally consider in everyday life. That does not mean we do not respond to the presence of such impurities. For example, an eyelash is roughly one billionth of our body weight, but if it gets into the eye socket, we are extremely well aware of it. In a modern material, the influence of an impurity that affects so many other atoms in a solid is initially hard to comprehend, but nevertheless, if we take social examples then many key people have major effects on the lives of billions. It is easy to find analogies of one person influencing many others. The 'one in a billion' effect was certainly appropriate from, say, a political leader such as Chairman Mao. He strongly influenced and controlled the lives of the population of a billion Chinese. A decision by an American or Russian president can also influence a billion people worldwide. Policy stated by a religious leader, such as the Pope, will similarly modify the lives of many millions of people. These are clear examples, but for the technologist the problems, or benefits, of imperfections in a material may be better paralleled by the decision of bankers or generals who trigger a sequence of minor events which still lead to changes on a global scale, such as a financial recession or a war.

Impurities in familiar materials

Our normal experience with 'high' purity materials is often very modest. For example, school chemicals may be only pure to a few per cent. In most aspects of our life we are very tolerant of unwanted impurities and tend to ignore them. Examination of labels on any packet of food, or bottle of drink, will happily tell us that our 'pure' natural product contains several per cent of additives to control the flavour, extend the storage life, or things which just happen to be included. Many impurities are never even discussed, or have names which are unfamiliar to anyone who is not working in the mass market food industry, so we ignore them or tacitly accept their existence. Once again, many TV programmes have pointed out that the trace additives can have very unwelcome side-effects for our bodies, even though they represent only a billionth of our body weight. Examples with herbicides that occur in foods at modest levels well below parts per million may seem innocuous, but once in our body they can concentrate in different organs to parts per thousand, or higher. Many herbicides unfortunately fit this pattern.

This self-denial of ignoring them may be a defence mechanism as we may not wish to think about the fact that our daily bread (or rice) is made not just from grain. Farmers and bakers are well aware that the grain storage silos inevitably include soil, weeds, mouse droppings (or even dead mice). Such rodent contamination is definitely an imperfection we would prefer to avoid, but it is a real problem, as rodents worldwide are thought to contaminate or eat 20% of our food supplies. Perhaps we hope that cooking sterilizes the products, but it does not remove the impurities. The rules governing acceptable levels of inevitable contamination vary between countries, but as an example one can consider a popular item such as chocolate. The cacao beans are left to ferment for a time before processing into chocolate and so are the focus of attention for many animals and bugs. This includes creatures which spend part of their life cycle within the cacao beans. Bug removal would be feasible with very large doses of insecticides, but the chocolate would be unpalatable and dangerous for humans. Accepting some contamination is therefore the only solution. Guidelines for 'purity' of our chocolate range from a maximum of 50 to 75 insect fragments per 100 grams of chocolate. Other items such as hairs from rodents (e.g. mice and rats) should similarly be kept below about 4 hairs per 100 grams. Just ponder on this with your next piece of chocolate, or any other food!

In general, we have a natural reluctance to consider such problems and so have glossy advertising portraying 'pure' healthy products. In our mental rejection of these impurities we can also forget the bonus side in which the 'natural additives' provide us with a diet containing important trace elements. For example, childhood exposure to soil and dust will reduce many problems of adult allergies. The focus on 'perfection' in advertising is equally a distortion of reality which makes us aspire to something which may be quite unattainable, in everything from consumer goods to the shape of our body.

Recognizing non-standard situations

Rather than feel limited to imperfections in crystals, we can recognize similar effects in human behaviour. Widespread effects can exist in a community from the presence of a non-standard inhabitant, such as a doctor who will treat many hundreds of people, or a foreign restaurant owner who can change the eating habits of an entire community. One person eating garlic, or the sound of a mobile telephone ringing, is evident far from the source. Choosing an example based on our sense of smell is a little delicate as it often produces emotive responses and many people react very nervously to any such discussion. City dwellers have suppressed their sensitivity to smell compared with those living in rural societies. Nevertheless, with good motivation our pheromone sensitivity is high. We certainly respond differently to perfumes which contain pheromones that are intended to attract either male or female friends. (Do not buy the wrong type because it is cheaper.)

Our noses do not have the sensitivity to match the sense of smell from animals, such as dogs. This is not unexpected as a dog has some 200 times more sense receptors than we do, but even dogs are far less sensitive to smell than moths. A male moth can apparently detect the scent of a female at pheromone concentration levels of one part per billion of the local molecular concentration in the air (i.e. equivalent to detection distances approaching one kilometre). This is spectacular, but the moth must also detect the concentration gradient so it knows in which direction to fly. Similarly, salmon return to their original home river to spawn when they are four years old. In the case of Pacific salmon, they then die. To find the home river they use the scent characteristic of that river, and this requires signal processing of the smells in the ocean to separate the effects of one part per billion of impurities characteristic of 'their' river. Human smell detection is normally inferior by comparison with these two animal examples, as both animal examples show how the one in a billion impurity effects are well within the range of influencing major events. In some cases, as for hydrogen sulphide (H_2S), as given off by bad eggs, we can detect it at levels of approximately five parts per billion. Sensitivity to pheromones and smells is much poorer for humans who live in cities compared to country dwellers and, since this sense is important in social interactions, one may speculate that this is a considerable loss. The scales of long range interactions (discussed later) for crystal imperfection examples are, therefore, neither unique nor particularly impressive.

The social analogy can be of wider value as, if we increase the number of outsiders (equivalent to a large concentration of dopants in a crystal), they will frequently segregate into a cluster of immigrants (or a foreign army base) with different language, customs or religion. The stability of the enclosed population increases with size, but it cannot be independent of the host environment. The inclusion and host population each have significant influence on each other. This type of behaviour is closely mirrored in the way impurities are incorporated into a crystal or glass. Our impurity atoms frequently produce a precipitate inclusion. This can have a different crystal structure from the host, but nevertheless it exerts very long range effects on the host, as well as being influenced by it. Precipitates were called colloids, but this has gone out of fashion and a currently fashionable term is a 'nanoparticle'.

Small nanoparticles differ in their behaviour from those of the bulk properties, and their cohesive strengths and interactions will be noticeably different. This is one reason why there is so much interest in the unexpected properties of nanoparticles. For example, small metal clusters, either as free-standing units or as inclusions in insulating materials, have totally different melting temperatures compared with the solid metal. The melting temperature of the metal nanoparticle progressively decreases with smaller units. Bulk metals such as iron, gold, silver or copper normally melt in the region of ~1,000°C (~1,800°F) but, in the case of very small nanoparticles, the melt temperature can fall below 400°C. Chemical properties similarly differ from the bulk. Interest in nanoparticles is equally exploring improved semiconductor optical and phosphor properties, plus a range of biological uses.

The financial advantage of imperfections

In a society driven by economics we should not underestimate the absolutely essential need for imperfections, even if they only develop later in the life of an apparently 'perfect' product. Despite our very best efforts to make high quality materials that perform as we would like them, there would be a very serious problem for manufacturers if the goods continued to do this without ever developing problems and failures. Some items, such as clothing, will easily be destroyed by usage and, additionally, changes in fashion force people to buy new products at very regular intervals. In the food industry, 'additives' are used, not just to increase shelf life, but to actively induce us to eat more. In part they interact with normal food processing so we develop a craving for more, or suppress the stomach to brain responses saying that we do not need more. Effectively this is intentionally inducing drug addiction, and it is a conscious commercial drive which results in obesity and all the associated health problems.

By contrast, the stringent requirements needed to make high quality electronics could well result in components that survive and function successfully for tens of years. The makers have lost any future market unless they can put on psychological pressure to buy differently packaged identical items, or offer such major advances that people consider their electronics are obsolete. Mobile phone makers are particularly adept at this, and some manufacturers are known to have deliberately remotely recoded older phones to make them run more slowly. Judicial fines are minor for the companies, and there is no recompense for the customers they have duped.

In the motor industry life is simpler as steels and components rust, drivers have accidents, materials in seats wear out and fashions change, so most cars will steadily be replaced after a few years. Other modern products, such as double glazing, pose more difficult challenges to the industry. The units are costly and in major buildings, such as skyscrapers, replacement costs are enormous. Therefore, in such markets the product must retain the insulation of the double glazing, airtight seals that no moisture penetrates between the layers, colour integrity for the nice architecturally selected bronze or green toned glass and stability of the heat reflective coatings. Damage will inevitably occur from UV sunlight exposure, weathering and flexing in storms, but skyscraper double glazing ought to have a predicted life performance of tens of years. By contrast, double glazing for home use does not demand such high performance standards, as replacement is much simpler and people move house and they are likely to change many of the features. In this case the manufacturers may actually welcome some imperfections and failure of the products. Since most companies offer ten year guarantees, then skilful engineering must have optimized the life expectancy to, say, fifteen years, which means replacement will keep the manufacturers in business without upsetting the purchasers.

Development of imperfections in physical properties or peer group driven changes in fashion and style are essential for economic survival. If the initial products are too well produced and lack serious imperfections then the manufacturers have to resort to other ways to encourage replacement. It is rare to find a manufacturer who admits to a deliberate effort to include an in-built life expectancy feature. Such events are normally not advertised, even if they are widely practised. A friend of mine, who had a canal boat, was buying some new rope fenders for the side of the boat. They looked particularly fine and rugged and my friend asked if people ever needed to come back for replacements. The rope maker said they would not need to unless he made some effort. He said, 'So I throw a shovelful of lime into the middle of the fender when I am making it'!

More subtle methods also exist. For example, many of us suspect that computer upgrades in software are only partly driven by new features, which may be un-needed and may never be used, but mostly because the upgrades require more computer power, or make older machines obsolete (particularly in terms of our social image). If imperfections do not exist, then they need to be created. Such artificially generated defects are not confined to tangible products, but are equally an essential part of society as the introduction of changes are a guarantee of employment for administrators, health and safety officers, or the legal profession. Commercially driven changes, and in-built imperfections, are therefore economic manufacturing priorities, even though they drastically undermine the natural resources of the planet.

4

A short log of technology from wood

Part I The importance of wood

I am always surprised that when we talk about ancient technologies, we happily start with the Stone Age, and then add various versions of Copper, Bronze and Iron Ages (but not necessarily in the right order). For me there is a major oversight, as before any of these developments we had items made from wood. Indeed, wood was essential and fundamental in our first steps towards tools and civilization. It was used for fuel for fires, fishing rods, and bows and arrows. It was equally important for travel with rafts and boats. Wood has been a basic starting material for the later technologies. Even today we quite literally use wood from the cradle to the coffin. Our need, fascination and interest in wood have certainly not gone away. We could survive without silicon electronics, but not without wood.

This is a salutary fact which we seem to ignore as we are steadily destroying the rain forests which are a very minor percentage of the surface of the habitable world, yet contain the greatest diversity of flora and fauna, and are a major generator of oxygen. The motives are variously financial gain from crops to feed cattle, or production of palm oil, etc. This is not a new pattern, only the destruction is very visible because of the scale. In reality, we have always removed the natural forests for agriculture, buildings and industry. Even in the semirural south of England, the Kent forests were used for fuel for iron works or for ship building. The bare downs of Sussex are the result of deforestation for sheep grazing.

Why is wood so multi-purpose?

In the twenty-first century, timber is still highly prized as a construction material, as well as for highly decorative surface variations. In terms of an idealized technological material it is, of course, far

The Power of Imperfections. Peter Townsend, Oxford University Press.
© Peter Townsend (2022). DOI: 10.1093/oso/9780192857477.003.0004

from perfect, as rather than being uniform, smooth, featureless, and homogeneous, it has all the 'flaws' of a living material. Non-uniformity of grain, and shades of colour, are desirable for decorative usage. They give each piece of wood individuality. Since I am unashamedly trying to show that imperfections can be desirable, wood and its imperfect properties seem to offer a range of positive examples.

Grain, texture and colour make wood attractive for floors, furniture, bowls and wood carving. We like the appearance so much that we happily try to simulate it in cheaper plastic versions of similar products. Decorative objectives and fashions have followed access to more exotic woods. Wood turners have a particular fascination with random patterns seen, for example, in burr wood. The burr wood is yet another example of desirable imperfections as it is formed by the problems of grain patterns near the bole of a tree. Burr wood (also called bur or burl) develops as a tangled mass of knots of dormant buds which are caused by a variety of stresses during growth, such as mould or insect damage.

Timeless usage of inlays and marquetry

If variation is the spice of life then we want differences of grain and colour and, to offer more extreme changes, we combine a range of wood types with inlays and marquetry. This is an extremely ancient craft in which a vast range of wood types are contrasted in solid or veneer versions to form geometric patterns, pictures or abstract designs. Two modern attractive examples are shown in Figure 4.1. (The simpler repetitive geometric patterning, as used on floors, is termed 'parquetry'.) In sophisticated marquetry, the defects of wood non-uniformity are turned to advantage, or deliberately enhanced. Experts not only use colour differences of the natural veneer, but also introduce more variations (imperfections to some, improvements to others) by staining wood or dipping veneer into hot sand so that it scorches slightly. Such variations enable gradations in colour that can be used in the construction of pictures and design. Scorching provides an effect of shadows and shading to give more pictorial interest in a two-dimensional wood pattern.

Whilst marquetry has been popular for several thousand years, the styles and techniques have developed with time. For example, much of modern marquetry is no longer hand cut by a skilled craftsman with

(a)   (b)

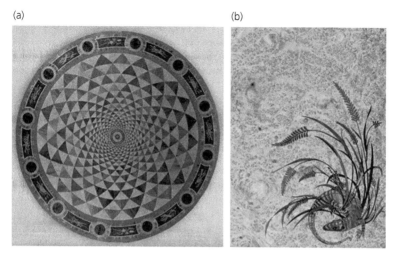

Figure 4.1 Examples of superb modern marquetry, provided by Aryma Ltd of Llandrindod Wells, UK

a very sharp knife or scalpel, but instead cutting is with extreme precision by a computer-controlled laser beam. A low power carbon dioxide laser can make a clean cut without fraying the edges of the wood grain. This bonus of making a laser cut with self-sealing and non-frayed edges is also beneficial when using lasers in cutting cloth for mass market tailoring, or in surgery. In surgery, a laser can be preferable to a scalpel as it reduces bleeding. In both cases laser cutting gives precision, and the cut edges do not fray, either because the laser power melts the boundaries or forms scar tissue. Computer-controlled laser techniques are ideal for low price mass production of marquetry, whereas skilled craftsman will adjust the cuts to benefit from random changes in wood grain and colour.

As a violinist I see some overlap between the marquetry skills and science of violin making. Decorative aspects have been used on many instruments from lutes, to mandolin, guitars and violins with inlays of coloured woods to embellish the instruments. There are some particularly ornate examples by Stradivarius. The functional inlay (purfling) around the edge of the back and the belly (Figure 4.2) is decorative, but it is really a technological necessity designed to inhibit cracks propagating along the grain across the length of the belly or across the back. The figure includes a less common fashion for a double line of purfling from makers such as Giovanni Maggini (1580–1632).

The central image of Figure 4.2 shows that the smooth f holes have relieved the stress at sharp corners. Nevertheless, some wood grain cracks are visible. Violin making has used this stress relief for 400 years but, unfortunately, the knowledge has not always carried into more sophisticated products. A very unfortunate failure resulting from this occurred in the first commercial British jet plane, the Comet, in the early 1950s. The design used square windows and a relatively thin outer skin on the plane. Because the plane went to new heights there was immense thermal cycling, as well as the normal expected stresses and pressure changes. This resulted in crack generation and splitting at the corners of the square windows with resultant crashes and many deaths. By the later Comet 4, the windows had become oval and the material

(a)   (b)   (c)

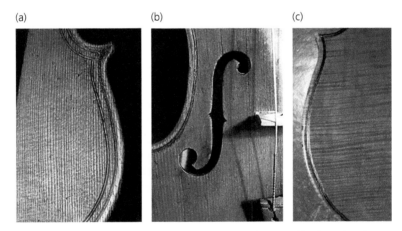

Figure 4.2 **Images of wood grain, purfling and violin f holes.** The photograph on the left shows part of the belly of a copy of a Maggini violin. Around the edge the protective inlay (purfling) has been made with two sets of inlays. On the figure the nearly parallel grain lines of the wood are clear, but the spacings vary as the result of annual growth rates in tree ring thickness and spacing. The middle picture is of a Beretta violin from ~1773. This has just the normal set of single purfling but demonstrates a smoothly rounded f hole which helps to inhibit splitting of the wood along the belly. The photograph on the right is of the back of the Beretta. The spacing of the annual growth ring patterns depends on the climatic history so it can give an estimate of the age of the violin. Normally the pattern from the belly is easier to interpret.

was stronger. A better understanding of the violin wood technology might have made an immense difference to the British aviation industry.

Early wood technology in stone axes and archery

Wood played a key role even in Stone Age tools, as a stone axe can be improved with a handle and to make one of wood is simple. Blocks of stone were tied into Y-shaped fork sections from a tree to make simple hammers and stone axes. A fairly major advance in skill was needed to progress to the construction of a good bow and arrows, but the rewards from a longer range hunting weapon were, of course, a significant stimulus. For arrows the material was preferably very straight grained and uniform (i.e. as perfect as possible). However, for the bow some degree of 'imperfection' was essential. During tree growth, the wood develops differently between established heartwood and the living outer sapwood. If archery were the means of hunting for lunch, it would have soon been obvious that one needed wood which not only bent smoothly, but also recovered the shape as rapidly as possible to cast the arrow on a flat trajectory over a good range, with enough power to penetrate the prey. The boundary zone between the soft flexible sapwood and the more rigid heartwood can offer an answer to this demand for better bow performance. A good bow was possible by working with material at the boundary of these two regions. In the classic English longbow, made from yew wood, the back of the bow (i.e. nearer the target) includes the flexible sapwood, which will stretch as the bow is bent. The belly side is the heavier heartwood which can stand compression and will rapidly spring back when the bending force is removed. The net result is a two-layer bow structure which matches the archer's requirements of flexibility and rapid recovery (Figure 4.3). To a physicist, one describes the stresses on the two parts as being a compressive force on the belly side and a tensile stretching force on the softwood section. This may sound more scientific, but the words and understanding have come several thousand years after the technology was introduced. Modern bows exploit the same concepts but do so with laminates of different types of wood or wood and polymer composites.

(a)   (b)   (c)

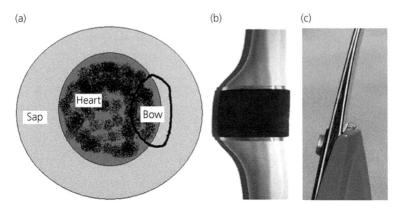

Figure 4.3 Composite wood types of an archery bow. High efficiency longbows are made by cutting the bow from the region of the junction between heartwood and sapwood. The figure on the left indicates this as it would appear on the original tree, and the centre image shows the change in colour between the two types of wood in a classical yew longbow. The boundary between heartwood and sapwood allows bending. Heartwood compresses and acts as a return spring, whereas the outer layer can stretch and will pull back to the original shape. The image on the right shows a modern design using laminates of two or more different woods. The same principles apply but more material layers are used and there is better uniformity over the length of the bow.

Is archery good for the physique?

Modern archery no longer needs high power bows so that arrows can penetrate mammoth hide or chain mail. Instead we think of it as a healthy, body-building, enjoyable sport. It is, although it is somewhat asymmetric in the way muscles develop. The modern modest development of the muscles is in strong contrast with the earlier military use of bows as weapons, which resulted in considerable muscular asymmetry and skeletal distortions of the archers. With twenty-first century composite bows, a high velocity arrow is normal. In earlier bows one had to hold the bow at maximum power whilst aiming, but via a cam and pulley systems, one can reduce the maximum force needed to hold the bow at full draw. Designs with reductions of 70% exist. This gives greater stability, less strain and higher accuracy. The new designs do not look like Robin Hood bows. The effort needed to pull back the arrow is called the draw weight, which is typically 30 to 40 pounds (approximately 15

to 20kg). Medieval archers needed high penetration power to force arrows through armour and draw weights of 120 pounds, or more, were
not unusual. These historic estimates were revised from bows found on
the Mary Rose (a ship of the time of Henry VIII). These medieval examples appear to have been nearer phenomenal values of 150 to 160 pounds
draw weight, and possibly as high as 200 pounds. They were also longer
than had been assumed.

Benefits of distortion during tree growth

Trees are beautiful and varied, but they have branches and knots so
are not totally ideal in terms of modern manufacturing materials. The
earlier approach was to fight against these features, instead of turning
them to advantage. For example, in ship building and timber houses,
wood was a key component but, rather than try to make angled joints,
it was common practice to use natural branching sections of the wood
to retain strength where sharp curves and knee joints were required.
Hence forests of oak were grown for the framework of ship building,
as well as a vast range of other woods such as elm and beech for keels,
deal (pine) for masts and a variety of woods, including elm, lignum vi-
tae, box and other hardwoods for the blocks and other tackle needed
in the boats. Whilst many of these features are not truly wood 'imperfections' they underline that the use of characteristics of non-uniform
homogeneous natural materials can be exploited in a wide range of
applications.

The demand for high strength, slow growing trees, particularly oak,
for shipbuilding was so great that in Britain, by the 1720s, there was a
major shortage of timber. The scale of the problem was obvious for
a ship named the Victory (the predecessor of the Victory which was
Nelson's flagship at Trafalgar). This first Victory was designed to be the
largest and most powerful warship ever constructed at that time. When
started, in 1728, it consumed some 5500 oaks as well as wood from 500
other tree species. There was such a shortage of wood for this vessel
that the timber was not properly seasoned. Naval historians have suggested this is one key reason why it later sank in a storm in 1744. The
only relevance to imperfections is to demonstrate selective historical
presentation. As a schoolboy I only read of Nelson's later 'glorious' Victory, and there was no mention of an earlier version that was actually

poorly designed, top heavy, took a very long time to build and sank in a storm.

Naval history was further skewed as our school text said we 'triumphed' against the Spanish Armada of 1588. In fact, a major reason for success was high winds that drove the Armada away from the coast. Not even mentioned was the English Armada that went to Spain in the following year, 1589. It was poorly organized and a disaster.

Natural bends in tree formation used in ship building are still seen and used in many other structures where supportive arches etc. are needed for roof trusses. These growth 'imperfections' offer greater strength than provided by supports of similar thickness manufactured from straight-grained wood sections. A typical example is shown in Figure 4.4.

Wood is still a key part of the building construction industry and is processed into many familiar variants. A recent addition to components for buildings is a mechanically very robust form where lignin has been removed. It is some ten times stronger than normal wood. It has a structure that scatters much of the visible light and, more crucially, strongly emits in the infra-red. The net effect is a material which is always cooler than the ambient temperature. Cooling of buildings is

Figure 4.4 Curved wood in building joints. The photograph shows how, in early building construction, curved timbers (crucks, or crooks) were used as roof and arch supports. In boat building the natural form of timber, as a branch developed from the main trunk, was utilized in places such as the hull. The cruck example here is of the Lych Gate of St Mary's Church, Ashwell, UK.

expensive and inefficient, but this new wood-based material requires no additional power source. As a cladding it can reduce electricity demands for cooling systems.

Part II   What are the hidden benefits of forests?

At a time when we are starting to realize that tree growth is beneficial to the entire world in terms of providing valuable resources, as well as consuming carbon dioxide, we are happily destroying the very same resources at an incredibly fast rate. Forests can vanish rather quickly but take millennia to re-establish. It is salutary to note that much of the Sahara Desert was forested not more than 5,000 years ago, but vanished extremely rapidly as the result of a minor climate change. Pacific and South American jungles are being depleted, which may well trigger a similar desertification over a short time scale. Before blundering into an irreversible catastrophe, we need to understand the resources that exist in forests.

Trees are alive and survive by adapting to attacks from weather or parasites, and part of their defence mechanisms mean they contain a vast range of chemicals and ways to respond to surface damage. Harvesting the chemicals, or parasites, is therefore possible without destroying the trees. An immediately obvious example to cite is the production of medicines from tree bark. Antiseptics were derived from witch hazel; yews yield a substance called taxol, which has anti-cancer properties; aspirin from willow; and many other tree derivatives are effective in disinfectants or syrups etc. Until the end of the nineteenth century it was probable that most medicines had origins in trees and plants. Overall, several hundred familiar chemicals and medicines were derived initially from trees. Many defensive tree resins are toxic and the prime example is the South American resin called 'curare'. It is an extremely powerful drug that attacks the muscles and is used as a poison and in hunting prey.

The anti-malarial substance quinine has been used for many years and it is extracted from cinchona bark. Malaria treatment has made a major advance with the drugs artemisinin and dihydroartemisinic, developed from Chinese herbs by the Chinese pharmaceutical chemist Tu Youyou. The work earned her a 2015 Nobel Prize. Interestingly, her source material had been cited as beneficial in several ancient Chinese

documents, but modern duplication of the process had been unsuccessful as modern technology used different chemical separation strategies.
She realized imperfections in modern equipment sampling technology
(including hot water) had not been used in the processes described in
ancient Chinese literature. The fact that some 240,000 compounds had
been unsuccessfully tested worldwide in the fight against malaria could
well imply that automated testing methods may have equally failed in
sampling for natural drugs applicable to other diseases.

The number and strength of protective toxic substances in trees increases towards the equator, as the tropical regions are more prone to
attack by bugs and disease. This is one reason why tropical rain forests
deserve more careful preservation, whilst we search to discover what
natural chemicals are present. An unexpected downside of the defensive toxins means that wood workers, especially those turning objects
from various tropical woods, can be exposed to aggressive pathogens,
particularly when working with tropical timber.

Valuable tree parasites and bark damage

Causing damage and imperfections to trees can have a wide range of
positive benefits for us. These range from parasitic damage as a source
of positive patterning of burr wood for use in decorative items, to parasites such as mistletoe, which features in the mystical rites of Druids,
and Christmas decorations. Because damage to the bark of a tree results in seepage in attempts at natural healing and scar formation, this
type of man-made damage has been widely used. The historic leakage of
sap produces the yellow substance that congeals into amber. Not only
is this considered attractive in jewellery but it has encapsulated a variety of insects which give historical records of long dead species (and
gave inspiration for the film *Jurassic Park*). Tree resin is still used for violin bows to provide friction on to the strings, and resin and vapour from
cedar are used as moth repellents. It is also the basis of incense, particularly in Eastern European churches. Other types of seepage from cuts
in trees have many uses. One such is maple, which provides the very
sweet substance of maple syrup, and there is an annual production of
some hundreds of tons in North America.

In the eighteenth century, expeditions to South America gathered
the sap from the tree now known as the rubber tree. Seepage of latex
was the basis for natural rubber production and is of immense value as a

water repellent and a flexible seal. Seeds secretly taken from Brazil were then spread to other countries to break the initial monopoly of Brazil and Portugal. The name of 'rubber' is unrelated to any native name for the plant but comes from the fact that it was used to remove pencil marks on paper. There are many examples of how we are reluctant to accept new ideas and products, and the same problem existed for latex. It was used to waterproof cloth and poor François Fresneau, who first demonstrated this water repellent effect, was charged with witchcraft.

Causing controlled damage to the surface of trees is also used in the production of bark re-growth in the production of cork. More drastic attack on trees is coppicing: to take saplings for burning, or production of flexible wood for fence materials or basket weaving etc. In the case of coppicing the plant actually undergoes an unusual beneficial transition in its growth, as coppiced roots can survive many times the lifetime of normal trees that have not been attacked.

Surface damage to tree bark caused by fire similarly can have a highly beneficial effect, as many species require a fire to germinate seeds and clear vegetation so that they can grow in an overcrowded forest environment. This may now seem fairly obvious, but initially it was counter intuitive (i.e. fire introduced damage and imperfection) so forestry management practice tried to block any forest fires until quite recently. Now the benefits of controlled fire clearance of undergrowth are better understood.

Historical and climatic records hidden in tree growth

Even for a non-scientist it will be obvious that tree and plant growth varies with changes in climate in temperate latitudes. We cannot miss the fact that most trees have very clear annual growth rings which mark the changes in density between winter and summer growth. The width and spacing patterns of these growth rings is therefore a historical record of our climate. Estimating the age of a tree merely means we need to count the number of tree rings. This is the science of dendrochronology. Not only are the variations (imperfections for the present purposes) in reproducibility of the tree rings tracking recent climate changes but, if we can construct the pattern, we can follow the changes back several thousand years. We are fortunate that the pattern of wood growth is relatively consistent for wood right across the northern hemisphere (i.e. the same patterns exist in Canada, Europe or Asia). It is possible to start with modern trees and progress back through older wooden planks to build up a very long time scale reference pattern. In Europe, and from the earlier Mediterranean civilizations, one can match the age of wooden objects from the simple tree ring counting methods with ages from historical records. The cross referencing of historic ages, even back to artefacts of the early Egyptian dynasties, is a reliable approach. An entire tree ring pattern over such a time scale, of say 4,000 years, can be monitored in a single specimen of several types of tree. In North America, there are examples of very slow growing ancient bristlecone pine trees which allow a climate calibration pattern over this time scale in single specimens. Ring sequences covering at least 8,000 years have been built up particularly well from both the small and shrivelled bristlecone pines and from the enormously impressive sequoia trees.

In the Southern hemisphere, there have been detailed studies of dendrochronology based on the slow growing, but immense kauri trees. These can live in excess of a thousand years and have existed for more than 50,000 years. The ring counting methods extend back to more than 4,000 years as the result of trees that have been found in bogs.

Tree rings and violins

There is a myth that the great violin instrument makers of northern Italy succeeded because they had a 'magic' varnish whose formula has been lost. This is rubbish. Their secret lies in their location and their skill. The high altitude and cold winters of the Italian Alps produced a short growing season that favours a closely packed wood grain structure, which is ideal for the acoustics of violins, as made in Cremona and the surrounding district. The violin photographs of Figure 4.2 show that, even in wood which has been carefully selected for violin making, the grain pattern is not uniform. From a violin it is not possible to say the exact date at which the tree was cut and violin making started, but since early violin makers matured wood for similar periods of time, it is simple to make an estimate of when an instrument was constructed. This means that dating a wooden object, such as a violin, to distinguish between an original and a later copy, should normally be fairly straightforward. Nevertheless, some makers have reused old violin plates in rebuilt instruments, so a degree of caution is still needed.

Among my violins, I have one with a good tone and a maker's label of Guarnerius, which stylistically is probably from near 1900; however, dendrochronology dating of the halves of the belly actually suggests the wood is from the period of Guarnerius. I like to think that maybe it has recycled Guarnerius components (if verified, the value would change dramatically).

Part III Other information hidden in the wood grain

Trees consume large quantities of carbon dioxide as well as other trace elements from the atmosphere and the soil. Not only do the tree ring patterns refer to specific years of growth but they also contain a historic record of the climate and atmospheric chemistry of that year, which has been trapped in the grain. Chemical and radioactive analyses of the components of each ring can reveal what was around at the time. Material from volcanic explosions, or products of nuclear weapons detonated in the atmosphere, are all clearly encoded with their date of origin. Even evidence for long distance wind-blown particle transport of, say, sand from the Sahara to the UK, or agricultural damage of dustbowls in the USA, is written in the tree rings. (Similar patterns of annual deposits occur for lake beds and those are known as 'varves'.)

Subtle dating routes using radioactive materials

It is visibly simple to count tree rings but, with more effort, it is also possible to follow the changes of atmospheric radio isotopes formed because of changes in the activity of the sun. For those requiring a little more scientific depth I will explain the method, called carbon 14 dating. The principle is very simple, although the actual measurements require sophisticated and accurate equipment and techniques. To summarize the science, the sun emits radiation (not surprising as it is an immense nuclear fusion reactor) including neutral particles called neutrons. A few of these neutrons strike nitrogen in the atmosphere and produce the alchemist's dream of converting one element into another. Nitrogen transmutes into a heavy version of carbon, with a mass of 14 units instead of the normal 12 or 13.

Nuclear theory says the carbon 14 atoms should not exist and, indeed, these heavier nuclei are unstable and, on average, half fall apart within the course of about 5,730 years. This is the radioactive decay half-life of an unstable isotope. Once formed, the heavy carbon in the atmosphere reacts to form carbon dioxide. This is absorbed into trees. The ratio of the numbers of carbon 14 (written as ^{14}C) to normal carbon is very small, but it is a measure of the solar activity for that growth ring year. Once the tree is formed, the carbon is stabilized in place and the radioactive decay steadily drops the fraction of ^{14}C in each tree ring. In principle one has only to measure the ratio of the radioactive ^{14}C to stable ^{12}C content of the wood and, knowing the half life, we can calculate the year that the tree ring was formed. Additionally, measurements of the ^{14}C changes between specific tree rings allows us to assess how solar activity has fluctuated with time. Very long time scales are feasible and, for example, from the kauri trees climate information goes back tens of thousands of years.

Experimentally it is challenging, but it works. The crunch problem is that the sun only forms about 7.5 kilograms of ^{14}C per year in the atmosphere. There is a background of normal ^{12}C measured, which is trapped in various ways in the environment and in the atmosphere. This total carbon store has been estimated to be about 42 million million tons. Of which a small fraction is in the form of carbon dioxide in the atmosphere. This background reservoir of 'normal' carbon dioxide is slowly increasing in the atmosphere. The concentration has been steadily rising as the human population has rapidly exploded, and we have started to burn fossil fuels of coal, oil and natural gas. This carbon dioxide increase is an easily measured effect which has been progressively, and rapidly, increasing over the last 200 years.

The ^{14}C method is strongly favoured by modern archaeologists but will be less useful for future generations as, not only is the background carbon content steadily changing (from our attempts to pollute the atmosphere with fossil fuels), but even worse is that the use of atomic bombs and testing of nuclear weapons in the atmosphere has added around an extra 2 tons or so of ^{14}C since the 1940s. This input can either be viewed as a disaster or a bonus, depending on whether one wishes to date the age of an old or a new carbon-containing object. Carbon 14 dating with nuclear-weapon-generated carbon will actually make life somewhat easier for future archaeologists studying the 20th century, so long as we do not use more weapons in the future. If we do, then archaeologists may be an extinct species.

Archaeologists and ^{14}C dating

In general, the practitioners of this carbon 14 scientific art produce dates which are in good agreement with other methods and historical records. Records go back some 6000 years from sites in Egypt and China. The method means we are able to monitor the radioactive carbon that entered the wood in different centuries. In routine usage the technique is effective but, unfortunately, there have been the inevitable over enthusiastic and incorrect initial claims with problematic high-profile samples, that have received more press attention than the hundreds of successful measurements. In the following two examples, egos and a desire for self-publicity seem to have intruded into rational science. The first dubious example was the attempt to date the Turin shroud, the second was to try to prove that the 1908 Tunguska explosion over Siberia had been from an alien spacecraft! At first sight the Turin shroud should have been an easy problem to tackle. Initially the question appeared to be to decide if it were a shroud from the time of Christ or a brilliantly executed medieval forgery. The date differences were so large that this should have been a trivial problem to resolve. However, only minute quantities of material were provided for the dating method. Further, the sample was taken from an area that was possibly contaminated by a medieval repair to the section used. There was a further possibility that modern handling, or cleaning, had contaminated the tiny fragments. Unfortunately, because of the religious content of the provenance of the cloth, the emotive input and consequent impact on scientific reputations, the claims and counter claims are somewhat excessive. Having seen many successful results from the technique I believe the currently available Turin data can be ignored, but a new and properly controlled measurement would be interesting.

By contrast the Tunguska incident is well documented, so there is no doubt about the date of the event in the morning of June 30th 1908. (The cited date depends on source as the Russian and Western calendars differed.) A large swathe of forest in Siberia was flattened by an explosion in the atmosphere. A later expedition to the area had eye witness evidence for the destruction of some 80 million trees. By the mid-century the obsession with nuclear bombs said the energy release involved was perhaps 1,000 times more powerful than the Hiroshima bomb (and in the range of energy of, say, 15 million tons of TNT). It had also produced an earthquake, now calculated to have been about five on the Richter scale. At the time, models for the cause of the event ranged from impacts in the atmosphere of a rocky meteorite, an icy comet, or the explosion of an alien nuclear powered spacecraft! For such a large explosion the option of a nuclear-powered alien space-ship attracted immediate media attention, and the estimated burst of ^{14}C from a 1908 nuclear spaceship explosion might have been as high as maybe 7% above the normal background. In the 1960s, experiments were made with Siberian tree rings dating from 1909. Using such well identified growth rings initially led to claims of a 1% abnormality in ^{14}C. This 'evidence' was rushed into the press and caused some consequent excitement and publicity. More systematic later estimates reduced this figure to less than 0.3% (i.e. at the upper end of the possible experi-mental errors). This killed the alien spaceship idea, but of course it was less newsworthy so, even in current discussions of Tunguska, the alien spaceship is still cited. The details of the non-alien explanation are not unanimously agreed by all the 'experts'. Proponents of impact from a rocky meteorite, or those favouring an icy comet, are still in con-flict. These are quite 'normal' events. In 2013 at Chelyabinsk in Russia, there was an extremely well documented, and photographed, high at-mosphere explosion from a small meteorite. The size was around 20 metres in diameter. The shock wave damaged some 7,000 buildings and 1,500 people were injured from the debris. We now know that there have been similar events around every sixty years. A similar airburst of a meteorite in 1490 in Ch'in-yang is thought to have caused some 10,000 deaths.

The Turin shroud and Tunguska examples underline that scientific data can be clouded by experimental errors and poor interpretations. The positive side of science is that, over longer time scales, the emo-tive input fades and interpretations become more rational, reliable, and acceptable, even if they are not cited in the media.

Future new ideas using wood

Our discussions so far have centred on using wood as it has grown, but that is, of course, just a simplistic view and there are numerous prod-ucts that exploit fragmented timber, or are the basis of new chemicals and medicines, plus items as diverse as rubber and maple syrup. Timber products include composites such as chipboard, multilayer floorings with sandwich layers oriented in different directions, and flooring with expensive woods as outer layers.

Less familiar will be examples of new methods of making violin bows with Pernambuco wood from Brazil. This was once the top grade of bow wood but it is now endangered and restricted. However, there are considerable quantities of the wood in fragments that were unsuitable for sections as long as a violin bow. New technologies have combined this 'scrap' material with modern glues to make composite instrument bows that are claimed to match the performance of conventional, and expensive, Pernambuco bows. Their price is much less than top grade originals (approximately £10,000 in 2020).

Summary on imperfections related to wood technologies

Wood does not have the popular image of a dynamic and exciting modern technological material; rather we accept it as an aspect of everyday life. Scientific glamour is associated to new inventions such as electronics or new metals etc. Nevertheless, the reality is that we used wood before any of these other materials and it will still be around when modern civilization has collapsed and faded away. We rely on wood products throughout our lives and have no inhibitions about the fact that woods are very varied and have many 'imperfections'. We should use this brief chapter to remind ourselves that a full technological perspective should contain materials other than those that hit the headlines for their novelty or commercial impact.

The physical basis for radioactive carbon dating

Since I have taken examples for radioactive carbon dating, I will conclude the chapter with a rapid, but simple, explanation of the process for those interested in how it operates.

As already mentioned, the sun is an immense nuclear fusion source of energy and it emits particles called neutrons. These neutrons can hit nitrogen nuclei in the atmosphere and the impact can cause a change to the nucleus of the nitrogen, which then partially reforms and turns into a heavyweight variant of carbon. Normal carbon has a positive charge from 6 protons and an additional 6 uncharged neutrons to assist in the stabilization of the nucleus, i.e. ^{12}C is (6p + 6n). It is chemically carbon because of the 6 protons. The entire atom is neutral because there are 6 negative electrons orbiting the nucleus. The outer orbit electrons define the chemistry that makes the atom behave as the material we call carbon. Similarly, normal nitrogen is ^{14}N which contains (7p + 7n) units within the nucleus, and 7 orbiting electrons.

The reaction with solar neutrons is as follows. The neutron interacts with a nitrogen nucleus and ejects a proton (i.e. a hydrogen nucleus) and leaves a heavy version of carbon:

Neutron (n) + ^{14}N (7p + 7n) combine, and then fragment to become ^{14}C (6p + 8n) + ^{1}H (p).

The new nucleus of ^{14}C is slightly unstable, so there is a radioactive decay in which half the nuclei fall apart in about 5,730 years. Overall, the sun generates a tiny quantity of some 7.5kg of ^{14}C per year in our atmosphere, so there is quite a limited amount of ^{14}C.

Carbon dating of wood only requires that we measure the relative concentration of ^{14}C to ^{12}C in a piece of wood. During plant growth, the ratio was fixed by equilibrium with radioactive carbon in the atmosphere. Once the wood is formed then the ratio falls as the ^{14}C

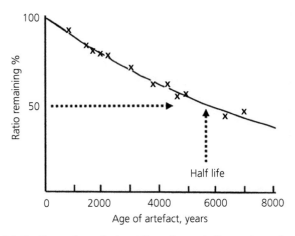

Figure 4.5 Radio carbon dating. The schematic figure shows how the radioactive content of ^{14}C decreases with time, so that in older samples there is a smaller percentage of the original ^{14}C. The age predicted by the decay matches well with historically-dated artefacts and tree ring dendrochronology.

steadily decays. We know the half life for the decay, so the ratio tells us the time since the tree ring grew. A half life of nearly 6,000 years means the progress is a good measure of the age of artefacts over several half life time scales. Carbon 14 ages match very well with dendrochronology tree ring values and ancient wood, such as coffins in early Egyptian tombs, and other historic records of age. When combined with dendrochronology, the ^{14}C offers both a record of solar activity and other major environmental events, such as volcanic explosions and modern nuclear weapons. Figure 4.5 shows how the radioactive content of ^{14}C decreases with time. The marker dots relate to early data with the technique where there were both carbon 14 values, and other evidence from tree rings, or historic written records. There is good agreement between the alternative methods. The half life value is indicated. Modern improvements in the measurements have extended the useful time scale to several half lives.

5

Reader beware

Part I Separating certainty from opinion

In the initial chapters, my examples of good technology from inherent imperfections in materials were invariably very familiar and easily appreciated. Even the models of simple atomic structures that I will use later are no more than children's building blocks. Additionally, we can draw parallels between atomic arrangements and human behaviour in terms of different communities living in a city, as these will be helpful. Structural models of lattices and their imperfections have existed for over a century, and so major conceptual errors will probably have surfaced and been corrected. This is not always the case, and deeper in the scientific literature there are many concepts, interpretations and experimental results which are still being challenged in terms of their detail. My personal guiding rule is that if a current model has survived more than one generation, say 25 years, then it is probably a reasonable starting point.

In materials research subjects, few people move far from their initial field of training. I have seen claims that in subjects, such as physics and chemistry, 90% of academics only expand around the subject area of their original doctoral thesis. This concentrated effort means their knowledge base and expertise can become ever more tightly focussed. In terms of progress it is a valuable luxury as they can stay at the forefront of their subject. The downside of this narrow focus is that, once established, one trains students who may also take an easy option and remain in the same field.

To some extent I have a similar limitation in that I rarely stray from an interest in optical properties of imperfections, which was at the core of my thesis. However, this is a sufficiently generic concept that it has been useful for me in areas as different as archaeology, mineralogy, engineering and cancer detection. It also has the bonus that at meetings outside of my main topic I am assumed to be an expert. In reality,

The Power of Imperfections. Peter Townsend, Oxford University Press.
© Peter Townsend (2022). DOI: 10.1093/oso/9780192857477.003.0005

experts are quite rare and when attending conferences my personal assessment is that topics are typically driven by say 10% of real gurus, who
understand in depth. Another 40% are good followers, 40% are variable
quality and the rest are dubious.

I assumed that such profound influence on a research area from a few
people was just a feature of a well-focussed science, but then realized it
is a far more general pattern. Power has always been concentrated in
small minorities, whether via monarchs, aristocracy, religion, or dictatorships. In the twenty-first century there is an emerging field of
'sociophysics' which examines the power of small minorities and their
influence on political decisions and voting. They assert that minority views will become accepted if they are from determined advocates,
and certainly appear to succeed once they reach around 10 to 20% of
the population. Political supporters seem eager to follow even extreme
views that are promoted by these minorities. This is evident in modern politics with many clear examples, as seen from the rise to power of
Hitler in the last century. One can equally cite other dictatorial rulers,
both past and present.

This simplistic pattern is typical for science and technology of materials, but definitely less clear in the case of multi-faceted disciplines of
medicine and biological sciences. Both are extremely complex, and our
knowledge and techniques are expanding incredibly rapidly. This leads
to conflicts in medical opinion and/or misinterpretations of results. For
example, doctors will have been trained in the views current during
their time before graduating. Medical opinion and treatments evolve
remarkably quickly, and there are very many cases where opinions totally reverse. All such new data and discussions will exist in medical
literature, but it is difficult to bring this to the attention of a community who are working with extensive workloads, covering a vast range
of diseases and medical conditions. To offer some clue to the scale of
the problem it is worth mentioning that there are some 30,000 medical
and scientific journals publishing a total of at least 2.5 million papers
per year. This does not include articles in magazines, newspapers, blogs
or web sites. It is not surprising that the easiest option is not to bother
to read anything. Nevertheless, there is a true benefit in the wealth of
literature which one can consult before going to see a doctor with a set
of symptoms. One can focus on them and see if they are described freely
in open web literature (and in a form you trust, not merely hidden
advertising and promotional literature).

The consequence of this overload is that many doctors inevitably continue with the opinions they were trained to accept. A very strong inherent human characteristic is that we are suspect of change, especially if it is contrary to something that we had previously believed. This bias is amplified if we are frequently dealing with problems where it appears that we are more knowledgeable than, say, patients at a surgery.

There is a further distorting factor in medicine, not just that there can be excessive and disparate workloads, but there is immense pressure from pharmaceutical companies to market all their latest products. To maintain knowledge of which drugs are truly effective, have serious long-term side effects or are not needed at all, is a major challenge. In terms of a general UK practice I am sure that, in the allowed ten minutes or so per patient, it is easier to prescribe something which may work, rather than persuade the patient to change their life style and have no need for any drug. Any search for side effects of even the most common of drugs can be disturbing and I have found examples for extremely common items with over 100 side effects, some rare, some serious. This is inevitable, as our genetic make-up and lifestyles are immensely varied.

The electronic search engines for science do have benefits but, equally, some downsides. When I was a graduate student one could pick up a single book that cited all the physics journal abstracts for the year. Such access is now totally lost in a mountain of articles from an internet search. Less obvious is that in the book of abstracts version, one was often attracted to totally unrelated articles, and this was excellent for broadening input of different ideas. The dedicated web search has suppressed this spontaneous stimulation for a diversity of scientific literature. Worse is that the top listings can be occupied by commercial backers, and/or the list will start with the most cited items, which in turn both hides originality and perpetuates old concepts and errors.

If these are problems for trained scientists and medical experts then it is an even worse challenge for the general public. It will, effectively, be equally impossible for our policy makers, who may be career politicians who have had minimal work experience, and totally lack any advanced education involving science or mathematics. They therefore lack the training to understand even simple presentations of data in the form of tables or graphs. Results which are clear cut, but need descriptions in terms of averages, probabilities and statistics, may be completely misunderstood, or consciously presented with bias.

Scientists are not immune to even simple errors and I recall, at a recent international meeting, an argument between a European and an American as they were both using the term 'billion' and did not realize that there are two definitions. The American is a thousand million (10^9), the older European meant a million million (10^{12}). In the UK we somehow switched to the smaller value. I wonder if UK and European politicians know which they mean when discussing budgets and financial commitments.

Some examples of confusing presentations

The potential list of poor presentations and misinterpretations is long. Some are deliberate, but many articles assume the readers will have the same background as the writers and thus will not be misled. Equally, it is easy to assume that something written in English (or French, or Spanish etc.) will have the same nuance to all others who speak the same nominal language, and that meanings are independent of country. For example, if I write a reference and say the person is 'quite good', in UK English it is meant as a compliment. In the USA it is distinctly negative.

Much later on, I will discuss an imperfection in behaviour which leads to obesity and diabetes. An article I saw said that, in the UK, we have no problem as we were ranked only 38th in the world obesity tables. This was totally misleading as the table was set out in terms of total numbers of people who are obese in each country. So those with enormous populations (China, India etc.) were at the top. Small island populations were very low down. The ordering changes dramatically if quoted in terms of a percentage of the population in each country that are obese.

It is often claimed that statistics are untrustworthy, but I suspect that in reality the mistrust should be aimed at those who present the numbers and information, not necessarily the basic data. As an example of such an imperfection in presentation, politicians will talk about average values of wealth and wages and cite highly reliable numbers from honest sites. Nevertheless, it is clear that what they (and the general public) understand by average can be very misleading. Here is an example of an imperfect presentation, which we believe we understand, and which will appear to match our particular biases and/or political viewpoints. The difficulty is that we may not recognize that 'average' in statistical terms might refer to a 'mean' or a 'median' value. These are very different.

Definitions of wealth are variable as they will depend on the type of work, lifestyle and expectations. Wealth and financial hardship therefore mean different things. Even using data such as 'median household disposable income' is only a simplistic guide since the value is shifted upwards by all the highly paid. In a big city there are many company executives and directors. In 2019, their UK average salary was above £90,000 per year, with a top end group in major companies exceeding £3.9 million. Bonus packages in addition to the salaries can be ridiculously large, with a recent example of £75 million. Top football stars will also have salaries that can be in millions (in 2018, the top ten were between about 1 and 7 million).

Although factually correct, I feel quotes such as that in London, in 2019, the median salary was around £30,000 can be deeply misleading, as *median* is the point where there are 50% of the people below, and 50% above this number. By contrast, the number is even higher if one counts the total income, including the many very-high-income people, divided by the total population. In London, the top salaries can be enormous so the *mean* value jumps up to £40,000. Focussing on the mean, or the median, draws our attention away from the fact that there are vast numbers of people with far lower income. The peak of the lower wages is termed the *mode*. Figure 5.1 indicates the pattern and the

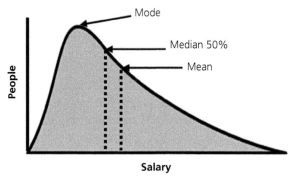

Figure 5.1 Comparisons of definitions of median and mean salaries. 'Median' is the 50% value with equal numbers of people below and above it. 'Mean' is higher because of a small number of very-high-income people. The peak income ('mode') is far lower and it has a significant minimal tail that is poverty.

relative terms. Statistics are not incorrect, but they need cautious inter-pretation. If the distribution is as in Figure 5.1, the *mean* is a comfortable £40,000 per year, but the peak of the most likely income (the mode) is only £20,000, with a significant number of families earning even less. I have almost fallen into the trap of language, where I am using salaries for the median and mean values, but was tempted to write 'wages' for the poorly paid.

Misinterpretation can arise from careless reading, or failure to ap-preciate that there can be a very wide distribution in income. Equally it may be a conscious effort from a company to disguise disparity be-tween workers and directors. Assume a hypothetical company that needs many low paid workers (e.g. a warehouse distribution for goods). They may employ 1,000 poorly paid workers at, say, £15,000 per year. By contrast the ten board members may draw salaries of £1.5 million each (i.e. precisely the same total sum). In an annual report they will correctly say the *mean* salary is almost £30,000 per annum. To a casual reader the worker/management split will be hidden. Such iniquity of income is typified by 2019 data that FTSE-100 chief executives received £901 per hour, and their employees just £14–37.

Presentation of statistics related to population and climate change

Here we are moving to topics where presentations may deliberately ob-fuscate the same data, depending on which political viewpoint is held by the writer. There have been estimates of world population which stretch back some 10,000 years. Inevitably these are not of high precision for historic examples, but there is good agreement on the broad pattern that is discussed. Global human populations were as few as 5 million some 10,000 years ago, but are currently around 8 billion. This is a range of over a million times increase. On a simple linear figure of population number versus date, the older data is totally hidden next to the axis of the graph. For a mathematically competent scientist, the solution is to use logarithmic scales as this can encompass the entire range and show the trends. Figure 5.2 shows a log-log axis plot. Broadly it is a smooth curve. To a non-mathematical observer, population growth is not a problem as the modern period, the top-right end of the graph, looks fairly flat. Note also that the figure uses a logarithmic scale counting

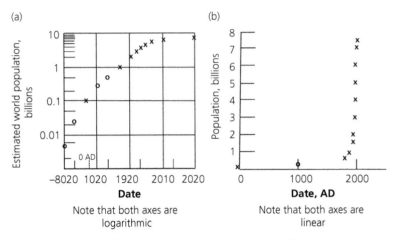

Figure 5.2 The SAME data are plotted on alternatives of logarithmic or linear axes for changes of world population with time. The logarithmic time axis compresses past data to accommodate a very wide span in time of some 10,000 years. The values marked are the same numbers on both plots. The aim of this figure is to show how our perception is very biased by the layout of the available information (see text for discussion). Values before ~0 AD are too small to display on the right-hand plot.

backwards into the past, so from 1 to 10 years ago is the same length of axis as from 10 to 100, 100 to 1,000, or 1,000 to 10,000 years. This is sensible, but somewhat unfamiliar, and thus easily misunderstood.

Next to it is a graph of PRECISELY THE SAME DATA on linear scales. Old data of small populations have vanished against the time axis. Numbers from more than the last few centuries are tiny, but they suddenly rocket skywards within the last 100 years. It is impossible not to panic once we realize that population growth is totally out of control. There is no way in the future that we can have the resources to sustain such a population explosion.

On the pair of graphs, the data are the same, but I doubt that even mathematicians will not be influenced differently by the two presentations of the data. My example demonstrates imperfections in the brain processing and personal bias, and these factors override the figures and statistics. Indeed, rather than merely adding confusion, many people will interpret these curves as totally opposite trends.

The recent population figures rise so steeply that Table 5.1 lists years at which we crossed each billion number in world population.

Table 5.1 Estimates of world population with time.

Year, AD	Population, billion
2023 ?	8
2011	7
1999	6
1987	5
1975	4
1960	3
1925	2
1825	1
0	0.1

Projecting forward is interesting as some government bodies suddenly assume a greatly reduced rate of growth on the assumption that there may be a dramatic reduction in fertility rate. Indeed, there is a small trend, but this may be offset by better health care which implies greater longevity and, therefore, an increase in population with a higher percentage of old people. (I will return to a discussion of these factors in Chapter 18). If one looks in fine detail at the original population data, some of these graphical estimates have a tiny downward blip at the end of the First World War from deaths during the conflict and the subsequent influenza epidemic (which caused more deaths than the war). A similar feature happened at the time of the Black Death plague. In both cases the numbers were more than one third of the European population at the time. Nevertheless, population recovery was rapid in both cases and it continued to rise. Indeed, this is absolutely standard human behaviour that a major drop in population is followed by a greatly enhanced birth rate. The implication of this is that from our current population of nearly 8 billion, even a major global, or regional, catastrophe would not have any long-term effect in reducing the world population and its demands on agriculture, land and natural resources.

We are a relatively new species in the history of the planet and may turn out to be a very short lived, transient one. Our demise could easily be self-generated, or result from a sufficiently major event, such as a large asteroid impact. A major rise in global temperatures, to the levels that have existed in earlier eras, would be equally lethal to us. The coronavirus offers a warning and a reminder that pandemics can

suddenly emerge and cause both high death rates and economic collapse. Two other features are that, firstly, the linear plot of Figure 5.2b
looks remarkably like the pattern of explosive growth of bacteria. Secondly, human population growth is conceptually not much different.
However, when making such analogies it is worth considering some
consequences. In southern England in recent years there has been a
rise in a disease called 'ash tree die back' which spreads through the
atmosphere and specifically kills ash trees, extremely rapidly. Totally
removing infected trees has helped to reduce the spread, but the prediction is that 98% of ash trees will vanish within a decade. Maybe 2%
are predicted to survive if they are dispersed sufficiently distant from
the sources of infection.

Who has difficulties with interpreting information?

One should not castigate non-scientists for incorrectly interpreting
graphs with logarithmic axes. Part of the difficulty is that, when presented with complex and/or detailed information, our brain looks for a
rapid and intuitive interpretation. Figure 5.3 shows data I once recorded
on how a luminescence signal changed with temperature. The dynamic
range of the system was considerable, so I used a logarithmic intensity
axis. I clearly see six peaks, but even though this was my own data, I do
not subconsciously recognize that the signals vary by more than a thousand to one. Instead, the impression is that they vary by maybe about
three times!

Some errors occur when we lack depth of knowledge. A trivial
example would be to ask what is the weight of a water molecule, in
terms of their atomic weights. Water has 2 hydrogen atoms and an oxygen (H_2O), and most people will assume ^1H is just a central proton
and so has unit mass, and most oxygen is ^{16}O (8 protons and 8 neutrons), so overall the obvious number is $1+1+16 = 18$. This is normally
fine, but with a little more knowledge, we may know that there are
other isotopes with different numbers of neutrons. Deuterium, ^2H, has
($1p + 1n$), giving weight 2; tritium, ^3H, is ($1p + 2n$ and weight 3). Less
familiar is that oxygen has three stable isotopes of masses ^{16}O, ^{17}O and
^{18}O. So, overall, there are a possible 18 different water molecules with
weights from 18 to 24. A nuclear physicist might say I am forgetting that

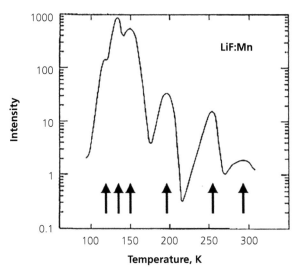

Figure 5.3 An example of a luminescence signal that varied with intensity as a function of temperature. In order to cope with the large dynamic range, the intensity axis is logarithmic. However, the impression it may give is that the signal levels of the peaks only differ by, say, a factor of three, instead of 1,000. (K is absolute temperature where zero centigrade is 273.4 in Kelvin.)

there are some very unstable oxygen isotopes with the 'wrong' number of neutrons ranging from [11]O to [26]O. Easy question, but a non-obvious answer.

Do we see what is there?

Examples of misinterpretation from complex images, drawings and pictures are very numerous. We believe we see what we expect to see, not what is there. This brain processing error is exploited in optical illusions such as the drawings of the Escher staircases, or *trompe l'oeil* paintings where we see a face which is actually a collection of vegetables or bodies. There is truth in the saying of 'Beauty is in the eye of the beholder'.

An example of concern is that medical cancer diagnosis from X-ray images is made very rapidly and thus prone to error. One underlying imperfection is that if one tracks eye movements during visual inspection then, once a cancerous region has been spotted, the brain controls the eye to keep returning to the site, and only poorly examines the

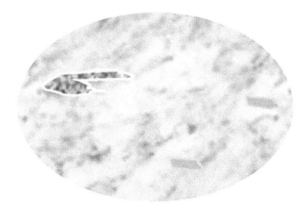

Figure 5.4 An image with more than one type of texture

rest of the image (even if other cancerous areas exist). This human imperfection has been partially addressed by artificial intelligence (AI) automatic scanning techniques, which label where cancers may exist. For the labelled image the major improvement is that the viewer spends a somewhat more equal time looking at the various target areas. Figure 5.4 shows a simpler, and much more obvious, simulation with two types of texture. Is it apparent which area has been changed in terms of texture?

Most viewers will immediately recognize the unusual area at top left (near 10 o'clock) which differs from the background. However, did you notice the two other regions? Do not be surprised if you missed one as normal mass screening views with X-rays have an extremely high failure rate (often 30%), either by missing cancerous sites and/or assessing normal tissue as cancerous. The consequent health risks and/or unnecessary operations have a significant impact on the lives of thousands of people, and are costly. In this instance AI is valuable.

Part II Data and interpretation of changes in atmospheric carbon dioxide

Discussions of climate change and global warming that are linked to human activity are incredibly contentious, not because of the data, but because of the predictions of the changes, and the fact that to reduce the emission rates would affect all aspects of our lives, from

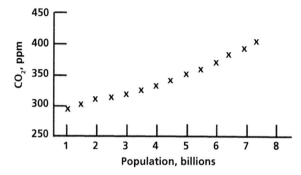

Figure 5.5 A comparison of world population with atmospheric concentrations of CO_2 (in parts per million) at different years. The pattern extends from ~1800 to the present day. The CO_2 values are rising slightly more rapidly than the population growth.

agriculture to industry and population. Here there are so many self interest groups (from both sides of the contest) that few will admit that there is a straight correlation between the patterns of growth for population, carbon dioxide in the atmosphere and climate changes. However, the correlation between increasing population and increases in the concentration of atmospheric carbon dioxide is unequivocal (Figure 5.5).

The measurements of CO_2 in the atmosphere have become very accurate, and include data that are obtained at places remote from industrial sites (e.g. Antarctica and/or a Hawaiian mountain top). In Figure 5.5 the correlation in growth of the CO_2 and population spans the time range of the last two centuries. What is clear is that the two sets of data are so similar that it is unlikely to be fortuitous that they are not linked. The link is simply explained as population growth since the start of the industrial revolution means we have burnt increasing quantities of fossil fuels, and released into the atmosphere carbon that had previously been locked into coal for the last 300 million years. Some oil sources may be somewhat younger, but they will still be on the scale of at least 70 million years old. It is these sources of carbon that we are reintroducing to the present atmosphere, and hence a correlation with the rise in CO_2 concentration and industrial activity is unsurprising.

Carbon dioxide is unlikely to be driving the birth rate, so the alternative conclusion is that human activity is driving the steady rise in CO_2. Industrial pollution is very visible in many modern industrial and major cities, where one is overwhelmed by the atmosphere, just as was apparent from images of nineteenth century Britain, where industries were blackening the towns and countryside. Interestingly the coronavirus lock-down was matched with extremely clear UK skies, no vapour trails, low pollution levels in the cities and, for people with breathing complaints, an incredibly improved level of life. The same features have been noted in many cities across the world.

Some articles have claimed that humans, and the animals we eat, are a major source of carbon dioxide in the atmosphere. Whilst it is true that more humans and cattle etc. will exhale more CO_2 this is irrelevant, as it is merely part of recycling plants that we, or the animals, had already eaten. The same untouched vegetation would have decayed and emitted precisely the same amount. Arguments for a vegetarian diet to reduce CO_2 are similarly flawed. Data from before and after the major attempts of American settlers to wipe out the bison in the late nineteenth century do not show up in the measurements, despite the killing of many million animals. One should also note that cattle are not the major source of meat across the world (surprisingly, it is goat). A genuine argument for eating less meat is quite different, as that involves health issues from overeating.

Major sources of pollution are not all obvious and openly discussed. For example, the electricity used for internet and social media produces *more* CO_2 than commercial airlines. Bitcoin computer processing uses as much electricity as a small country. Neither aviation nor shipping are covered by the Paris Agreement on climate change. Shipping is crucial for international trade with some 90,000 major ships operating (in 2018), that burnt some two billion barrels of low-grade fuel oil. Low grade oil is cheap, but rich in sulphur and other pollutants, as well as producing CO_2 at around 3% of the world output of greenhouse gases. To put this in perspective, the fifteen largest ships match all the emissions from the total number of cars on the planet. Some ships use 'scrubbers' that partially reduce emissions, including sulphur (and hence sulphuric acid emission), but they dump the pollutants and heavy metals into the ocean. Scrubbers also increase fuel consumption by around 2% (and so generate yet more CO_2).

Predicting the scale of climate change

To a scientist, or indeed anyone interested in gardening, the concept of a greenhouse (or a sunroom) is very familiar. For our world the greenhouse ceiling is not glass, but a mixture of carbon dioxide and methane, both of which trap heat in the atmosphere. Better greenhouse performance therefore implies a hotter zone beneath it. It follows that raising the carbon dioxide concentrations (and potentially vast quantities of methane, which are currently trapped in frozen soil) will imply an increase in global temperatures. Not least, as methane is an even more effective greenhouse gas than carbon dioxide. The 'only' difficulty is to predict the scale and speed of the change (not that there will be one). Other factors which might enhance, or reduce, atmospheric temperatures are linked to pollutants such as those from major volcanic eruptions. These have often caused cooling, crop failure, famines, deaths and civil unrest. There is a further pattern from other factors in our orbits and/or solar output that have driven a periodic pattern of ice ages (Figure 5.6). In this figure, with data deduced both on Antarctic temperatures and concentrations of CO_2 trapped in ice cores, there is a remarkably strong correlation that with more CO_2 the temperature is higher. One can even sense some natural periodicity of the pattern (roughly 100,000 years), together with substructures. Climatologists understand some of the many causes of both the periodicity and structure, as in part they relate to minor changes in the orbit of the Earth around the sun, and periodic changes in the tilt of the axis of rotation. Focussing on the year axis, one might claim we are near the end of a natural rising phase of warmer climate, or for some reason have inhibited a decline into the next ice age. Nevertheless, there is no evidence from the last million years suggesting that the current world geography has experienced the extremely high CO_2 concentrations that are now being measured above 400 ppm (parts per million of the atmosphere). Much earlier, when higher values did exist, the temperatures were more extreme, and would not have supported the plants and animals of this century. A positive note is that we have never experienced the CO_2 atmospheric concentrations seen on Venus (i.e. 96.5% of the atmosphere). There the greenhouse effect and the closer proximity to the sun combine to give a surface temperature of the planet around 470°C (780°F). The temperature is less on Mercury even though it is

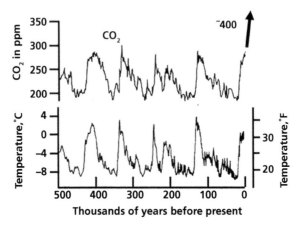

Figure 5.6 The upper sketch shows data obtained in places remote from industrial sites for atmospheric CO_2 concentrations, and the lower one has Antarctic temperature values. Broadly they are very similar patterns. Note the CO_2 concentrations were always below ~ 300 ppm, whereas during the last century they have shot up above 400 ppm, and are rising.

even closer to the sun, but it does not have a CO_2 atmosphere and it is a smaller planet.

Imperfections and bias in scientific discussions

Discussing CO_2 concentrations and the consequences is an ideal example for my caveat that our interpretation, understanding, and bias for or against particular scientific models, is a major imperfection in making progress. There is excellent and reliable data that CO_2 in the atmosphere has increased beyond the range experienced in the last million years. The pattern of increase over the last 200 years matches the population increase, and we understand how CO_2 and methane can act as a heat retaining component in the atmosphere. The difficulties and uncertainties are in predicting what are the immediate and longer-term future consequences. These become confused because data, such as that of Figure 5.6, show we have a long-term cyclic pattern of warm periods and ice ages. Since we are possibly near the top of a warm cycle, it is not feasible to claim if the natural contribution to the temperature should now be rising or falling, as the natural speed of temperature change is quite slow. In Britain and Europe, the last ice age peaked some 22,000 years ago and ice vanished around 11,000 years later. However, the predictive problem is that, since our atmospheric greenhouse gas concentration is now so much higher, it is unclear how this will perturb the natural temperature cycles and we must be concerned for the very near future, measured in decades, not ten millennia. In many models we will destroy our ability to survive, and definitely we cannot sustain the current population growth. This is such a serious issue that we must question why there is such a reluctance to even admit that the situation exists. When we really face up to the danger it is still unclear how we can react, since a key factor, and the only guaranteed solution, is that we need to make an immense reduction in world population. This is so unwelcome that we hide behind suggestions of changing agriculture, stop excessive eating, maintain tropical forests etc. All are valuable and sensible proposals which certainly should be followed, but they do not address the real villain which is that the population explosion is the basic driver of our problems.

I therefore will comment on reasons and views from countries where I have lived and worked as I can appreciate why some regions will have no understanding that there is any threat, and/or do not wish to admit it, since it will imply immense economic and social changes. I have worked in different parts of the USA and am well aware that, with such a very large country, there are considerable variations in climate and seasonal temperatures. One consequence of this is that the bulk of the population, who are city dwellers, have isolated themselves from the variable weather with central heating in the winter, air conditioning in the summer, and make most of their travel in air-conditioned vehicles. Changes in the seasonal weather patterns are therefore not readily appreciated. This will not be the case anywhere for farming communities as they will recognize that seasonal dependencies for planting, rainfall, temperatures, etc., have clearly altered within the lifetimes of many of them.

The second factor for the USA is that the general public still use the temperature scales of Fahrenheit, whereas in Europe (and even the UK) the Centigrade temperature system is used. Changing is not easy and I know that my older UK friends and their parents have no concept of numbers in Centigrade. Scientific data is almost always cited in

Centigrade and so, on reading the literature, a large percentage of older UK residents, and most of the USA, have no idea what the numbers mean. In discussing global warming etc. this is a disaster. Worse, is that scientists often use absolute temperate numbers (in Kelvin), as I deliberately did in Figure 5.3 above. This immediately cuts themselves and their data off from the general public. Zero Centigrade, or Celsius, is 273 K, or 32°Fahrenheit.

Errors in converting, or understanding, different units are not limited to temperature and, for example, accurate astronomical distances, to say Mars, have been measured in metric units of metres and kilometres. For such a very large distance, conversion to units of miles needs very high precision. Failure to do so apparently resulted in a space probe crashing into Mars as it was fractionally closer, by a few kilometres, than predicted from the inadequate conversion. The minimum distance from Earth to Mars is about 33.9 million miles (approximately 54.6 million kilometres) so the conversion factor needs to be much better than 1 part in a 100 million (i.e. more precise than the arithmetical conversion on a calculator or many computers).

The second difficulty in discussing any topic which impacts directly on the quality of our lives, wealth, or religion, is that we automatically (and subconsciously) initially reject any arguments or facts that might conflict with our previously held ideas, and/or would result in a reduction is some way of our wealth or lifestyle. This is a universal problem but, in the present context, I will include one example that certainly has applied to Australian politics. Australia is a very large country with a tiny population of around 25 million people. In terms of wealth they have a substantial income of around £25 billion from the export of coal. Current volumes are about 400 million tonnes. In terms of pollution of the atmosphere and additional CO_2, this translates to almost 1.5 billion tonnes of CO_2 etc. (carbon atomic weight 12, oxygen 16, produces CO_2 weight 44). The economic dilemma is that coal generates around £1,000 per head of population, but at a cost of some 60 tonnes of pollutants and CO_2 per individual (just from the sale of coal). This makes them one of the largest CO_2 polluters.

One additional factor that must be considered is that Australia has always had a major problem with wild fires (bush fires), but in recent years the highest recorded temperatures have been increasing (in 2019 by more than 1.5°C (2.4°F) above the long term average) and rainfall has decreased in parts of the country. The rise and scale of the fires has thus soared to around 6 million hectares (about 15 million acres), which is comparable with the land area of England and Wales. There is a quantifiable cost in terms of insurance, lost business, homes destroyed, firefighting and health problems, etc. Estimates for the various negative economic factors number in the range of £5 billion, or more, for 2019. Many people sensibly link the increase in such natural disasters to higher temperatures caused by rising global temperatures. The overall effect is a diversity of opinion between the political ministers seeing the economic gains, and those who are suffering immense losses of homes and living from the consequences.

Long range effects are apparent as, 2,000 miles to the east of Australia, the New Zealand glacier called Franz Josef became a picturesque brown colour from the ash. This will help absorb summer heat and melt the surface.

My simple examples indicate why apparently clear scientific data do not always produce acceptance and an unprejudiced discussion of the consequences and actions that are required.

Understanding subtle effects

I have only focussed on problems of well documented examples, such as measurements of temperature and atmospheric concentrations of carbon dioxide. Unequivocally this implies a warmer Earth greenhouse. The more difficult challenge is to predict the type, scale and speed of the consequences of such heating, and the speed at which they will impact on all our lives. This requires a far greater understanding of a vast range of interacting factors than merely predicting that warming will occur. Even with the best of intentions the details of how these climatic factors will develop is extremely complex and, for the majority of us, we can only guess at the consequences, and almost certainly we will have a mixture of predictions that could be correct or could be wrong. Tracking links between a rise in air temperatures and changing climate is incredibly complicated and, even as a non-expert, one can suggest a dozen major climatic factors that are involved. For example, over the last thirty years so much of the Arctic ocean has rapidly warmed and progressively become open water during summer, that there are changes in movements of icebergs and current flows. Glaciers have vanished in both Alaska and Greenland. This has resulted in a cooling of the waters of the North Atlantic.

By contrast, the near surface temperatures of virtually all the other ocean areas in the world have steadily warmed over the last century, by more than one degree Centigrade (around 1.5°F). Such small surface changes not only influence marine life and fish stocks, but also control our climate. The UK and eastern US weather is powered by heating and surface water evaporation in the central Atlantic near the Caribbean. Historically, the surface had been around 23°C. This has now increased. With a minor heating of 2 degrees the water vapour pressure increases by 13%, and for a 5°C (8°F) rise it leaps upwards by 35%. Such changes match temperatures that are now being recorded. The obvious consequences are an immense upsurge in warm moist air and energy that powers the hurricane season, as well as the rest of the North Atlantic climate. Indeed, there has been an increase in both the number and power of storms and hurricanes in the last few decades. These are matched by extreme precipitation, floods and highly unseasonable weather. Far more subtle, and therefore difficult to model and predict, is the shift in patterns caused by cooling of the North Atlantic and open water in the Arctic. All these changes further modify the patterns of the high atmosphere jet stream which drives a great deal of our weather patterns. Extreme changes are guaranteed, but modelling their scale and timing is dependent on the assumptions fed into the computer programmes, even with excellent current data.

Astonishingly, in 2019, 2020 and 2021 there were unprecedented temperatures recorded in Arctic regions of Siberia, with highs up to 38°C (around 100°F). Numbers unmatched for many hundreds of millennia. Yet some political views reject even this as evidence for any sign of global warming.

I would like to have concluded this chapter with very definitive comments, but I can only outline our difficulties when handling complex interactive mechanisms, such as climate. Unlike the comments on world population, where it is unequivocally rising far too rapidly to unsustainable levels, and must somehow be drastically reduced (preferably not by war or plagues). I have no doubt that we are altering our climate, not least by increasing our protective greenhouse with industrial atmospheric additions of carbon dioxide. The greenhouse may become even more effective once surface trapped frozen methane is also released. However, the details and speed of change, together with predictions of future climate are immensely complicated, and thus model dependent. The only clear conclusion is that we must minimize the rate of change by lowering pollution levels, so that the plants and creatures on the planet survive, or have time to evolve to cope with the new conditions.

The really hard question is how to dramatically reduce the world population. This is the root cause of the problems, but the ongoing explosion in population is something we seem afraid and unable to discuss and address.

Despite our difficulty with multi-parameter problems, we have acquired an enthusiasm for computer-controlled artificial intelligence to make decisions for us. The good news may be that it relies on data, and may not have been processed with deliberate in-built bias. If it is truly impartial, and asked to tackle problems such as 'Directions and Limitations of Earth's Kinetics' it may first choose an acronym for the project. Older sci-fi fans will see this will be an acronym of DALEK, and recall that the consequent solution will be a command of 'Exterminate!' Reader beware.

6

Key features of chemistry and solids

Part I Background to ion sizes and chemical bonding

In this chapter I will offer some simple background, and many examples, related to simple, or historic, uses of glass products. The next chapter will focus on more modern applications. Since I am assuming some readers may have minimal relevant scientific background, I will rapidly summarize how atoms combine into molecules and solids as this will lead smoothly to understanding the imperfections that are likely to appear.

An atom has a heavy core region of positively charged protons mixed with some neutrons (that have no charge). The positive core attracts an equal number of negative electrons which circulate around it. No two electrons can have precisely identical energies so there is a staircase of energy levels. The major 'stairs' variously accommodate 2, 8 etc., electrons. For simplicity, we can think of them as being in a series of orbits around the core. School texts may use the sun and planetary orbits as an introduction idea but, in reality, the electron density patterns are far more complex. Moving from hydrogen to helium, and all the rest of the elements in the periodic table, means electron shells steadily fill up the levels at each step, and then progress to higher level shells. 'Happiness' for an atom is a completely full outer electron shell.

For common salt, sodium chloride, the sodium (Na) has 11 protons, so the electron staircase has filled inner shells with 2 and then 8 electrons, the eleventh is alone in the third shell. Similarly, chlorine (Cl) has 17 electrons which fill the inner two shells (2 + 8), leaving seven more in the next shell, which has space for 8 electrons. Chemically, the Na to Cl bonding is then ideal if one electron moves from Na to Cl, as both will have full shells. Figure 6.1 sketches this behaviour.

The Power of Imperfections. Peter Townsend, Oxford University Press.
© Peter Townsend (2022). DOI: 10.1093/oso/9780192857477.003.0006

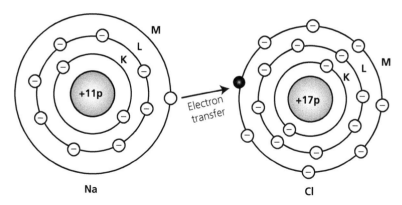

Figure 6.1 Electron transfer to make ions. The sodium nucleus has 11 protons, whereas chlorine has 17. The associated electrons fill energy shells (labelled K, L, M etc.) which can accommodate 2 in K, 8 in the L shell, and 8 in the M shell. The solitary M shell electron moves to the chlorine and then both ions have complete outer shells. The picture is greatly simplified by sketching orbits as circular, and evenly spaced away from the nucleus.

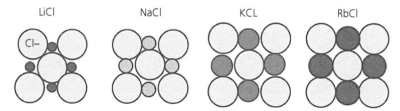

Figure 6.2 Ion sizes in compounds. The sketch shows the relative sizes of metal and halogen ions for chlorine alkali halides from lithium to rubidium. Despite the range of sizes, these fit together in the same structure. For even larger disparities, such as caesium chloride (CsCl), they switch into a different cubic based structure.

The simple example is fine for molecules, but to build a solid we need to see how all the ions might pack together. This will depend on the sizes of the various atom types that are in the structure. The situation will vary if the ions are of dissimilar sizes. For historical reasons we discuss chemicals and draw pictures of their relative positions in a crystalline model, but subconsciously often imagine that all the ions of the group are of comparable size. This is misleading. Figure 6.2 shows examples of how a few different alkali halide crystals are stacked together. The sizes of the neutral and ionized versions of a few elements are listed in Table 6.1. The Figure 6.2 is of one plane, but this same chessboard-type pattern is replicated in the third dimension in a crystal such as NaCl (salt). One consequence is that crystals of NaCl grow with cube-shaped facets. The atomic packing of crystals is frequently reflected by the natural crystalline shapes, as for pyramid facets of calcite and amethyst, star patterns in sapphire, or cubic shapes for salt crystals. The differences in structure are always detectable from X-ray and other crystallographic methods, even when other properties are similar.

Atoms and ions do not have an exact size since the effective average size depends on the chemistry of the neighbours. In reality electron orbits are diffuse, so 'radii' merely indicates the most probable distance from the nucleus. The table shows some values of this variation for positive and negative ions. The silicon examples emphasize the wide scale of possible ion sizes between positively (4+), and negatively (4-), charged versions of the same element. Inclusion of a very heavy element, uranium, indicates that, although the nuclear mass is much greater than

Table 6.1 **Examples of nominal ionic radii.** The effective radius of an atom, or ion, depends on the chemical interaction with its neighbours. The examples here indicate the spread which is possible and, more importantly, how the ion size varies with how many electrons have been added or lost. The units are in nanometres (billionth of a metre, 10^{-9}).

Element	Neutral atom	Ion	Apparent ionic radii
Lithium	Li	Li^+	0.059 to 0.092
Sodium	Na	Na^+	0.099 to 0.139
Potassium	K	K^+	0.133 to 0.164
Caesium	Cs	Cs^+	0.169 to 0.188
Magnesium	Mg	Mg^{2+}	0.057 to 0.089
Barium	Ba	Ba^{2+}	0.135 to 0.161
Aluminium	Al	Al^{3+}	0.039 to 0.054
Hydrogen	H	H^-	0.208
Fluorine	F	F^-	0.133
Chlorine	Cl	Cl^-	0.181
Iodine	I	I^-	0.220
Oxygen	O	O^{2-}	0.0132
Silicon	Si	Si^{4+}	0.041
		Si^{4-}	0.271
Uranium	U	U^{6+}	0.045 to 0.086

for the other examples, this does not have much influence on the size of the outermost electronic orbits surrounding the heavy nucleus. Density depends on the tiny central nucleus, and metal densities of lithium to uranium span a range from 0.53 to 19 grams/cc.

Pictures of the relative positions of ions in a crystalline model are normally made as though ions are spherical instead of the reality of deformable electronic shells. Nevertheless, it helps to imagine what might happen as atoms are packed together. For glass, which is primarily based on silicon dioxide (variously called silica for the glassy version, and quartz for the crystalline one), Figure 6.3 sketches the ionic arrangement. There is a pyramid of oxygen ions, with a silicon ion trapped in the middle of them. The size of ions depends on their charge state. A neutral oxygen atom (i.e. O^0) will increase once it gains extra electrons and becomes a negative oxygen ion (O^{2-}). The oxygen ions are large and the silicon ion is small. In units of nanometres (nm, one billionth of a metre) the radius of the O^{2-} ion increased as it gained two extra electrons and it is 0.132 nm. By contrast, the positively charged silicon Si^{4+} has lost four electrons, and it has shrunk to just 0.041nm radius. In volume terms, this is like a tennis ball trapped at the centre of four footballs.

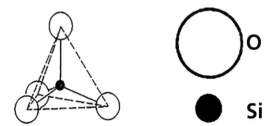

Figure 6.3 The basic unit of both quartz (crystals) and silica (glass) is shown in which the silicon (Si^{4+}) is a small ion surrounded by a pyramid of four larger oxygen ions (O^{2-}). Note that they are ions, not neutral atoms as, during the bonding, electrons move from the silicon (so it becomes smaller and positively charged) and the electrons are added to the neighbouring oxygen ions (which increase in size and are negatively charged). The pyramid is sketched in terms of the positions of the nuclei as, if one includes the electrons, the silicon ion would be hidden in the middle of the oxygens. The ion sizes are approximately as sketched, but are not spherical.

An historic oddity of physics texts and lectures is to cite ionic radii, because we draw two-dimensional pictures of circular ions. In reality, ions are slightly deformable three-dimensional objects. The key factor for mixing ions together in a solid is not their radii, but their volume. The ratio of oxygen to silicon ion radii is $0.132/0.041 = 3.22$, whereas the ratio of their volumes is $(3.22)^3$, which is far more dramatic at almost 33 times. Once we recognize such facts, we have a totally new intuition about the design of molecules and crystals. We are then far ahead of many science students (and many scientists and lecturers) by doing so.

Chemical notation

The chemical formula reflects the relative number of atoms within the building block of a compound. In inorganic chemistry, there is a metal atom coupled to another element, such as a halogen (e.g. chlorine, Cl), or to a cluster which are behaving as a single unit. Typical familiar clusters include a carbonate (CO_3), a sulphate (SO_4) or a niobate (NbO_3). The relative numbers of such partners, or units, are denoted by subscripts. If there is not a subscript it means just one unit of that component. So salt is sodium chloride, NaCl, (one sodium and one chlorine make the basic molecule); silica is SiO_2 (one silicon and two oxygen); calcite is calcium carbonate, $CaCO_3$; and lithium niobate is $LiNbO_3$. Water is a hydroxide (OH) of hydrogen. Hydrogen can also act like a metal, so it could be written as HOH, but the more familiar shorthand is H_2O.

Chemistry has analogies with the behaviour of people. People associate into small groups or families, chemically these are the molecules. If they gather together in large tribes, or groups of football supporters, they are the solids and liquids. Bonding in small groups produces stability. Bonding between humans can also lead to some stability. As tangible proof, car insurance premiums are higher for single people than they are for those with partners. The premiums reflect the actuarial accident statistics, so confirm which state is more stable.

Part II Glass: the epitome of an imperfect structure

The historical background of glass goes back 4,000 years, so in technological terms it is as ancient as our use of copper or iron. Both metals and glass would have appeared by accident from sand being heated in a fire. Naturally occurring examples of glass are quite rare and the most cited exception is obsidian. This is a hard, dark, volcanic material and obsidian was used as a reflective surface in small mirror beads, and it was valuable in making very sharp tools and arrowheads. The origin of such obsidian arrowheads is of interest to archaeologists. This is because the impurity composition of obsidian glass is not universal, and it will contain characteristic impurities from different areas in the world. These can be identified by shining ultraviolet light (UV), X-rays or a beam of electrons to emit luminescence. Such impurity-controlled differences in 'colour' of the luminescence spectra are excellent news for an archaeologist, as analysis of the obsidian luminescence spectra will identify the source of a particular type of obsidian arrowhead. The most surprising result of such studies is that the obsidians were transported over incredibly large distances at a time when we imagine people were just going from their cave to the local hunting ground. The obsidian data support claims that the early homo sapiens were trading over distances of hundreds of kilometres.

Although metals and glass may have appeared at a similar time, it is obvious that they are two very different types of material. Metals are good conductors of heat and electricity, whereas glasses are poor in both properties. We now term them metals or insulators. The mechanical properties (basically determined by a mixture of the atomic structure and the types of imperfections in the material) are also different for insulators and metals. For example, obsidian, or flints, could be fractured for their use in tools, whereas this would not have been feasible with metals, and is still not. Semiconductors fall midway between these extremes, but were essentially ignored until about 100 years ago. However, since they are the basis of modern electronics they are probably better understood than either metals or insulators.

Glass is so familiar to us in our daily life that we rarely consider it and, because it is clear and transparent, we probably assume it is pure. Most of us will fail to appreciate its importance or be ignorant of the high level of technology that is involved. Understanding the problems of glass manufacture is much easier if we understand even a little basic physics and chemistry (or cookery). Once this is included then the reasons for the colour of various glasses, pottery glazes and enamel will be much clearer. The basic ideas offer some insights into the way glass surfaces are toughened and the science behind the extremely high technology glass products used for optical fibre communication. I scarcely need mention that all these aspects of glass are totally dominated by impurities and other imperfections. Whilst some remain as undesirable defects, the majority can be used to our advantage.

Characteristic glass properties

In terms of their atomic structure, materials tend to fall into two types defined by the way their atoms are packed together. There are very simple units of building blocks, such as the silicon-oxygen combination of Figure 6.3. This same pyramid structure occurs in both the glass (silica) and the crystalline version (quartz). The fundamental difference is how atoms sit relative to their more distant neighbours. In an amorphous glass, the building blocks are arranged in an ordered pattern only over a very short distance, but appear randomly spaced over long distance. By contrast, crystals have both short and long-range order in their layout. (My earlier social analogy was a random crowd (glass) or a military parade (crystal)). Figure 6.4 sketches the arrangements of building blocks in the two alternatives in just one flat plane, but randomness, or ordering, extends into the third dimension as well. More closely packed crystals tend to be denser than their glass variants, The SiO_2 of quartz crystals is approximately 10% denser than silica glass.

Figure 6.4 Crystalline and amorphous patterns. In the crystal there is an ordered pattern and the average density of atoms is slightly higher than in the glass. When there are larger building blocks, as in silica, the local arrangement between the neighbouring silicon and oxygen bonding is very similar, but the separation of the groups is random over longer distances and there can be variations, even in the local density. Variations will depend on how the glass was made or subsequently heat treated.

Materials used for glass production

Most familiar types of glass are made from high quality sands that are mainly composed of particles of quartz, and related silicate compounds. Very white sands exist that are mostly quartz and these are favoured by high-grade window glass makers, as they give clearer material. As for a SiO_2 molecule, in a glass every silicon is sitting in the middle of a pyramid of oxygen ions. Silica glass only softens and melts at a very high temperature, but experimental cookery with additions of other compounds lower the melting point. Some of these dopants result in colouration of the glass, and others are needed to remove the unwanted colouring.

In nature there are very many types of silicate based materials, with a common group of minerals called feldspars. These are a mixture of silica with other elements, such as aluminium, sodium and potassium, and therefore are called aluminosilicates. Feldspars are one of the most common minerals on the surface of the earth.

Additives not only change the melting temperature, but also control optical, mechanical, and chemical properties. Various metals give a range of colours. Materials incorporated by accident in window and household glass include iron from the original minerals, and chemicals from cullet (recycled broken glass). Another impurity is tin from the production process of making very flat window glass via the 'float' process. To emphasize the variations, Table 6.2 lists some typical compositions.

The bulk composition and the outer surfaces are often very different (just like the edge and the bulk of a Samurai sword) and some changes occur by chemical reactions with the atmosphere, or are added deliberately to give a coloured glass. This is a positive usage, but equally chemicals, such as the salts used in dishwasher compounds, can attack and cloud some types of glassware. Compositions have varied greatly throughout history and, in archaeological terms, contaminants and compositions can pinpoint the location of a source of material. This is valuable for uses as different as dating the age of Roman artefacts, or deciding on the original location of a modern wine bottle. The methods can distinguish between genuine high price vintage wines and those that have been falsely labelled from regions where they were not manufactured. The benefit of the method is that the surface of the glass bottle can be analysed without interfering with the wine (e.g. with X-ray excitation).

Table 6.2 Chemical composition of simple types of glass. The table indicates the weight percentage of typical components of some familiar types of glass. The details are not important, but the length of the list indicates that many component ions are involved.

Material	Formula	% by weight		
		Modern window glass	Older glasses	Glass containers
Silica/quartz	SiO_2	66 to 74	variable	74
Sodium oxide	NaO_2	14		13
Calcium oxide	CaO	9	15 to 20	10
Magnesium oxide	MgO	4		0.2
Sulphates	SO_3	0.2 to 0.3		0.2
Iron oxide	Fe_2O_3	0.1		0.04
Potassium oxide	KO_2	0.02	Up to 10	0.3
Lead oxide	PbO		25 to 35	
Aluminium oxide	Al_2O_3	0.15		1
Titanium dioxide	TiO_2	0.02		0.01

As is apparent from the production of enamels and pottery glaze, the final colour of the glass layer depends not only on the atmosphere used during the kiln process, but also on the temperature and the rate of cooling. This is because the elements (e.g. iron, manganese etc.) can exist as ions with different numbers of electrons involved in the chemical bonding. Heating establishes different chemical compounds from the same mixture of materials, as electrons transfer between ions, and the structures relax to form new compounds. These changes define the colour of a pottery glaze.

Far less publicized is that heat treatments are routinely used to alter the colour and boost the value of minerals and gem stones. Understanding the result is simple, but predicting the outcome and controlling it in a commercial process can be very challenging, and invariably the normal optimization of the manufacturing process is by experience, rather than theory.

Tricks needed to make a glass

To make a glass we need to melt the components together. This is certainly not easy as quartz melts around 1,722 °C (about 3,131 °F). If the melt is then *slowly* cooled, the little tetrahedral building blocks have time to shuffle around to all fit together neatly in an ordered arrangement. This, by definition, is a crystalline structure. There is more than one way the molecules can pack together (Table 6.3) and the first structure formed on 'freezing' (!) is called cristobalite. On further cooling it switches into another ordered crystalline packing at 1,470 °C (trydimite), then yet again at 867 °C to a new structure labelled β-quartz. Finally, on cooling to 573 °C it goes into the α-quartz crystal structure, which is the version we normally handle at room temperature. However, if the liquid is cooled too quickly there will only be short-range structure at the scale of a few basic pyramids, as the building blocks do not have enough time to properly line up. They are randomly oriented, so there is no long-range order. This version is silica glass. Since there are several variants of the crystalline arrangements which are stable in different temperature ranges, it will not be surprising if the glass version has a mixture of the possible local silicon-oxygen possibilities. Scientists are not as neat and tidy as one would wish, as even in different parts of Europe some countries call the crystal 'quartz' and the amorphous material 'silica', whereas others use phrases of 'crystalline silica' or 'quartz glass'. Often users are very careless and one may have to guess which material they mean.

The examples in the table are for crystals which form under normal atmospheric pressure. Increasing the pressure leads to very different crystal formations of SiO_2 that are not totally stable, but can eventually convert to normal quartz (i.e. they are metastable). The first of these examples is termed Coesite and it forms at temperatures above 700 °C and under high pressures above 2 or 3 Gigapascals (GPa). The second is Stishovite and is named after a Russian scientist who fabricated it in the laboratory prior to it being discovered in nature. Stishovite is an extremely dense version of quartz that forms above 1,200 °C and under pressures of more than 10 GPa. Since we probably have no intuitive feel for a Gigapascal, consider it as a pressure of 10,000 atmospheres

Table 6.3 Typical phases of crystalline silicon dioxide during cooling.

Temperature °C	Structural phase name
Above 1,722	Liquid
1,722 to 1,470	Crystalline cristobalite
1,470 to 867	Trydimite
867 to 573	β-quartz
Below 573	α-quartz

(i.e. about 70 tons on the area of a small matchbox). Both minerals are formed under the conditions of high temperature and pressure occurring during meteor impacts, and so appear in meteor craters, and in extrusions of rock from deep within the earth's crust.

From treacle to sticky glass

Liquids have effectively the same arrangement of short-range packing of molecules, but no long-range order. Hence modern descriptions consider glasses as very viscous liquids. Once again, we will lack any feeling for the numbers used to describe the relative viscosity of materials but, to offer a crude comparison, water would be about 100 units, treacle somewhat more, and glasses at room temperature maybe at least 10^{12} times larger (i.e. more than a million million times stickier). I lack a feel for numbers which are so diverse, but at least I appreciate that room temperature glass does not flow very well.

Most of the glasses we are familiar with, such as window glass, soften by about 500 °C (a red heat) and then can be moulded and shaped. At that point their viscosity is around 10,000 to a million units. (Old units were called 'poise' whereas the modern naming is Pascal-second.) At room temperature, glass is so viscous that it does not flow under gravity at a rate that we can visually appreciate, but with skill it is just about measurable. At room temperature, window glass is many million times more viscous than liquid glass.

There has been an unusual viscosity experiment in progress for the last ninety years, since 1930, to measure the viscosity of pitch. Pitch is a naturally occurring black tar, which in many ways behaves like a brittle glass. An experiment set up by Thomas Parnell in Australia has been recording how fast and how many drops of pitch fall from a funnel. The white heat of excitement was when the eighth drop fell in November 2000 and the ninth in 2014. Fascinating experiment, but not an ideal one for a doctoral thesis! An even longer-term experiment on slowly flowing materials was started by Lord Kelvin in the nineteenth century and it continued for nearly a century.

Glasses with different compositions can be more or less viscous. Some people suggest that very old church window glass sags with time, as does the lead in leaded glass windows. This is unlikely, but glass compositions are so variable that it might be feasible. Our view of permanence and stability is coloured by our short life span. On a grander time scale the rocks and mountains, which look so fixed to us, will steadily flow, distort and show obvious folds. This wider view has been adopted by the American Society of Rheology as 'panta rhei' (i.e. 'everything flows').

What use is a very viscous liquid?

Since, in many respects, the methods used today still have similarities with those used a few thousand years ago by early glass makers, the development of glass technologies has not caught the public imagination in the same way that has happened with semiconductors. For my objectives, a focus on glasses is ideal as nearly every property of interest and application is squarely based on some attempts to control impurities and imperfections in the glasses. To justify this, before I meander through some of the modern uses of glass products, I want to very briefly underline the diversity of products that now exist. As from the earliest times coloured glass and glass beads are attractive and one cannot minimize this appeal. Gifts of beads have worked, and not only for early cave men looking for someone to keep their cave warm. Later explorers negotiated land transfers, and gifts of gold, in exchange for glass beads that have shaped the destiny of many countries through development (or invasion). Glass beads have also been fashionable as an alternative to the making of woollen tapestry. In hindsight, bead work is preferable to the woollen tapestry as the colours of the glass have not faded.

Coloured glass has been developed for uses in tableware, ornaments and windows. Such materials are cited as examples of advanced cultural progress with an abundance of references to stained glass windows in churches, decorations on the exterior of many types of temples and houses, as well as drinking vessels, vases and many other objects. Figure 6.5 shows some familiar examples of glass products and their decorative features. Tiffany lampshades are widely collected and examples of chemically etched Lalique glass can command immense sums of many thousands of dollars. Our attraction to coloured or etched glass has definitely not gone out of fashion, and instead such articles are viewed as valuable antiques.

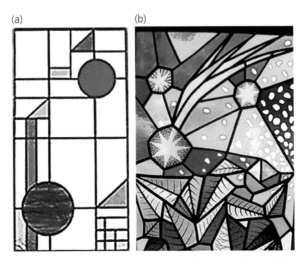

Figure 6.5 Examples of modern coloured glass windows. These are two glass windows produced by Werner Haas (a renowned Swiss glass artist). He has employed several of the glassmaker's skills to make images and pictures. He uses not only bulk colour, but also layered glass so the colour layer can be chemically etched to reveal the clear underlay. This is very obvious in the snowflakes and clouds in the right-hand image, which is part of a winter scene in a panel depicting the four seasons. The windows include shading and fine lines formed by chemical paints which are then heated and diffused into the glass, both in bulk zones and in fine lines etched into the glass. Surface texture on the glass (as in the mauve disc on the left-hand image) gives further depth and liveliness to the windows with sunlight coming through them.

In all these decorative examples, a major skill is to add colour in a controlled fashion. This is not easy and, in order to help keep their proprietary manufacturing practices confidential the thirteenth century, Venetian glass makers moved to the island of Murano. Basically, the options for colours include adding impurity ions to colour the entire glass uniformly, to add metal droplets which produce colour by an alternative physical process, or to make layer sandwiches of different coloured glass which could be variously cut back to give cameo type surfaces. A different approach is to use surface treatments or coatings which add colour or other special properties. The same methods are used in making glass glazes on pottery and in enamelling of metals. Many people do not even associate such glaze or enamel surface finishes as examples of glass technology. I suspect glass is more important to us than we might first realize.

Impurities and glass colouring

Impurity additions in glass melting are essential in order to lower the melting temperature of the mixture, but in all cases the impurities add optical absorption. Each impurity will preferentially absorb light in a particular part of the spectrum of the incident light. It will therefore change the colour of the transmitted light, and similarly modify the reflected light. Our visible spectral range is only a very small part of the spectrum that we can measure electronically, and not all colouring effects appear within our optical view. Pure quartz and silica are transparent over a far wider spectral range than is the case for window glass. The quartz/silica materials are excellent windows, even for short wavelength ultraviolet light. They absorb virtually none of the sun's visible light spectrum, and the only losses are by reflections at the surfaces (and any surface dirt). By contrast, once silica is doped into the form of the multi-component silicate window glass, the material strongly absorbs the UV light. This is the reason that it is not possible to gain a suntan when sitting inside a room with normal glazing. Double glazing is both thick enough, and has coatings on the surfaces, that it not only removes UV light, but it even absorbs the blue tail of the visible spectrum. Relative to window glass, silica is extremely transparent in the very limited spectral region which we can see, as humans.

Our vision is not the same as that of many other creatures. Many insects and birds have far better UV vision than we do. For them this is useful for finding pollen in flowers and it may be a factor in choosing a mate. For example, starlings are very varied and colourful when viewed with detectors that record UV colours. Less obvious to us is that the view of the world for a caged bird, or in a glass aviary, where UV has been blocked, will be very 'coloured' compared with their real-world image. The equivalent distortion might be for us to live in a house with windows that block blue and green light. Unfortunately, few people who keep caged birds, or build large glass aviaries, realize this.

Glass impurities have different colour effects and, for windows, we do not mind if there is boron or sodium in the glass, as their absorption effects are not important for visible light. Inclusion of iron is more of a problem as it tends to give glass a green tinge. Iron is often an impurity found in sand, so it can contaminate the original glass. It may also enter the float window glass, as some of the 10% of the material added into the furnace is cullet. Cullet is broken glass and recycled material. It includes the trimmings of the float production line edges that were pulled by the rollers. If recycled glass has included green bottle glass, then iron will have been in the melt. The absorption from the iron impurities is not serious when viewing the world through a few millimetres of window glass, but if you look along the length of a glass window there is a very definite green colour to it. Even for a few millimetres of thickness the iron introduces ultraviolet absorption, which is beneficial in blocking light that would cause fading of fabrics and pictures. This is a bonus feature by default, not by intent.

To add strong colours in the visible part of the spectrum, a number of impurity ions have become fashionable, both in glass as well as in pottery and glazes on ceramics. Impurity colouration depends in part on the choice of ion, or ions, the total concentration of impurities in the layer and, quite critically, on the way the material was thermally treated. This is because the chemistry of the impurity ions can be altered to bond in different ways into the glass matrix. This is not easy for glass makers and even more difficult for glazes on pottery, where several layers can be applied sequentially. After each addition the pot has to be re-fired at a different temperature. The temperatures which cause one colour, because of the charge state of the ions (e.g. an impurity sitting as a 2+ or 3+ charge state), may result in the wrong charge state for other ions. The sequence of temperature processing with glazes is a skilled art. Even from good manufacturers there are many 'seconds' in the products of a kiln because reprocessing steps have altered the colour or shade of the tones of glazes in undesirable and different ways.

Unfortunately adding one impurity ion and processing it to a specific temperature does not give a unique colour. There are physical and chemical interactions between *all* the components of the glass so the end result is complicated, and difficult to predict theoretically. This is a classic case where controlling the imperfections is essentially only possible from the expertise gained from earlier empirical (experimental) results. The glass maker has a role like the conductor of an orchestra who knows how the performers sound individually, but when all are added together, they do not always make the sound that was intended.

It is, therefore, important in the choice of the impurities that are used to colour glass to recognize that their chemistry depends both on their ion size, and the other ions present in the glass. For example, in borosilicate glasses (i.e. boron and silicates), additions of cobalt will give a pink colour. This is because in a borosilicate the cobalt is surrounded by six neighbouring oxygen ions. This defines the chemistry and the directions and strengths of the electronic bonds between the cobalt and the oxygen ions. The packing geometry is influenced by the size of the boron ions in the borosilicate glass. Boron is a very small ion which allows an oxygen packing arrangement where cobalt finishes up in a building block of CoO_6. By contrast, a cobalt impurity in a potash glass (one which contains the larger potassium ions) sits at local sites where there are only four oxygen neighbours. This change in chemistry results in different electron energy levels for the cobalt ions and there is then a mixture of CoO_6 and CoO_4 groups. Overall, the changes in the energy levels give a purple colour to the glass.

My simplistic overview, of merely listing factors which cause trouble, should be easy to understand, but for the glass maker the problem is extremely complex and one for which it is hard to make useful predictions, so experience and empirical data are essential. Detailed chemistry and quantum mechanics may be fascinating, and probably will be helpful in hindsight, but rarely in predictions of new treatments.

Similar problems occur with additions of copper. The copper can generate a range of colours from a blue in soda-lime silicate glasses, to green for the same concentration of copper in borosilicate glasses. The same type of colour variations may be familiar to us from school chemistry or geology. The copper in copper sulphate crystals is an intense blue, whereas in malachite the mineral has a strong green colour. Many metal impurities are, like copper, able to bond in different ways in a glass (chemically this is described as a change in the valence state, or number of outer shell electrons). In, say, a soda-lime glass, Mn^{2+} gives a deep purple colour, but Mn^{3+} gives a faint shade of pink. Chromium doping can go from green when sitting as Cr^{3+} to yellow as Cr^{6+}. Yet a further option occurs in a crystalline aluminium oxide host. Here the chromium Cr^{3+} replaces the aluminium Al^{3+} ions and slightly distorts the lattice. The new chromium energy levels absorb both blue and green light, so the crystal becomes a familiar red colour, and with this impurity doping it is called ruby.

This may seem an unnecessary comment but I have heard people say that ruby is *made* red by the chromium. This is poor phrasing as the chromium ions absorb both blue and green light, and the only signals easily transmitted are the residual red parts of the spectrum that has been provided by background lighting. However, when we look at luminescence of ruby the situation changes and the higher energies of the absorbed light (UV, blue, green) *produce* the low energy red light. Indeed, the first type of laser was made with ruby precisely by this type of excitation.

Copper behaves quite variably in glass to produce no colour when bonded as Cu^+ ions, but a blue colour when as Cu^{2+}. Glasses with iron can vary between blue-green, yellow or dark brown. Amber and brown, as seen in beer bottles normally, are the result of a mixture of iron, sulphur and some carbon. The skill and art of glass colouration is challenging and totally dependent on the impurities and imperfections, often in ways not totally appreciated even by the glass craftsmen.

Particle inclusions in glass

A further complexity arises when the added impurities are in too high a concentration to dissolve into the glass, or their size or chemical misfit precludes them bonding into the glass network. In these cases, they drop out of solution and form little metallic or other chemical clusters. These clusters are now termed 'nanoparticles'. Metal particles, such as gold or copper, are particularly effective at producing strong colours in the red and are the basis of the early 'ruby glass' colouration, as originally made in Venice.

For those interested in the detailed physics, there is a major difference in the mechanism of separated ion impurities and the clusters. For separated individual impurity ions, the absorption occurs as the energy of the light at a particular wavelength (colour) lifts electrons between possible energy levels. These energy levels, as in a staircase, are well defined and localized by a particular impurity site. Incoming light just lifts electrons from the lower to upper levels and the energy of the light is only absorbed if it matches the height of the step. It selectively absorbs specific colours from a white light source. By contrast, the mechanism of light absorption for metallic nanoparticles is very much a team event. In the case of the metallic nanoparticle, all the outer electrons of the particle are interacting. The energy jump is not localized for one electron, but is a feature of the entire system. Physicists call this process a 'plasmon absorption' and it means that the frequency of the light matches a collective electron resonance oscillation of the metal particles.

In the nineteenth century there was a great fashion to have a yellow/green-type glass. It was interesting because the colour never seemed as simple as for normal coloured and stained glass. Indeed, even at night with no illumination, the glass would still faintly glow a yellow/green colour. The key impurity which gave both effects was from inclusion of uranium in the glass. At that time radioactivity had not been discovered, so no harmful effects were considered. Many people still like the appearance of this Victorian yellow/green glass. I have a German friend who, as a student, had a supervisor who was frequently unwell and used to lie on a settee in his office to recuperate. My friend happened to pass his office with a Geiger counter and picked up a strong signal, which she traced to the supervisor's collection of Victorian green uranium glass. It was stored under his settee. She never subsequently sat on the settee!

Part III Imagine life without glass

Instead of considering glass as a low technology product, we should ask ourselves a different question as to how our world would change if we did not have glass. From this perspective it suddenly becomes far more important than semiconductor electronics. The more obvious consequences of no glass products would mean no windows, spectacles, microscopes, telescopes, drinking and kitchen glass, electric light bulbs or vacuum valves. Effectively no civilization as we know it. Removing silicon electronics would set us back seventy years, but a loss of glass products would set us back several thousand years.

Where glass is used for optical purposes, such as spectacles, microscopes or camera lenses, many compositions are used. For example, dense glasses have a high refractive index which allows designs with a thinner glass for spectacles than is possible with low index glass type compositions, or from plastic. High index glass is therefore particularly valuable in production of very strong spectacle lenses or in microscope objectives etc. Many lenses use combinations of different glass lenses to correct for the problems of colour aberrations that are unavoidable for a single glass lens element. High grade camera, telephoto and telecentric lenses are all constructed with a number of lenses of different glass composition.

A less obvious variation which emerges from inclusion of materials is the possibility to make photochromic glasses. In this case the transmission of the glass is reduced by the UV content of sunlight. The photochromic effect is ideal, not only for self adjusting sunglasses, but even in toys such as self-tanning dolls. There is an anecdote that the development of such glass was funded in an attempt to protect soldiers from the flash of a nuclear explosion. As anyone who uses photochromic spectacles will realize, they are quite slow to respond, so would have been useless for that application, but the military money has resulted in a valuable product.

A faster, but achievable, example of automatic self-darkening glasses is needed by people working with welding systems. They have the obvious problem of wishing to see what they are doing under normal lighting conditions when they align the components and the welding arc, but need an instant reduction in intensity as soon as the arc is struck, so that they are not blinded by the intense light. Photochromic glass compositions cannot cope with such extremes of intensity nor such a rapidly fast time scale. So modern designs use a glass material which responds to an electrically driven mechanism to change the transmission of the welding goggles (this is termed an 'electrochromic' glass). Similar electrochromic layers have been put on glass windows to achieve night time privacy without the need for curtains. The addition of indium to glass is critical in making transparent, but conductive, windows for smart phones and tablets with touch screens. Very few dopant ions offer both transparency and electrical conduction. Indium is a key industrial material but, unfortunately, it is relatively rare and with the current production rates of conductive touch screens the supply may well be exhausted.

Finally, the mechanical and chemical weaknesses of glass can be manipulated by carefully controlled surface cooling and other surface treatments. Hence modern safety glasses, as used in car windscreens (or kitchen glassware that bounce when dropped), exploit these surface stress 'defects' of the glass. Somewhat less familiar is the use of glass bonded to other material to add additional strength or specialist properties. Glass composites were used in early car window glasses to minimize fragments of glass during an accident. Initially the glass fragments stayed glued to a plastic layer. Modern glass/plastic composites are bonded differently and have more exotic properties. These materials work not only for car windows, but also in the production of armour. In somewhat specialized car windows, I have seen not just simple bullet proof glass, but also one-way bullet proofing. The composite material has a glass outer layer and a softer plastic interior. An incoming bullet loses energy in shattering the glass and the plastic zone flexes and absorbs the energy of the fragments. This blocks incoming missiles. Bullets fired outwards easily go through the plastic, explode the glass outwards, and carry on their trajectory towards a target. This allows people to fire through the glass from the inside towards the attackers. One-way bullet transmission glass may not be everyone's priority for their car, but apparently there is a lucrative and substantial market. We have come a long way from making glass beads.

The good and the bad news of stresses in cooling of glass

Controlled heating and cooling of glass is definitely an art as, not only can the material make a nice uniform transparent material but, under unfavourable conditions, it may cool faster in some regions than others, or even start to grow crystallites. For transparent window glass both effects are bad news, not only because they scatter light, but also because there are stresses set up in the glass. Inbuilt stress influences how it will fracture. By now it will be obvious that, if we can identify and control an imperfection in a material, then there is always some situation where we can exploit it. This is certainly true of stress effects in the surface of a glass which cools more rapidly than the interior. Fast cooling of the surface, often made with jets of cool air on the hot glass object, causes a fast contraction which compresses the surface layer. This surface skin toughens the glass and makes it less likely to break. It was also used on lightweight bottles.

The science is that normal glass has a limited range over which it can be stretched before it breaks. So small changes of surface and volume distortions from dropping it cause fracture. Less obvious is that glass is quite elastic once it is compressed. If we can make a compressed skin on the glass there is a bigger elastic range for it to relax and flex if it is dropped (or hit) before it reaches a breaking point. Surface compression effectively gives the glass a range of elasticity before it breaks. There are many examples of surface-compressed kitchen and glass tableware which now exist that usually survive being dropped. However, if they do fracture the result is spectacular as the object breaks into a multitude of tiny fragments. A minor danger of using such glass is that we can become careless with normal, unstressed glass objects. It may not be realized that microwave cooking with toughened glass vessels can modify the glass. The microwave energy can allow chemical bond relaxation within the glass, and so the benefits of the surface toughness and resilience are undermined. I have a set of glass bowls that were initially circular, which have deformed as the result of use in a microwave oven. This relaxation process is taking place at a very low temperature (i.e. during cooking of food) compared with softening in a furnace at a red heat. Instead, the microwave energy is coupling directly into the lattice bonds of the glass.

Stress effects seen with polarized light

A more modern example of controlled surface stresses is obvious in car windows, where a pattern of air jets non-uniformly cool the windscreen during the shaping phase. This sets up a stress pattern which under most conditions is not visible to us. However, if you drive a car wearing polarized sunglasses then the stress pattern is revealed as the stress fields also polarize light that is transmitted. Being aware of the pattern is irritating, but the stress pattern means that in a major accident the glass fractures in small pieces rather than as sharp shards. Estimating the number of serious accidents avoided by glass damage to passengers is difficult, but certainly the numbers run into millions.

The stress effects may be visible even without the need for polarized sunglasses if the incoming light has been polarized in some way. This happens with light reflected off a wet road or the sea. It is obvious if we are looking in the direction of the source, such as the sun. Light energy comes in packages called photons, which can either be described in terms of waves or particles. We can imagine that the photons are equivalent to flat stone discs and the polarization axis is along the plane of the disc. When sunlight hits the sea the flat discs parallel to the surface bounce (just like bouncing stones across a pond), by contrast the stone discs which have their axis perpendicular to the water crash into it and sink. Reflections from the sea surface change the light from being unpolarized (equal numbers of discs randomly aligned, including vertically and horizontally), to a set of photons polarized in one plane. Polarized sunglasses use stretched molecules so only one axis of the light passes through. When the transmission axis is for light that is vertically polarized they reject the glare of light bounced off the sea or the beach.

We may assume that using polarized light is some recently exploited effect. This is not so. Not least because it is part of the detection system used in navigation by bees. Polarization is equally obvious if we look at sunlight reflected off a polished metal surface. An historical example that exploited this was to find the direction of the sun in misty, overcast weather. In conditions when the direction of the sun was not obvious, it is claimed that Viking sailors looked at the intensity of light reflected off a knife blade. Even in overcast and misty conditions, when facing the sun, the polarization effect is much more obvious than when looking at light from other directions. Viking sailors also used crystals of Iceland Spar (a pure form of calcite, $CaCO_3$) as the crystallography gives it two refractive indices which split different polarizations into two displaced images.

If you have polarized sunglasses, try looking at your computer screen and rotate the glasses. Polarization effects are then extremely obvious. Here on my computer there is zero transmission when I rotate the sunglass axis by 45 degrees in one direction, and a maximum intensity in the opposite 45 degree direction. A less obvious, but fascinating example, is to view a rainbow and rotate the polarization axis of the glasses. The polarizers reveal new rainbow features, and because it rejects some of the normal background light, it makes the rainbow far more obvious. It therefore helps to see the secondary bow at a higher angle with a reversed set of colours.

Has glass making progressed in 4,000 years?

The answer to this question must be a definitive yes and no. Certainly in the technology of furnaces and temperature control, modern equipment is excellent. There is more reproducibility in mass products and the underlying physics and chemistry are well appreciated. Nevertheless, the scale of understanding such a multi-parameter problem with strong interactions between the detailed heating cycles and glass compositions is, basically, extremely hard. Despite 4,000 years of production the expertise relies on experience. The earlier analogy with cooking is valid as we enjoy the food even if the chef does not understand the role of all the ingredients, or the chemistry of the reactions in the glass cookery.

Surprisingly, I will include a comment on cosmology. Experience with glass making is that we can assemble billions of atoms to produce a semi-randomly ordered array of silicate, or other building blocks, and do so to have a consistent and nominally uniform product in terms of composition and density. Despite this, we also know that more detailed inspection will show there are localized density and stress variations. For air-jet-cooled car windows these are large zones, but even in the most careful manufacture there are still variations at the microscopic level. The density and composition changes occur over short-range order and reflect the many different crystalline or packing alternatives that could be formed from the starting products. I quoted examples of SiO_2 in the forms of silica, quartz, coesite and Stishovite, where the densities range through 2.2, 2.65 and 3 to more than 4.28. Stishovite is more than twice as dense as our familiar versions of silica and quartz, despite having the same composition. Local inclusions in the glass can potentially trigger a wide range of local densities that depend on the processing history. Although there is clear experimental evidence, we tend to happily ignore this reality and say we have a uniform randomly ordered glass system.

Part IV   Some digressions from materials science

Imperfections are unequivocally central to materials science, but I thought it might be useful to realize they are likely to be just as important in other disciplines. Therefore, I will offer two potential examples.

Perhaps this tangible evidence for glass localized density variations should be considered by cosmologists who are trying to understand the properties and formation of the entire Universe. In their models they assume a constant average density and composition of the cosmos. They currently have severe problems to explain the apparently increasing rate of expansion of the Universe, but they lack a suitable driving force. To this end they introduce a new quantity which is termed 'dark matter'.

Beware. Like the Big Bang, it is a theory and, although it may be broadly correct, the trend is that after much publicity and twenty or so years of teaching, it is then discussed as being a proven fact. Similarly, dark matter is a highly fashionable topic with very large research efforts and lots of funding (so it attracts extremely brilliant scientists who are good at raising even more funds), all seeking this mythical material and the associated prizes for the discoverer. Some recent critics have suggested that if the Universe is less uniform than imagined, and we happen to exist in a region of somewhat lower density, then some of the problems go away, and dark matter may not be needed. My instinct is to see some parallels between the billions of atoms in a 'uniform' glass and the billions of stars and galaxies in the 'uniform' Universe. The empirical and quantitative data for glass unequivocally says that glass only has a constant density on a large scale and we can always find zones of locally reduced, or increased, density. Perhaps our local region of space is not at the overall average density. If so then the analogy for cosmology may imply dark matter is just a passing concept, and in the same category as many of the earlier all-pervading proposals for matter, such as phlogiston, or the ether, that were once assumed to exist throughout the Universe. They were useful ideas in their time but gently faded away with more experiments and greater understanding.

Imperfections within the stars

The chapter started with brief comments on how atoms are constructed. On Earth this is sensible, but many atoms in the Universe cannot survive as they are exposed to the immense pressures and temperatures that are found within stars. Indeed, stars are the source of all our elements, with lighter ones aggregating into heavier nuclei. However, the heaviest nuclei may only originate in massive or cataclysmic stellar events. When discussing atoms, we focus on a tiny nucleus at the centre of any atom which include protons with a positive charge (written as p^+). To make a complete neutral atom there are very low mass electrons (e^-) orbiting around this core. For convenience at an introductory level we can think of electrons as planets orbiting the sun, but in reality, their movements are far more complex and are more appropriately discussed in terms of waves. The simplest atom is a neutrally charged hydrogen (p^+ + e^-). There are also particles of similar mass to the proton which have no charge, called neutrons (n^0). The hydrogen nucleus can bind to one or two of these neutrons and these different versions (isotopes) are called deuterium ($p^+ + n^0$), or tritium ($p^+ + 2n^0$). The tritium is unstable and has a limited survival time with a half life of around 12.3 years.

For all the other elements neutrons are essential, as attempting to make a nucleus of two protons is impossible because the electrostatic repulsion between them would break them apart. However, gravitational attraction and other nuclear forces between particles provide bonding. Hence adding neutrons allows the construction of heavier nuclei. The next lightest nucleus is helium three ($2p^+ + n^0$). In all other elements with stable isotopes there are at least as many neutrons as protons, with more needed for heavier elements such as uranium. Away from the Earth, in the core of stars such as the sun, their immense mass means the pressure at the core can be greater than 250 billion atmospheres! Under these extreme conditions, atoms cannot survive, and nuclei directly interact with one another. For example, a hydrogen nucleus can combine with a helium three nucleus, to form a conventional helium four nucleus. However, this would be an imperfection in terms of charge, and so it requires conversion of a proton into a neutron plus some major associated nuclear reactions. The key consequence of the combination converts excess mass into energy according to the famous expression $E = mc^2$, where c is the velocity of light. This simple example is thought to generate about 40% of the core energy in our sun, and produces an inner temperature towards 7 million °C. An alternative process is the fusion of four protons (hydrogen nuclei) into a helium nucleus ($2p^+ + 2n^0$), which we also call an alpha particle. Various other nuclear fragments (positrons and neutrinos) are released in the event, as well as enormous quantities of thermal energy.

Our sun is a modest star but, between the central core of intense nuclear activity and the cooler surface where 'familiar' atoms can exist, there is a confused mixture of vastly different conditions. This implies that there are regions where neither our familiar science nor the zone of pure nuclear physics exist. Hence a region full of imperfections and complexity in terms of structure and dominant science.

7

Examples of new glass technologies

Part I Why glass has become high technology

Glass manufacture in the twentieth century included a real leap forward with, firstly, the manufacture of float glass, and then optical fibres for communications. Float glass gave us very large flat windows and revolutionized building construction and architectural styles. It was also the enabling technology for the widespread use of double glazing and sophisticated heat control windows. Impressive though these advances may be, they are still improving. One only needs to look at the Shard building to realize that glass can be a highly effective cladding material, on one of the tallest buildings in London. Other uses range from housewares and decorative figures, and all benefit from the imperfections that colour and stabilize the materials. In Kew Gardens, in London, there has been an exhibition by Dale Chihuly emphasizing what is now feasible in large scale glass sculpture.

With iron and bronze, outer layers react with the atmosphere and form iron rust or a bronze patina. Although far less visible, glass surfaces can also react with the atmosphere and current advances address the difficulties linked to problems of modifying and stabilizing glass surfaces, including water vapour, rain and all the other air borne chemicals and pollutants that come into contact with it. This is particularly challenging where there is deposition of metal or other materials to give surface colour for architecture, or layers to offer selective reflection or transmission, including climate control, or surface toughness. Any surface layer can be viewed as an imperfection of the bulk. For glass we are at the stage where there is a classic battle between the 'good' and the 'bad' glass surface properties. In this chapter I will give a glimpse of both sides of the character of glass surface treatments. The situation is encouraging and I am confident that better, more stable, surfaces will continue to appear.

The Power of Imperfections. Peter Townsend, Oxford University Press.
© Peter Townsend (2022). DOI: 10.1093/oso/9780192857477.003.0007

Millennia of glass science has resulted in clear and coloured glasses plus skilled cookery in furnaces, subsequently followed by experience and understanding of the chemistry. In principle, developments of iron and steel advanced in an identical fashion, but they were more willingly researched as they often had military applications, and hence financial support. There can be an elitist view that glass is not appropriate for study by high level technologist and theorists. This view has hindered progress both at the research level and in industrial manufacture (e.g. as described in the next chapter for optical fibres).

The good and the bad of float glass technology

A brief description of the float glass process will immediately show both why it is successful and why there is a potential problem of surface contaminants. As with all types of glass, a mixture of materials is included to gain some control over the melting and softening temperatures and/or to try to give some physical or structural stability to the glass. The aim of the float glass process is to create a glass in the form of an extremely wide, flat ribbon. This is a continuous process where pure silicate sands are poured into a molten bath of glass at the same rate as the ribbon of uniform sheet glass is being pulled from it over a small weir by rollers. It is cut into lengths as it flows along. Subsequently it can be further processed in other ways to add bronze or other colour tints for buildings. The viscosity has to be just right to maintain the ability to draw a glass ribbon from the melt.

In addition to fresh white sand, scrap glass from trimming the ribbon and other cullet from recycled material are included. This 'rubbish' glass is economical as it may constitute 10% of the total, but it is limited, as brown or green cullet is not acceptable in clear glass production.

Sheet glass is relatively thin, so there is a need to have a bath of molten glass deep enough that the components can mix, and a mechanism that does not allow it to stick to the furnace, or the 'weir' over which the ribbon is pulled by rollers in a continuous stream. The flowing river of glass moves faster than walking pace. The ingenious solution is to arrange for the glass mixing to take place on top of a bath of molten tin. It allows the glass components to melt and mix together in the furnace and this liquid glass flows smoothly across the tin. The underlying tin gives the glass an extremely flat base, and the furnace is large enough that the thickness is constant.

This all needs careful temperature control so that the glass is sufficiently liquid that it can flow, but with enough viscosity that it can be pulled by rollers. I have seen such factories which run continuously for many months at a time at a rate of, say, 600 tons of glass per day in each furnace with two metre ribbons of glass. If it is intended for surface tinting for buildings, then it is moved to a vacuum chamber where the air pressure is greatly reduced and a metal coating is evaporated on to the surface. Alternatively, it goes to furnaces where it can be bent and toughened for car and other vehicle windscreens and/or other types of processing.

For vehicles, the hot glass of the windscreen etc. can be shaped and then rapidly cooled with jets of air. As already mentioned, rapid cooling of the surface produces a compression of the interior because the thermal conductivity of glass is poor, and the interior remains in the expanded hot state. This forms a slightly different density of a compressive surface glass region even when the bulk eventually cools down. The big bonus is that the surface stress means the material has some elasticity and will be much tougher than the bulk glass. In a car accident it will shatter into tiny fragments rather than into sharp splinters.

A minor problem of floating molten glass on a bath of tin is that during the time in the furnace there is some diffusion of tin into the lower glass surface. The upper face will also have some tin contamination from the tin vapour in the furnace. For windows, tin is not obvious as it does not produce any colouration for visible light. Nevertheless, the tin causes changes in surface chemistry and it interferes with subsequent coating and bending treatments. Problems may appear immediately in terms of coating adhesion, or only appear at a later stage when the material is exposed to the weather. In one unfortunate example, a skyscraper with beautifully bronzed glass needed replacement after a few years. That was a rare event, but difficulties of bonding solar and reflective coatings are bound to exist.

Applications for tin-doped glass surfaces

I am trying to show that imperfections can have positive properties and, therefore, it is worth asking if the accidental tin contamination of the float glass surface could have any useful purpose. There are thousands of hectares of low-cost float glass produced each year, so finding a new use would be intriguing. Tin is a heavier element than the silicon that it replaces in the glass. This means it is denser and has more electrons with which light can interact. This causes the light to travel somewhat slower than in normal glass. School level physics describes this by saying the refractive index is increased. A raised index in the tin-doped layer can act as an optical waveguide, where light injected into the tin rich zone is trapped in the surface layer and is guided across the surface by bouncing within the layer without escaping. This is an interesting solution in search of an application.

As a personal digression I want to mention that many people have attempted to make surface waveguides in glass by chemical reactions to raise the refractive index above the bulk material. The experimental attempts were typically on a small scale of a few centimetres, the processing was expensive, and often not uniform or reproducible. Many years ago, it occurred to me that float glass should inherently avoid such difficulties, and so I bought a sheet of over a metre length from the local glass shop. I introduced light into the tin-doped surface. As I had assumed, it acted as a waveguide and I measured the low losses and scattering as the light beam travelled across the entire sheet trapped in the tin rich surface layer. The result was excellent. To demonstrate this large-scale uniformity, I plotted a graph of waveguide performance with a scale in metres to emphasize how brilliant float glass is. When I published the result, imperfections took their toll, as the editor, or printer, obviously thought I had made an error by using 'm' on the length axis and carefully changed it to 'mm'. A correction was hurriedly made!

Changing the surface properties of glass

Glass has many desirable properties, but there are significant reasons to introduce surface changes and coatings to the original surface which far outweigh the costs and difficulties of the processing. The types of outcome are very varied and I have listed just a few in Table 7.1 to offer a flavour of possible opportunities and benefits.

Tough surfaces

At least in terms of giving stronger surfaces, which inhibit glass fracture, both stress and impurities can be very successful. Our normal experience with glass is that it is a brittle material and so if we drop a glass

Table 7.1 Examples of benefits from coating or changing glass surfaces.

Product	Function
Changes in surface chemistry and density	Toughen the glass and/or add chemical resistance
Heat reflective	Temperature control of cars and buildings
Self-cleaning /Water repellent	Windows with less maintenance
Anti-reflective layers	Spectacles, paintings and computer screens
Multilayer coating	Mirrors which select specific colours
Float glass surface with tin	Windows for buildings and vehicles
	Optical waveguides – in search of an application
Photochromic materials (e.g. that darken in sunlight)	Sunglasses and windows. Most sunglass examples are based on bulk glass effects

object, we assume it will break rather than bounce. Surface treatments do not directly alter the toughness of the interior but stress compacts and reduces the number of micro-cracks which act as pathways from the outside to the interior. For many glass products the penetration of water or other chemicals is a significant factor in causing the glass to be weakened. Making a toughened skin is therefore an excellent piece of technology, not only for glass, but also for metals. There are several easy routes to do this. The first is to rapidly cool the surface to form a prestressed surface layer (e.g. the use of cold air jets during shaping of the car windows). One bonus of surface toughening is to allow the use of thinner glass and more elegant, but rugged, glass tableware.

The second approach is to do some chemical changes which replace small surface ions with larger ones that put the layer into compression. The chemistry of the glass means that in addition to the silicate building blocks there are many other types of ions that were added. These help to reduce the melting temperature, or improve the way it can be moulded. For example, most glasses contain some sodium ions. Sodium is an alkali metal, but there are other alkalis ranging from small lithium ions to progressively bigger and heavier ions of potassium, rubidium and caesium. If we can do some surface chemistry and change the surface composition by replacing sodium ions with larger potassium ions, then the surface will be compressed and under stress compared with the glass interior. This is a routine chemical route using impurities to toughen glass surfaces. Adverse chemical effects include alkali ion exchange in a dishwasher, where the chemicals used in cleaning replace the surface ions of our best glassware and cause pitting or a cloudy appearance.

Coloured glass

As explained in the previous chapter, the chemical valence can vary with temperature so colour 'perfection' is tricky and expensive, and may be matched by many examples of second grade items. Colour can be added just on (or in) the surface layers of the glass. But surfaces are vulnerable to chemical attack or abrasion, and have problems of adhesion and survival. Many coatings have suitable colour properties, or reflect heat and offer temperature control, but their adhesion is weak. For float glass the villain in this role can be the residual surface tin from the float process. An extremely common example is the coating which reflects infra-red light to stop sunlight from overheating the room. The layer needs to reflect the long wavelength light (i.e. the heat) but it should not distort the spectrum of visible light or we would be looking through a coloured window.

For buildings there is a partial solution in double glazed windows, as the coatings (and tin side of the float glass) can be put on the inner faces of the glazing so that they are not exposed to mechanical abrasion during cleaning, and are relatively protected from atmospheric attack. During assembly of the double glazing, skilled glaziers can tell the difference between the two surfaces as the coated layer feels rougher than the uncoated surface. Sometimes, in extremely large double-glazed windows, gales will cause the two panes to flex and touch. This is very bad news as it normally rips off the coating and leaves a visually damaged section of glass surface. Glazing is now remarkably strong, as implied in a TV advertisement where a heavy weight is swung into a window without causing damage.

There are ongoing improvements in the types of material that are used and modern glazing can give climate control with blocking and reflective properties for incoming sunlight (to minimize heating from a very high temperature source – the sun), but at the same time trap radiation from the low temperature room (as in a greenhouse effect).

Some people assume that one of the commercial versions of heat reflective glass contains a potassium surface (chemical symbol K) as it

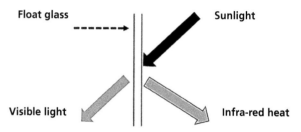

Figure 7.1 K pattern of heat reflection and light transmission of a glass surface.

was called K glass. Figure 7.1 shows the advertising-type sketch which caused this mistaken assumption.

The next item in Table 7.1 mentions glass surface layers which actively react, either directly, or in the presence of sunlight and water, to break up and reject dirt and contaminants on the outside of glass structures such as greenhouses and conservatories. I live in sight of the sea and this effective self-cleaning glass to remove the salt spray and output of seagulls is highly desirable. (The chemistry used to do this will be mentioned in a later chapter.)

Multilayer coatings

Single coating layers are low cost but as soon as one moves to technologies, that need either accurately controlled deposition of many types of layers, or gradations of a single type of layer, then cost rises. Nevertheless, multi-layer deposition has a host of applications in advanced technology and it has become routine.

The physics underlying multilayer coatings is conceptually simple. When light hits a glass surface a small percentage of it is reflected. The intensity depends on the refractive index of the glass and the surroundings; for a typical window glass the refractive index is about 1.5, but for air it is 1. In this case the glass reflects about 4% of the light at each surface. A really high refractive index materials, such as diamond (approximately 2.42) reflects far more light (around 18%) and so is ideal for glittery gem stones. The high index for diamond is unmatched by any other transparent natural material. Ruby, sapphire, garnet and alexandrite are all relatively high index at around 1.76. Some synthetic materials can be very good substitutes for diamond in terms of index;

and the structure called cubic zirconia, with an index around 2.16, is often used as an alternative to diamond.

Reflection can be reduced, for example if the light had gone through the boundary between, say, sapphire with a coating layer of a silica film, with indices around 1.75 and 1.46 respectively. The reflection at this interface is a mere ~0.8%. If we now consider making a stack of materials by alternating, say, silica and sapphire, then for each interface there will be multiple reflections as the light waves bounce back and forwards at each junction. Depending on the thickness of each layer the light path will occupy different numbers of wavelengths. These wavelets will either add together if the waves are still in step, or cancel out if they are exactly out of step. By controlling the thicknesses of the layers, we can choose whether the waves will add or cancel. Overall, this can make an extremely high reflectivity mirror for just one wavelength. The idea is simple, but it needs quite good technology. I will skip the deposition details as they are challenging. For a laser mirror, there may be as many as twenty-five layers, each of which must be controlled to within a small fraction of the wavelength of the laser light. In crude terms, each layer thickness cannot vary by more than, say, 1/1,000th the thickness of a human hair. In the laser case the multi-layer mirror is extremely reflective at the wavelength of the laser. There are two mirrors for a laser cavity and the exit mirror has to allow the laser beam to escape.

Once we have the control to put down many layers of chosen materials in precise thicknesses then several options emerge. The opposite of the laser mirror is to have a glass which has minimal reflectivity. This is ideal for the glass in front of pictures, anti-reflective spectacles, or, in principle, the glass of TV screens where we do not want any reflections from the room. A second option is a glass coating that adds a colour to the transmission or reflection. We can also make filters which either totally block one single colour, or only transmit a single colour. One example might be a filter to block yellow sodium light from a telescope lens, so as to be able to view the stars at night in an area with yellow street lights.

The advantages are obvious, but multilayer deposition is expensive, sensitive to layer bonding and easily damaged by surface chemicals and mechanical rubbing, as during polishing of one's glasses. For spectacles, and many other optical components, anti-reflective coatings are now routine additions to the glass. Since most people change their spectacles for new prescriptions every few years, any long-term failure aspects are not serious, but for scientific optics more permanent adhesion is a problem.

Simple ways to colour glass surfaces

Rather than use sophisticated multilayer coatings we would prefer simpler, and cheaper, options of adding colour to a glass. Surface metal film coatings can do this, as used in window glass, but there is a limited range of available colours. Adding a second glass is quite standard but, in order to reach a sufficiently dense colour, the layer will be of millimetre thickness, rather than a thin film. We need to look for alternative routes. The first of these is to consider a chemical exchange or ion diffusion into the glass (as used in chemical toughening). This is certainly possible in a few cases, but the speed at which ions diffuse into the glass interior tends to be quite slow. Also, very thin surface layers may not offer a strong colour. To some extent, the strength of the colour can be altered by the length of time that the chemical reaction is allowed to run. Diffusion rates will improve during heating so high temperature diffusion is a lot simpler and can offer stronger colours, and diffusion-controlled surface chemistry is useful for shading and tone to be included in the decoration of the glass.

Part II Ion implantation and glass surfaces

Coating a glass surface is a little like applying makeup on a face. There will be foundation layers to help with bonding between the face and the colour layer, and with more skill one can achieve better results. However, with detailed patterns the colours blend sideways and the effects are far from permanent, or will be removed by touch or contact with liquids.

An alternative route to skin colouring is to use tattoos. Here the dyes are driven below the skin surface by needles, so this is an energetic injection of the colourants. It has advantages that the effect is relatively permanent. The choice of dye can give good colour contrast and, because the needles are quite small, the colour can be tightly localized in a pattern and the colours will not run sideways. Clearly injecting dyes into a glass surface with needles is not feasible, but in fact there is a close analogy with the technique known as ion beam implantation, which was mentioned earlier for toughening metal surfaces.

Ion implantation is the absolutely fundamental step used in semi-conductor production (as discussed in a later chapter) in which all the electrically important dopants are added to silicon (or any other semi-conductor). The principle is crude. The element of interest is formed into a beam of ions and accelerated to a high energy, and accurately aimed via a mask to define the electrical circuitry. The penetration depends on the mass of the ions, the type of target and the voltage used to give them energy. Not surprisingly, the injection of such high energy ions causes damage and destruction of the local atomic structure of the target material. In many cases this is acceptable, or even the intention, but in other examples one has to use some heat treatments so that unwanted damage to the host lattice can be removed.

If the process is so effective with both metals and semiconductors then we need not be surprised that it also works in glass to add surface colour, or other properties. For glass, there do not seem to be many examples where this has been considered. The obvious reason why ion beam technology is not applied more widely to glass surfaces is that glass is basically a very cheap material and used in large area applications. Using expensive ion beam equipment to modify large area glass surfaces does not always add enough value to the product to justify the cost of the treatment. By contrast, ion implantation in semiconductors is absolutely central to semiconductor production, not only because of the very high added value to tiny components, but also because there are no alternative manufacturing options.

Mirrors and sunglasses

Ideally, car mirrors are preferentially reflective in the blue end of the spectrum. These are ideal for night time driving as we benefit from the blue sensitivity of the eye at low light levels, and we are not blinded by the green, yellow and red colours that dominate bright headlights which we see in the car mirror. They are normally made by an expensive multilayer deposition route. An agent for such mirror makers visited me, and asked if I could think of an alternative process. My absolutely instant reply was to propose making mirrors with ion implanted silver nanoparticles. I just happened to know that such particles have a strong collective response (termed a plasmon resonance), which for the correct particle size is in the blue. This would allow control of the reflectivity spectrum merely by selecting the size of the particles and the energy of the implantation to control their depth below the surface. It avoided the complexity and cost of the existing multilayer deposition technologies. He offered me a considerable contract and the process worked, plus several spin-off ideas, patents and publications.

It is interesting to compare this with normal fund raising, where one spends days writing brilliant research proposals that are not funded. I am eagerly waiting for more rich, stray visitors with a problem.

The concept is not limited to blue mirrors as, even for silver, if we change the particle size, the larger nanoparticles have their peak reflectivity at longer wavelengths. Figure 7.2 presents reflectivity data for both blue and green biased mirror surfaces.

With more skill, one can also convert the same implanted layer into a different wavelength mirror by modifying the size of the silver nanoparticles. Figure 7.2 explores this possibility using precisely the same quantity of silver implanted into the surface layers of the glass, but allowing it to cluster into different size particles. Subsequent treatments can vary the reflectivity curve. In this example between a blue and a green mirror, the difference is not in the quantity, or depth profile of the silver, but instead relates to how the silver is incorporated. To achieve the blue curve, we need to have small clusters of silver ions as little metallic nanoparticles. By contrast, if they gather together into large particles then these larger units have electronic resonances which give the green maximum. Control of a heat treatment can allow movement of the silver to cluster together into large units. When

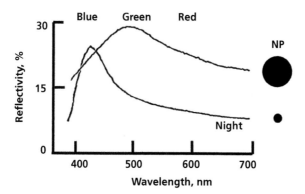

Figure 7.2 Reflectivity of different size silver nanoparticles (NP) in glass. The curve labelled 'night' is ideal for car mirrors.

the material is cooled these particles are frozen in place. This is somewhat like the science of cookery. However, to achieve the smaller size distributions needed for the blue mirrors, we could heat the nanoparticles to a high temperature so that they dissolve and disperse into the glass structure. This is simple in an oven, but the trick is then to cool the glass so quickly that this new dispersion of the silver is frozen in place. (Precisely the same concept as for sword blades.) That was fine for a thermally conductive metal, but glass is a very poor conductor of heat and so bulk rapid cooling is difficult, further, the thermal shock can fracture it.

At this point we moved to a quite sophisticated method of heat treatment using a high power pulsed laser beam. By choosing a laser with pulse durations of only a few nanoseconds (10^{-9} s) we absorb the power directly at the silver nanoparticles. They melt and heat just the immediate surrounding tiny volume of glass. The silver diffuses into these hot surroundings, and the heat rapidly vanishes into the mass of adjacent cool bulk glass, leaving a dispersed collection of small nanoparticles. By controlling the pulse rate and laser power, we can tailor the final size distribution of the nanoparticles.

More imaginative patterning of the laser treatments could include a coloured reflectivity pattern or images built into the mirror. Overall, we needed the modern high technology to achieve the results in a simple glass material.

A much simpler variant on the concept was to make mirror front sunglasses by implantation of metal ions to form nanoparticles reflecting across the spectrum. A flat spectrum mirror was achieved by selecting a different ion for the implantation. Viewed from the front they are excellent mirrors. But in our initial attempts we saw there was a problem of a strong reflection on the inside faces of the mirror, from light sources behind the head. The solution was to add a second, deeper implanted impurity layer beneath the mirror front as in Figure 7.3. This extra layer was chosen to strongly absorb light. The method works as incoming light at the front is only absorbed once. By contrast, light coming from the rear that is reflected back from the mirror layer, is absorbed twice. This solved the problem. (A similar type of approach had been made with TV screens to reduce reflected room light, even though this means it is necessary to start with a brighter picture.) The transmission and absorbance factors of Figure 7.3 are chosen for simplicity to show the effect by using a layer that absorbs half the light.

A bonus of our implant route to mirror front sunglasses is that they have minimal colour distortion (far less than is usually seen with brown or tinted sunglasses). We also made variations with graded strength sunglasses (dark at the top and clear at the bottom), and double graded glasses to cut reflections both from snow or sea, as well as from sunlight. This is not a new idea as I had seen wooden slit glasses used by Inuits designed for the same purpose. For a further example I used coloured glass with a transmission response that matched that of the eye. These gave excellent mirror front sunglasses with no detectable colour distortion.

A further example of exploiting imperfections caused by ion implantation in glass was to use the ion beams to generate damage and a reduced density of the glass network. This may seem odd, as most people have tried to avoid such damage production. Nevertheless, we produced a surface layer which was deep compared with the wavelength of visible light, and which tapered in density with depth as a result of the defects. This had the effect of steadily reducing the refractive index up towards the surface, and the graded index layer interfered with the normal surface reflection. The result was an extremely low reflectivity glass surface for light across the entire visible spectrum. Reflectivity values were near 0.2% instead of the usual 5%. Our method was also rapid and low cost. The downside was that the mechanical strength of the glass was lowered, so it is not a method suitable for windows or devices which will be rubbed, cleaned, or otherwise abraded. For optical components which are not exposed to handling it offers a

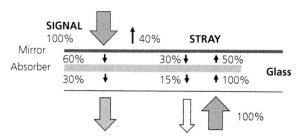

Figure 7.3 The layout of a mirror face sunglass. The front mirror face was implanted with metal ions and a different ion was used to absorb transmitted light. Note that stray light entering the glass from the viewing side will also be reflected, but since it passes twice through a 50% absorber, it can be sufficiently suppressed. The same concept has been used on TV screens to minimize room light reflections.

rapid, low-cost and useful alternative to the costly, complex, multi-layer depositions that are normally employed to give low reflectivity over a wide range of colours.

Insulator implants for optoelectronics

There are a number of other high technology areas, using transparent materials, where the unique control offered by ion beams is economic, especially in the production of optoelectronic devices. For many optical materials it is the only route to fabricate waveguides and lasers. Rather than following conventional wisdom of ignoring implantation into glass, my friends and I have successfully modified glass surfaces with ion beams, and fabricated a host of optoelectronic components. These will be mentioned in the next chapter.

Part III Glass processing and optical detection of cancer

In order to highlight how sophisticated glass processing can underpin a significant advance in diagnostic medicine, I need to offer some brief background of our objective. In conventional breast cancer mammography, X-rays are used to sense differences between healthy and cancerous tissue. It has been used for many decades and steadily improved, both in terms of reducing the X-ray exposure, and current new examples of artificial intelligence to sense the possible cancer sites. Despite their value, the downsides are that medical X-rays are cited as causing nearly 10% of cancers, and since many million women have mammograms each year, there are also a very significant number of cancers caused by the scanning. There are additional problems of actually imaging on to X-ray plates, and then relating the images to the position in the breast. Clearly an improvement is to use a totally non-damaging scan of an undistorted breast. This is feasible with laser light. If such optical scans reveal a possible cancer then an X-ray scan could be a reasonable second step. Note that there is a potential bonus by using several different wavelengths of light, as it is possible to separate cancerous and other inclusions such as cysts. This route could significantly cut the medically induced cancer rate, and indeed should be more accurate. It is sad to report that currently there are many failures to detect existing cancers and many false positives where operations and breast removal are conducted. Considering the difficulties of diagnosis with X-ray images, this is inevitable.

On near-surface skin cancers, one can typically illuminate them with a laser beam and observe a weak fluorescence at longer wavelengths for healthy skin (NB: energetic short excitation wavelengths re-emit via lower energy, long wavelength, emission). There can be very clear differences in spectra from healthy and from cancerous cells. 'Surface' is not limited to outer skin cancers, but is equally successful for internal surfaces such as the lungs, and Figure 7.4 offers an early demonstration of the differences recorded via an optical fibre for healthy and cancerous bronchial tissue. Note this totally avoids a difficult, high dose, X-ray investigation.

Moving to breast cancer diagnosis, the challenge is to optically excite internal breast cancers to produce luminescence that can be detected. Flesh is obviously slightly translucent in the red region as we can sense light through our eyelids. If our eyes had longer wavelength red sensitivity we would see below the surface of the chest and see strange images looking like those drawn by Leonardo da Vinci of musculature. Fortunately, there is a window of opportunity to explore optically beneath the surface which is set by the absorption curves for the main constituents of breast tissue, such as water and haemoglobins, etc., and

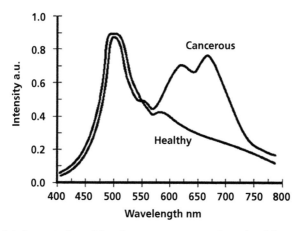

Figure 7.4 Laser induced luminescence spectra from healthy and cancerous bronchial tissue.

surface melamine (Figure 7.5). Excitation and luminescence of tissues in this optical region (say, 600 to 1,100 nm) offers a direct identification of the presence of cancer cells.

Even more sophisticated analysis is possible in a second approach by using very short pulse laser beams to give scatter patterns that define the position of the scattering site. This is equivalent to looking over a wide angular range with laser beam scattering from, say, a spoon in a jug of milk, to locate the site.

The principles are simple, but they require optical detectors that are efficient in this long wavelength range. For the echo type location such detectors must also be extremely rapid. The most sensitive light detectors are termed photomultipliers, and so our challenges were to both improve their efficiency and to increase the speed of imaging versions of the detectors. A schematic of a simple photomultiplier is shown in Figure 7.6. There is a glass, or silica, window on a vacuum tube. The inside of the window is coated with a semiconductor type material which, when excited by a photon, emits an electron. The electron is accelerated to gain energy and hit a secondary surface which, in turn, emits many more electrons. After a chain of such events the one original electron becomes a burst of a million or so, and is easily detected and processed. Photomultiplier tubes (PMT) are not new and were first used in the

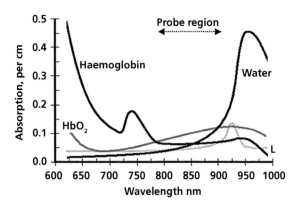

Figure 7.5 **Components of breast tissue.** This demonstrates a region of low optical absorption which is suitable for a laser probe study of breast tissue. HbO2 is oxy-haemoglobin, L is lipids. There is additionally some surface absorption from melanin. Normal visible eye response is minimal beyond about 700 nm.

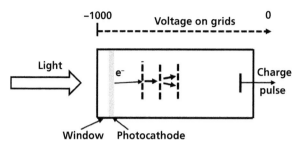

Figure 7.6 Light enters the window of the photomultiplier. In the vacuum, electrons leave the photocathode and are accelerated into grids producing more energetic electrons; this cascade can then eventually deliver a million electrons per incoming photon. Just three grids are schematically indicated.

1930s to record sound that was encoded in the blackening of a strip on the side of cinema film. The probe light was intense as the overall efficiency was extremely low with less than 1% of the light resulting in electron emission. Later tubes improved in the blue region, but far less so at long wavelengths.

At short wavelengths, PMTs are quite sensitive but their sensitivity falls steeply in the range of breast transparency as the incoming photons have lower energy. The challenge was to raise PMT sensitivity at long wavelengths. The inherent emission efficiency for normal incidence light is low and we realized that there could be significant improvement if we could replace the internal flat window surface with a redesigned shape with regions at different angles to the incoming light. The enhancements require different angular changes with wavelength and polarization, so for a wide spectral range compromises must still be made. With modern laser technology, we carved a truncated pyramid structure on the glass and then subsequently processed it into a dome pattern. The images of Figure 7.7 show results in two of the steps of this approach. The size marker is for 40 microns, which is around one third the thickness of a human hair. This is precision engineering that one would rarely consider for a simple glass window on a PMT.

Clearly this was a success or I would not have mentioned it. Figure 7.8a indicates how this restructured window pattern exceeded the previous steady developments in PMT units over the preceding seventy years. Both blue (i.e. near the normal peak performance at 400 nm), and red data at 750nm (where breast tissue is translucent) are shown.

Figure 7.8b displays an example of improvement factors compared with commercial tubes worldwide that were then available, in 2004. The medical advantages of these improved cancer detection systems are still only being used in a few locations, unfortunately this is typical human behaviour by people entrenched in the ways in which they were trained.

In conclusion, we improved the efficiency in the blue to approximately 50% for wide spectral coverage detectors (we reached higher values of approximately 70% in PMTs designed for a narrow spectral range). The key area for optical breast cancer detection was significantly raised and the PMTs became useable over an extended longer wavelength range. The improvement factors at the very long wavelengths are impressive, but this is of course because the initial performance was so poor. In a separate redesign we enhanced the response speed of the

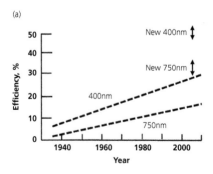

Figure 7.7 Images of a laser ablated glass structure. Left, as first cut with the laser, and right after thermal treatment. The marker bar is 40 microns.

Figure 7.8a The pattern of developments for commercial PMT performance worldwide in the blue and red regions with year. The arrows indicate the enhancements from the structured window surface.

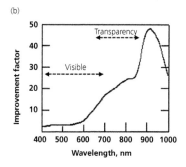

Figure 7.8b **A monitor of the improvement factor across the spectrum.** The arrows indicate the region of visible light and the zone of useful transparency for breast tissue.

photon imaging detector (effectively an incredibly sensitive camera) by fifteen times.

Part IV Photochromic glass and the photographic process

One of the classic examples of a highly valuable optical effect, which is entirely the result of a complex set of imperfections, is the photographic process. The mechanism which produces the photochromic effect is very similar and the concept was based on photography. However, there are two rather obvious differences in that the photographic effect produces a permanent change in the material, whereas the photochromic changes are reversible. Secondly, photography can be activated by a wide range of colours, not only with visible light but even with lower energy light in the near infra-red region. By contrast, typical photochromic sunglasses only darken with the high energy UV component of sunlight. They steadily fade back to a more transparent state when there is no UV light.

Photography may now seem an outdated technique for capturing images but it is an interesting example of exploiting imperfections. The effect relies on processes on the surfaces of silver halide grains embedded in a gelatine matrix. The photochromic sunglasses are similar. I will summarize the imperfection science of the photographic process. The roles of the various impurities and defects are subtle, with the result that the models for the mechanism only developed rather late in the history of the subject. Variations of photography have been around since the 1820s, but even now complete understanding is still not perfect. This slow and gradual development of our understanding of photography is interesting as it totally contrasts with the theory and practice of taking pictures with digital cameras. Photography preceded any theory, but for the electronic example the model preceded the production of the detectors.

A photographic emulsion contains small triangular grains of silver halide crystals embedded in a film of gelatine. Colour film is similar except that there are several layers with different dyes to make the film sensitive to different ranges of colour. The incoming light is absorbed by the silver halide grain and the energy of the light (the photon energy) is transferred to a silver ion in the grain. Surprisingly, the bond energy between the silver and the neighbouring halogen ions is quite small. A concentrated package of light energy, termed a photon, when absorbed at a single ion site, is enough to break the bonds. The silver ion escapes and moves around freely for a brief period of time until its energy has been dissipated. In a large block of material there is a strong probability that the silver will just return to an original silver site, and the light would have done nothing. However, we are using very small silver halide grains, so it is possible for the silver to reach the surface of the grain. These individual silver ions are partially trapped and appear to be social objects and, if they meet up with other impurities on the surface, they remain bonded to them for some time. This is not a permanent arrangement, but if the light is bright enough then other silver ions may also arrive at the surface from within the grain. These form more permanent bonds and several silver atoms can join together on the grain surface. When two or three are bonded together their survival rate increases, and by the time they build a block of four silver ions they are quite stable. This silver cluster is the basis of the latent image produced by the camera on the film. Exposed film can then be taken away for chemical processing to distinguish between grains which were illuminated and those which were not, and this is the chemical development stage. There is a need for four silver ions to be formed in the same grain in quite close succession. Failure to do this explains why, at very low light levels, even prolonged exposure does not build up a photographic image, as the grain surfaces never manage to reach the critical cluster of four silver ions.

In hindsight it is interesting, and extremely fortunate, that the first photographers used gelatine as the emulsion support for the silver halide as this has a crucial role in the entire photographic process. Gelatine is made from cow's hooves etc., but cows live in fields and chew grass and weeds, such as mustard seed. Mustard seed contains sulphur and this works through the processing chain to the hooves and into the emulsion. There it sits on the silver halide grain surfaces waiting to bond and stabilize any stray silver that appears on the surface. Without mustard seed the photographic process might not have worked. A modern chemist might have begun with a pure chemical support layer without sulphur, and the entire process would have been missed. Mustard seeds may be exceedingly small but they have played an immense role in our lives for two centuries, from Victorian to modern photography, and the entire cinematographic industry of the last 120 odd years.

The science of photochromic sunglasses

Basically, there are no local surfaces and sulphur equivalent sites in the glass to tie down the silver. When silver ions are freed and drift around, they only temporarily form small silver clusters. The clusters are not totally stable and in the absence of UV light, which was liberating silver ions, the clusters break up and individual ions drift back to their original type sites in the glass. Energy is needed for movement through the glass and this is provided by thermal energy. The process is quite sensitive to the thermal energy and so the speed of darkening and recovery of photochromic sunglasses differs very noticeably between cold and warm conditions.

Is the future clear for glass?

Unquestionably, glass and its uses are heavily reliant on empiricism, experience and experimentation. These are cornerstones in any type of industry using complex composition materials. In the examples outlined here, it is obvious that many of the important properties have been derived from skilful use of trace impurities, large scale concentrations of additives, and various types of thermal processing or surface chemistry. Nevertheless, many of my examples refer to familiar or relatively low-cost products, and applications could have been apparent to readers a century ago. Unfortunately, neither simplicity nor low cost glass materials inspire high technology kudos. The more positive view is that the glass industry will remain an important part of modern life and new and improved products will steadily appear. An even more revolutionary change in glass technology has emerged over the last fifty years to the point that it has totally changed our way of communicating. This is the introduction of optical fibres, which are some million times more transparent than normal window glass. The production problems were substantial but the benefits are enormous. The entire field is based on understanding, and benefitting from, control of imperfections and additions of dopants, plus the ability to bond different types of glass together.

8

Optical fibre communication

Part I Signalling with light

The largest earth-moving engineering project in the UK in the last seventy years was not the construction of the channel tunnel to France, but the installation of optical fibres throughout the streets of the country. Optical fibre communications have completely changed the way we live and relax by providing hundreds of TV channels into the home, access to internet and websites across the globe and visual links and remote office working. It has not improved the quality of the TV programmes, but it has definitely produced a revolution in lifestyle. The science behind the ideas was long established, but the current success of the methods came despite the widespread views of many of the leading communications industries that optical fibres were never going to be a realistic component of long-range communication.

Overcoming the technical difficulties has strongly depended on our understanding of the role of impurities and imperfections in glass, as well as designing the light sources and detectors to allow signals of light to carry information. In order to appreciate the reasons for the views that optical fibres were not valuable for communication, I need to consider the background history and science of optical fibres, and emphasize that the success was only possible because of the single mindedness from just a very few people, plus some good fortune in terms of the timing of medical and laser developments. The combination of these apparently unrelated factors allowed the fibre industry to advance. I particularly enjoyed a fascinating commentary on the scientific progress and personalities of those involved in the 1999 book *City of Light*, by Jeff Hecht.

Sending long range optical signals is not a new idea. It has been used by many nations over the last few thousand years. In Britain, beacons were lit by the Romans to help in road construction to give a long-range view of where to build a straight and direct road. This particular

The Power of Imperfections. Peter Townsend, Oxford University Press.
© Peter Townsend (2022). DOI: 10.1093/oso/9780192857477.003.0008

transport skill seems to have been overlooked by later British generations, and only reappeared a thousand years later with railway and
motorway construction. Early Chinese and Roman, as well as later civilizations, all used beacons built on high ground to signal attack by
invaders, for example a British beacon chain was constructed to warn
of the approach of the Spanish Armada.

Information has been sent, both by using smoke signals to control
puffs of smoke and by reflecting sunlight with heliographs. Coloured
smoke signals are still a feature of the state of progress of Papal elections at the Vatican. The more informative optical signalling methods
were coded via on or off pulses, which is exactly the modern method
of signalling in a binary code, with either a one or a zero. It is ideal for
data transmission in modern optical fibres. The idea may be the same,
but the pulse rate has gone up from a few pulses per minute with heliographs, to much more than a thousand million per second with the
optical fibres. Part of the improvement in the fibre case is that signals can
be sent with many different colours of light, so wavelength selective encoding can mean a further increase in signal capacity down a single fibre
with a hundred separate colour coded channels. The historical progress
of such ideas is summarized in Table 8.1.

Bending light around corners

Whilst Indian smoke signals could be seen across the American plains
for long distances and heliograph reflection of sunlight was also visible over many miles, their weaknesses are that both methods rely on a
straight-line field of view (and they only worked in daytime and clear
skies). To do better than this, and optically match the convenience
of electrical signals running in wires, needs three basic components.
Firstly, an intense directional pulsed light source; secondly, a system to
guide the light around corners; and finally, sensitive detectors. There
are signal processing electronics to encode and decode the information
at both ends of the fibre. These were all major challenges. Nevertheless,
it clearly succeeded as there are billions of messages and photographs
sent by fibre every day. Only with all these features, from signal encoding to recovery, can the high rates be reached.

As mentioned earlier we can make optical waveguides by bouncing
light off the inside boundaries of a glass sheet if the light is nearly parallel
with the surface. Such surface reflections are even seen when fishing, or

Table 8.1 Historic examples of signalling.

Inventor	Period	Method	Data rates
	~400 BC	Hilltop beacons	A few pulses/minute
Claude Chappe	~1792	Optical semaphore telegraph	3 charac- ters/minute
Abraham Edelcrantz	1794	Version with 9 flags	
Samuel Morse	1838	Electrical pulse code	~200 per minute
	1865	Atlantic cable	
Alexander Bell	1876	Telephone	Speech
Guglielmo Marconi	1895	Radio signals	
	1956	Atlantic telephone	72 voice channels
	1976	Better re- placement cable	8,400 voice channels
	1965	Satellite links	
KC Kao, GA Hockham	1965	First optical fibre trials	Short range
	1988	Atlantic optical fibre cable	
	1990s	Multiple wavelengths	
	2000	Fibre with 100 wavelengths	$> 10^{13}$ pulses per sec

swimming under water. The critical angle for trapping is defined by the refractive indices of the guide and the adjacent layer. Some typical index values are given in Table 8.2.

When a beam of light hits a surface at an angle, it changes direction because the speed has altered. School textbooks present this in terms of Snell's law, as shown in Figure 8.1, where:

$$n_1/n_2 = \sin\theta_2/\sin\theta_1 \text{ (i.e. } n \sin\theta \text{ is a constant)}$$

Since we assume the refractive index of air is very close to 1, if we discuss light hitting a prism or refracting at a water surface we tend to just label $n = \sin(\text{incident})/\sin(\text{refracted})$, however, with a glass fibre

128   The Power of Imperfections

Table 8.2 Examples and definition of refractive index. The refractive index is a measure of the speed of light in a material, relative to light travelling in vacuum (refractive index = speed in vacuum/speed in material). Examples of values are shown from vacuum (~air) to diamond, which is the highest index of transparent natural materials.

Material	Refractive index	Speed m/s	Miles/second
Vacuum	1	3×10^8	~186,400
Water	1.3	2.3×10^8	143,400
Glass	1.5	2×10^8	124,300
Ruby/sapphire	1.79	1.67×10^8	104,134
Diamond	2.4	1.25×10^8	77,700

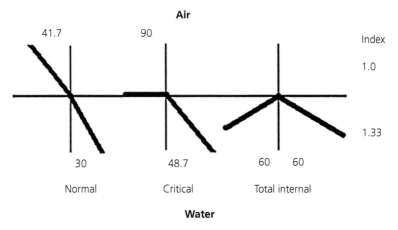

Figure 8.1 Refraction and total internal reflection. An example of light crossing a water-air surface. By an angle of 48.7 degrees a submerged swimmer cannot see through the surface as the light is reflected and trapped. Similar problems exist for looking into the water at a shallow angle. The sketch effectively describes how light is trapped in an optical waveguide.

and a lower index cladding, we need both indices to define the limiting angle at which the light is trapped and bounces along in the fibre.

Snell's law is sufficient for our purposes, but advanced work requires the more detailed mathematics of Maxwell's equations. The principle is the same but, for an optical fibre, the indices may be much closer at, say, 1.48 in the guide and 1.46 in the cladding. Further, within a guiding layer there are only a limited number of angles which sustain successful bouncing of the light along the guide with multiple reflections. These,

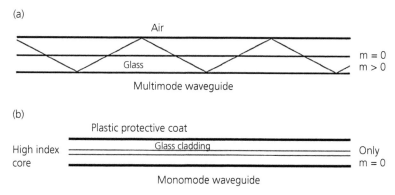

Figure 8.2 Modes of light trapped between two surfaces. Light can bounce inside a block of glass or an optical fibre. If the dimensions are large compared with the wavelength of the light, there are several allowed angles of bouncing, called modes, termed m=0, m=1, m=2, etc. The angles here are exaggerated to show the effects. For an optical communication fibre, the core region of high index is so small that effectively the light is going straight along the fibre in an m=0 mode.

the successful conditions, are called waveguide modes, as sketched in Figure 8.2.

A bare optical fibre would suffer from surface dirt and scratches, so the solution is to deliberately clad one piece of glass with another one of a lower refractive index. This is absolutely essential in an optical fibre system. This sounds like an obvious idea which any science student in school could have solved, even without the hindsight of modern literature. However, the reality is that even top scientists are imperfect and their failure to successfully spot this problem, and the simple solution of adding protective glass coatings, not merely hindered the development of optical fibres, but also meant the research was abandoned in some major industrial laboratories.

Light trapping in water fountains

To emphasize that confining light in a material is not a new idea, we can quote many early examples. Basically, the same approach to trap light was used in France in the early 1840s, by both Daniel Colladon and Jacques Babinet. Colladon sent a beam of sunlight into a column of water that curved from a tap and headed down to the floor, whereas

Babinet used a rod of glass and realized this could be bent to guide light around corners. The first general public demonstrations of such effects were made a few years later with a bright light source under a water fountain. Demonstrations were made in France and at the 1851 Great Exhibition in Britain. The light was trapped within the fountain jet of water, even when it curved. Light escaped and could then be seen when the water jet broke up.

There are many variations on this idea and I once designed a modern equivalent with a light emitting diode (LED) in a container of wine. In countries where waiters will pour wine into a glass from a great distance, the wine arcs across to the glass. In my device the light is trapped within the column of wine, but escapes as the liquid fragments and hits the glass, with a corresponding burst of scintillation. The colour is defined by the light emitting diode, and I picked colours which were not absorbed by the wine.

In the case of a glass rod, one can easily demonstrate the waveguide action as a small light source at the end of the rod will emit light at the far end, and if the rod is curved the signal will go around the bend, so a variation on the wine trick is to shine a green laser pointer underneath a glass of red wine. The red wine absorbs the green light so looking straight down into the wineglass reveals nothing, but light trapped within the glass follows the curve of the glass and waveguides round to the edges. The glass rim emits a beam of bright green light. An alternative party toy version of this example can be a light emitter at the base of the glass which turns on when the glass is lifted. These trivial examples just show that waveguiding of light is a very easy idea to understand, and for the first hundred years the uses were mostly limited to similar toys and party tricks.

Uses of short lengths of glass rods and fibres

We now manufacture optical fibres in lengths of tens or hundreds of kilometres, but a mere fifty years ago it was just a few metres. Technologically it was very difficult to achieve even modestly long lengths of fibre. Interest in fibre systems would have been completely killed if there had not been some potential goals which could exploit quite short lengths of thin flexible glass rods. Fortunately, a major incentive came from the medical profession as doctors wish to examine the inside of the body, as well as the surface. The first attempts to peer inside, say into the stomach, were by emulating a sword swallower and inserting a tube and light source through the mouth. Use of a rigid system with a large enough diameter to gain any image of the interior was a disastrous approach and numerous patients were injured, or died in the attempts. Significantly more medical patients died compared with sword swallowers. It is surprising that doctors persisted, as initially even if the medical problem was obvious, there was probably no surgical method available that could benefit the patient from the diagnosis.

Flexible and small diameter endoscopes were needed and a number of dedicated researchers attacked the problem, not only by using more flexible tubes and a series of lenses to transport the image, but also by using bundles of glass fibres. The lens systems were very bulky and unbelievably large diameter considering their use. The prospect of an examination for a prostate problem using a half inch (1.25 cm) glass lens system in a tube definitely sounds worse than the problem. Fibre bundles were more flexible and slightly smaller in diameter. If the fibres could be optically separated, each fibre sent a signal from just one viewing spot. A bundle of, say, fifty fibres would then offer a fifty spot image; dreadful by modern camera views with a few million pixels, but still an improvement on rigid tubes and lenses.

The pioneers realized that not only did they need very clear glass to transport the light but, also, they needed to add cladding materials which kept the light trapped within the fibre and avoided leakage if the fibre/rod touched one of the neighbours. Light losses were very serious if the glass had been handled, as grease from fingers offered surface scattering and decoupling sites. Coating the surface with metal seemed initially like a sensible way to confine the light, but as the light bounces many times, even up a rod from the stomach, most of the light will be lost as metal mirrors lose, say, 15% of the signal at each reflection. For a short length of bent fibre with just ten reflections, this drops the intensity by more than 80% (i.e. only 20% remains).

Rather than using metallic mirror surfaces, cladding was attempted using plastic type outer surfaces. These plastic coatings are less absorbing than metal, but poor bonding and rubbish trapped at the glass/plastic interface resulted in considerable scatter and loss of signal. The first real success came using a very clean ground and flame-polished glass rod. This was inserted into an equally clean and polished glass tube of lower refractive index. When this package was heated, softened and pulled into a rod, then the overall result was a length of semi-flexible 'fibre' which confined the light within the core glass. Fixing many fibres into an aligned bundle meant image information could be transmitted. The devices were so popular that they generated a significant medical market which kept up the interest in fibre production.

The medical profession and the general public do not always use the same language. Because the use of optical fibres to inspect the interior of the body does not require surgery, the procedure is euphemistically called 'non-invasive'. Friends who have had such experiences use quite different words (not quotable here) to describe the investigation. The non-invasive investigation also frequently results in subsequent infections, presumably because the fibre systems are difficult to fully sterilize.

Misaligned fibre bundles

Making aligned fibre bundles is not easy and if the bundle is scrambled then image information is lost. This idea appealed to the US government and considerable funds were offered to make a scrambled bundle which could be cut into two parts. The idea was that it might form a secure encoding system so a document could be recorded (photographed) through the scrambling part, but the image could only be decoded by a person with the identical descrambler piece of the initial fibre bundle. Unfortunately for this security application, but fortunately for fibre development, it was some time before it was realized that the use of an unchanging scrambling code meant it could be solved relatively easily, even in the 1950s.

Short-aligned fibre bundles were also used, and still are, as the face plates of very sensitive photon imaging tubes, as the fibres split up the image into a series of dots (i.e. pixels) and the pixelated image signal is compatible with the detectors that turn optical signals into amplified electrical currents (as in photon imaging tubes).

The market for compact fibre bundle coding continues to have considerable potential in terms of personal security coded devices, even if it is unsuitable for state secrets. One can readily imagine devices for number coded ignition systems in cars, houses or similar simple security systems.

Part II Crossbows and the first attempts to make glass fibres

Probably the earliest successful attempt to make controlled thin strands of glass, in the form of fibres, first emerged from the need to have very thin glass fibres for use in the torsion wires in nineteenth century electrical measuring instruments. One such instrument was a very sensitive moving spot galvanometer, to measure tiny electrical currents. Torsional twists on an electrical coil are made apparent by a mirror mounted on the system which deflects a beam of light. One possible mirror support 'wire' was a fibre of glass. Although a skilled glassblower could make a short length of large diameter fibre, the diameter would vary along the length and the method was not readily reproducible. A real technical advance was introduced in about 1887 by Charles Vernon Boys. He designed a miniature crossbow that had an arrow attached to a heated glass rod. By firing his glass arrow, a length of very fine molten glass was formed which cooled into a uniform and strong glass fibre. Mechanically it was stronger than steel of the same diameter. It was smaller than achievable by drawing fibres by hand, more transparent, and of nearly a constant diameter over several metres. The crossbow route only made lengths of tens of metres but it enabled some of the fibre properties to be appreciated and measured. Optical fibres had thus become a laboratory reality.

The sensitive moving spot galvanometer was an essential part of the transatlantic cable system for detecting electrical pulses of Morse code signals. In hindsight we see the spin–off technology totally overshadowed and replaced the original system.

From related methods of simple glass fibre production, the glass was even included in clothing materials to give an iridescent finish. This became fashionable and was widely used for decorative nineteenth century clothing. Clothing examples actually benefitted from light scattering as fibres touched other items, or fibres, and so for neither the torsion wires nor the clothing was there any need to confine the light with a cladding.

Longer optical fibres

By the 1960s, the next major effort for endoscopy involved cladding the fibres to stop signal losses where they touched. It slowly became appreciated that a proper and carefully controlled optical fibre could be made with a high refractive index for the core region, where the light was travelling, and an outer cladding region which had a lower refractive index. The cladding added strength, blocked surface reactions with water vapour and, more critically, meant the light did not escape when the fibre was bent.

The possibility of using optical fibres for long range signal communication was still totally rejected by most of the major US companies. In part, because the existing performance was so poor, but equally the companies were concentrating on microwave and radio links between signalling towers. They also assumed that it would be possible to build microwave systems within metallic pipes that could be laid underground. Microwaves were well understood and sources and detectors had been constructed as a consequence of wartime and military interest in radar. Their inherent difficulties were that (a) metal waveguides have extremely high losses and so the signal intensity fades, and (b) signals could not be bent around tight corners. To solve these problems would require many stages of detection, amplification and/or other ways to repeat and boost the signal every few hundred metres. The microwave waveguides were also sensitive to distortion, thermal effects and weather conditions that influenced the air (and water vapour) within the pipe-work through which the microwave signals were transmitted. In view of these very obvious difficulties it now seems very strange that the method was pursued with so much enthusiasm, conviction and funding.

Light loss in early fibres

Very few scientists had seriously looked at the problems associated with optical fibres or even believed they could carry long range signals. The most obvious problems being that glass fibres of that period reduced the signal intensity very rapidly, and it was unclear how to make long lengths of fibre and join them together. It was assumed the problems of making joints would be extremely difficult. It is. The total diameter of a modern fibre is not much larger than a human hair and the signal carrying core is only about one tenth the diameter of a hair. To accurately align two pieces of such glass fibre core with a tolerance of much better than 1% when working as a telephone engineer, whilst making repairs outside in bad weather, seemed impossible. It is still not easy, but methods of doing this are now reliable and routine.

Although the idea of guiding light in a fibre is simple, the reality, in 1960, was that in the first types of fibre that could be made there were very severe losses from absorption and scattering of the light. Even using the best quality glass materials, the loss was typically as high as a 50% decrease of intensity per metre of fibre. For a laboratory demonstration this was progress, but even for communication within a room using a 10 metre fibre, the input light was over a thousand times weaker as it emerged from the fibre. Technology to make kilometres of fibre did not exist, and it was irrelevant as no useful signal would emerge because of the losses. Basically, what was required was a material which is a million times less absorbing than window type glasses.

Window glass is functional as we view through a thin direction of a few millimetres thickness and probably assume the losses are merely the result of dirt and surface reflections. For a refractive index of 1.5 there is a reflection loss of about 4% at each glass to air interface for visible light (so, in the absence of absorption, about 92% of visible light should be transmitted). As previously mentioned, if we look at a pane of window glass edge on, then it is far less transparent and appears slightly green in colour (from iron impurities). The reflectivity loss has not increased but we are able, even for a few tens of centimetres of glass, to see absorption effects. For kilometres of fibre, as needed for signal transmission, this means the signal is destroyed.

More transparent materials than window glass (a silicate composition of numerous metals oxides) is pure silica (SiO_2). However, it was initially rejected as a possible fibre material as it has a very low refractive index (at about 1.46), even though it was the clearest material available. It was not obvious how to make it the core of an optical fibre, since it needed to have an outer casing of a glass with an even lower refractive index. Further, and a limitation for many experimentalists, was that to draw a silica fibre from a near molten piece of silica rod required a very high temperature. Silica melts near 1,713°C (~3,115°F). Furnaces and crucible materials in 1960 that could manage such temperatures were rare, and the only easy heating methods involved oxy-hydrogen gas torches.

Light scattering

It was unfortunate that the glass industry had no detailed idea what limited the transmission of glass, as it had never needed to address the problem. It was correctly assumed absorption came from impurities, such as iron or other metals, that existed in the sands used to make glass. Nevertheless, one advantage of a material such as silica is that it has just one very simple composition, and it is totally SiO_2 throughout the material. This implies that there are not likely to be serious variations in composition or density that could act as light scattering sites.

Scattering is linked to the wavelength of the light. Intuitively obvious, as adults can walk in nearly straight lines on stony ground, but a shorter-legged child will be scattered off the intended path. Optically the light scattering decreases as the fourth power of the wavelength, so blue light with a wavelength of 400 nanometres (400×10^{-9} metres) scatters some 9.4 times more that red light (700 nanometres wavelength). This difference in scattering by blue and red light gives the familiar observation and explains why the sky is blue. Direct sunlight is a mixture of all the wavelengths from the sun, and looking in the direction of the sun the light includes intense, and unscattered, yellow and red light. However, the light which comes from other directions only exists due to the fact that it has been scattered many times. Since the shorter blue wavelengths scatter more than the red, it means the rest of the sky has a blue colour.

Transparency at longer wavelengths than are transmitted by silica should give less scattering, and this resulted in studies with heavy metal fluoride glasses. Much effort was expended on making various compositions of materials called ZBLAN. The various ZBLAN glasses are very complicated mixtures of fluorides of zirconium, barium, lanthanum, aluminium and sodium etc. Whilst their long wavelength absorption is indeed better than for silica, there are severe losses as a result of non-uniform density/composition scattering. The objective for research into such complex materials was that, instead of operating at the present wavelength near 1.5 microns, a move to a much longer wavelength of, say, 10 microns would drop the intrinsic scattering loss by a factor of approximately 250 times. The reality was ZBLAN has highly variable compositions and is brittle, so we have stayed with silica.

Silica-based fibres and improvements from impurities

Hindsight is quite different from reality, and in the 1960s the focus was on how to make and use the most transparent materials. To reduce the absorption, one clearly needed to remove impurities of metals and water etc. The best candidate material for transparency was silica but, as mentioned, this has a very high melting temperature and the lowest refractive index of any glass. The emphasis on attacking the problem was to remove all impurities and worry about the melting and cladding difficulties later on. This was a sensible approach as it was possible to show better transparency could be achieved. The more important result, even for these limited successes, was that it meant that it was possible to maintain a level of research funding and support. As ever, impurities come in two forms, good and bad. The villains were ones which caused optical absorption, but a blinkered focus on them meant that one was overlooking the benefits of other impurities.

Window glass manufacturers use silicate glasses heavily doped with other oxides such as boron, sodium, calcium etc., (to lower the melting temperature and/or act as stabilizers to make a non-brittle glass) or lead (to increase the refractive index). These were well known facts, so a similar approach was therefore possible for the silica fibres. Surprisingly it was not instantly recognized. Not all standard glass dopants are compatible with fibre usage. For example, boron introduces optical absorption at the infrared wavelength used for many modern fibres (at 1.54 microns), but it was suitable for use with the earlier red laser signals near 800 nanometres (i.e. 0.8 microns). Physicists are sometimes their own enemy, as we mix up the units we use to describe wavelength. For visible light we use nanometres, so blue to red spans 400 to 700 nanometres. For longer wavelengths we use microns (one millionth of a metre). 1.54 microns is 1,540 nanometres, or about twice the wavelength of visible red light.

Because of the very large range of values, Figure 8.3 shows a logarithmic plot (i.e. in dB) of the changes in overall losses caused by absorption, scattering and bends etc. for an early optical fibre. The lowest loss values were clearly apparent in the minimum of the curve near 1.5 microns. Early fibres only used the transmission valley near 1.3 microns because there were problems in the choice of light sources and detectors. Modern material is somewhat clearer than the example used here. The absorption bumps between the two valleys come from excitation of residual water in the fibre. It is obvious on the figure, but we need to remember that the cause is water impurities, probably on the scale of just a few parts per million of the glass. It is still a very transparent glass, even at this peak wavelength, but we are looking at losses per kilometre, not just at window glass thickness.

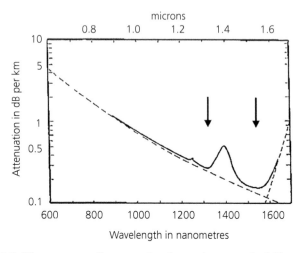

Figure 8.3 The pattern of attenuation losses in an optical fibre. Optical
fibre losses are the sum of scattering (the lower dashed curve) and absorption. There are two main valleys where transmission is good. The one near
1.3 microns was used initially, but now the light sources and detectors benefit from the better transmission near 1.54 microns. There are three absorption
peaks caused by tiny traces of impurities, such as water. The longer wavelength
absorption arises from vibrations within the Si-O bonds. The intensity of absorption falls by about 25 times between 600 nm (orange) and the 1540 nm
valley.

The silicon in silica can be replaced with other tetravalent elements,
such as titanium or germanium. Heavier elements have more electrons
and these cause more interaction with the light, which slows it down
and increases the refractive index. Early fibres from one US company
included some titanium to raise the refractive index of the glass. However, there is a better size and bonding match by using germanium to
substitute for some of the silicon ions. Germanium also raises the refractive index. Production of a germanosilicate core glass, surrounded
by a pure silica cladding, therefore has the ideal situation of a high refractive index core and a lower index cladding. We view the germanium
as a beneficial dopant, not as an unwanted impurity.

The melting point problem has been addressed in the same way by
adding low melting point materials. Their chemistry may differ from
that of tetravalent silicon, so not all the chemical bonds are correctly
satisfied. If one uses trivalent aluminium then it is necessary to put in

further compensators of, say, pentavalent materials such as phosphorus to redress the electronic state of the material (i.e. the average of a 5 and a 3 is a 4, which matches the silicon bonding arrangement). Errors in the bonding can result in the glass becoming coloured and absorbing. Details may differ, but the example indicates the principle of the methods used to reduce the melting temperature of a germanosilicate glass.

Removal of water and metal impurities, that cause absorption, is essential. The key fact to recognize is that one can add large quantities of impurities so long as they do not degrade the property of the glass that is needed for the application. For optical fibres, the 'bad' impurities are now down to parts per billion and this is the number quoted in publicity hype and marketing. The large quantities of germanium (for index enhancement) or sodium, aluminium or fluorine etc. used in adjusting the melt temperature are never mentioned. Further impurities, such as erbium, are also added in optical fibres which include optical amplifiers and lasers.

Whilst the preceding science can describe the construction of a long-range transparent glass fibre, the details and requirements rapidly escalate in terms of limitations and difficulties. For example, the indices of the core and cladding will be very similar at, say, 1.48 (core) and 1.46 (cladding). A large core makes it easier to couple in the laser light of the signal, but larger diameters will support many more optical modes which effectively have a range of speeds as the bounces increase the path length and slow down the signal. For pulse coded signals this limits the pulse rate that can be used. Solutions thus aim for narrower cores and/or graded indices at the boundary.

Summary of imperfections in the science of optical fibres

A summary of the role of imperfections is revealing in that in the development of optical fibre material, it became necessary to manufacture glasses which were at least a million times more transparent than previously existed. To a large extent, this meant exclusion of many metals and water vapour which all introduced absorption and attenuation of the light at wavelengths that were to be transmitted. Simultaneously, other impurities are added, (useful ones, so these are called dopants) to control the refractive index, melting and pulling temperatures and aid in high tensile strength glass formation.

A completely different set of social imperfections emerged in the failure of many eminent people and industrialists to understand the potential of glass fibre communication, severe prejudice for pre-existing and/or entrenched technologies, failure to provide funding, corporate competition, collapse and very destructive patent litigation (e.g. see Hecht's *City of Light* book). To me, working in academic-based research, such insights into corporate competition, deliberate attempts to undermine competitors, prejudice and obviously inept decisions were instructive. Such problems of course also exist in academia, but they are problems that one rarely mentions in standard literature or scientific teaching. The most positive conclusion that can be drawn is that if there are enough people with foresight, charisma, skill in salesmanship and sheer hard work, then progress is made, despite the efforts of the majority.

I often wonder how many equally good concepts have been lost by ideas not emerging at the opportune moment. I therefore suggest that there is potential value in reading the un-cited literature (both published and patent) of the last century. There will undoubtedly be many ideas which might now be fruitfully resurrected, and accepted, because of advances in technology and understanding, which could enable the discarded ideas to come to fruition. To emphasize such potential benefits, one may cite the predictive work of the Irishman Edward Hutchinson Synge, who published articles which, many years later, were re-invented as multi-mirror telescopes, laser ranging and near field microscopy. A slightly later example is that of the Hollywood film actress Hedy Lamarr, whose invention, in the 1940s, of frequency hopping for submarine signals was discarded as fiction, whereas it is now the basis of modern communication technologies such as smart phones and Bluetooth. Imagination of science fiction writers have equally moved from print to reality.

Since my aim is to show how to benefit from imperfections, I will quote a recent example from optical fibre transmission. When buried fibre cables are distorted, for example by the passage of a heavy vehicle, an earth tremor caused by an earthquake, landslip or impact on the surface, the tiny fibres can flex and this causes some signal loss as photons are reflected backwards down the fibre. Clearly this is an imperfection that interferes with the primary objective of optical communication.

However, cables frequently become obsolete and are replaced as there is a demand for greater capacity. Nevertheless the 'dead' fibre systems will remain in place. It was realized that these signal reflections can locate such movements. In one example, a geologist graduate student (Celeste Labedz) spotted the 'noise' in Alaskan fibres had detected some glacier quakes. Instead of Alaska having a single localized earth sensor, recognizing the fibre potential added 300 miles of sensors. Other examples have now mapped under sea fault zones and earthquakes with predictive information on the build-up of major earthquakes and volcanic eruptions.

9

Beauty from imperfections

Part I Why minerals and gemstones are attractive

We do not need to be scientists to appreciate the beauty of natural materials such as minerals and gemstones, and we may not have guessed that their shapes and colours are firmly linked to the physics of imperfections. Geologists and mineralogists spend great effort, and visit very inhospitable places, just to build immense collections and examine properties of variously coloured stones and pieces of rock. The lucky ones discover new commercial deposits, or they may identify a new type of mineral. New finds may be named after themselves, or the local region. This is fame. It is actually not that easy to achieve fame this way as, in practice, there are only between four and five thousand distinct minerals. By contrast there are well over half a million natural carbon compounds. To a non-geologist there appear to be an endless list of minerals, chiefly because, even for a single mineral, there are many variations in terms of colour caused by impurities. There are also many ways minerals combine into different strata. Therefore, five thousand is perhaps a surprisingly small number of natural minerals (it is fewer than the number of standard Chinese characters). Nevertheless, understanding natural materials has spurred the development and growth of many hundreds of synthetic crystals with desirable, or better controlled, properties. The main difference is that normally the synthetic crystals are purer. Scientists have also made a wide range of new materials which cannot form under natural conditions. A few colourful minerals are shown in Figure 9.1.

Growth of synthetic gems

The initial answer to this question is probably a mixture of profit, and a lack of natural stones of large size. A classic example in this field is the

The Power of Imperfections. Peter Townsend, Oxford University Press.
© Peter Townsend (2022). DOI: 10.1093/oso/9780192857477.003.0009

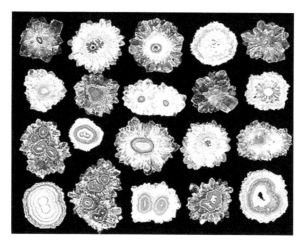

Figure 9.1a Examples of attractive natural minerals. This collection shows examples of the variety of colours and patterns that can exist from a very limited range of minerals. The samples are stalactites of agate and amethyst.

(b)   (c)

Figure 9.1b and c Two examples of less familiar minerals used in table tops and jewellery are the green variscite and the mauve charoite. These three photos have been provided by my friend Dr Javier Garcia Guinea of the Museo Nacional de Ciencias Naturales, in Madrid.

development of synthetic ruby and sapphire where, for more than 100 years, it has been possible to grow them by melting their constituents in an oxyhydrogen flame. The basic material is aluminium oxide, and the pure material grows as a very hard clear crystal. Slices from it make excellent scratch resistant watch fronts and, mostly, we will not realize it is not just a normal glass. If we add some chromium oxide into the mixture, the chromium ions easily replace aluminium ions in the lattice (they have the same bonding chemistry) and the crystal becomes the gemstone we name as ruby. The technique is fairly simple. From the heated region of the flame molten droplets fall and grow on to a tiny seed crystal. The seed sets the pattern of growth and, with careful control, one can then develop a very large sample, say 12 inches long (30 cm) and several inches in diameter. Stones can be cut from these immense blocks (called boules) and incorporated into jewellery, or used for lasers etc. Indeed, this was the material of the first visible laser in 1960.

Alternative dopants into the same host material produce other colours, and additions of nickel in the aluminium oxide structure give a yellow sapphire. Blue sapphire is made by titanium doping, but this is rather more difficult as the chemical properties of titanium require four bonds to oxygen ions, whereas the aluminium and chromium need only three such neighbours. Aluminium and chromium are trivalent, whereas titanium is tetravalent. The problem can be resolved by addition of secondary impurities and it is then feasible to grow blue sapphire samples of good quality.

The jewellery trade is probably not seriously concerned with competition between synthetic and large natural rubies or sapphires, as careful examination will show that the synthetic material is usually *too perfect* compared with natural gemstones. Jewellers are, therefore, one of the few groups of the public who welcome signs of imperfections. In non-jewellery applications, the real value of the sapphire structure is in the extreme hardness of the material, and its chemical inertness. The difference in quality between natural and synthetic gem material can often be very obvious and Figure 9.2 shows some amethyst examples. Amethyst is just an impurity-doped version of quartz (crystalline silicon dioxide).

Effects of cookery on the colour of gemstones

Heat treatments at a modest temperature, say a few hundred degrees, rarely alter the crystal structure, but they provide enough energy for electrons to move around in the crystal and new types of chemical bond can form that modify the optical absorption properties of the materials. If absorption bands occur in the visible region then we see coloured crystals. It has become fashionable to market many minerals with colours that did not occur naturally, and two prime examples are topaz and spodumene.

Topaz is named after an island in the Red Sea and, in Sanskrit, it meant fire. Topaz exists naturally in a range of colours from colourless

Figure 9.2 Comparisons of natural and synthetic jewellery stones. The photos show examples of jewellery made with natural and synthetic amethyst. The nominally flawless examples are the synthetic ones. Value is normally greater for the natural, and visibly flawed, material. Photo from Dr Javier Garcia Guinea.

to blue, yellow or red, and the colours depend both on the impurities and their electronic bonding in the lattice. Some colours are greatly preferred and so colour conversion is usually made by a variety of heat treatments and some 300 tons per year are processed this way.

Spodumene, lithium aluminium silicate $LiAl(Si_2O_6)$, is normally a fairly drab colour as it forms grey or opaque crystals. However, there are two types of clearer gem quality material with the names kunzite and hiddenite. These are nice classic mineral names as the pink or violet kunzite was named after a New York jeweller, whereas the green/yellow versions were named after Hidden, who discovered the material. Drab stones do not have a great market value but the colour often results from manganese impurities. In nature the manganese sits in the trivalent Mn^{3+} state, but if the material is X-ray or gamma irradiated, it switches over to the tetravalent Mn^{4+} charge state and the crystals become an attractive brilliant green – which can easily be sold.

Jade is a generic name which covers a range of similar materials of different colours from green to white or brown. The green versions tend to be more valuable and the colour is related to the presence of iron.

Once again scientific methods have been tried to change the colour, and so enhance the value. Chemical surface reactions can achieve this and allow ion exchange with iron into the surface. Ion beam implantation is also feasible. Such treatments are, commercially, not totally successful as only a very small depth beneath the surface is modified and the material can be easily scratched which exposes the true interior colour.

Mineral prospecting needs a good knowledge of minerals, not just in the form displayed in museums and shops, but also in the totally different form that they can have when still in the ground or in a mine. We know that valuable gemstones may be disguised within uninteresting exterior geodes (as for amethyst), or in combination with other minerals. Less obvious is that they can be a totally different colour in a mine compared with the gemstones we see later. A classic example of this is called sodalite. This is not a mineral that is suited to beginners of crystallography as, although the overall pattern is cubic, the detailed building block is a mixture of sodium chloride in a sodium aluminium silicate matrix. This forms as a chemical package described by $(NaAlSiO_4)_6(NaCl)_2$ (i.e. some 46 ions in the building block of the unit cell). The presence of the chlorine means it behaves in a somewhat similar way to the photographic halides, in which exposure to light allows atoms to move around. The consequence is that in the low light level in the mine the crystals look pale and colourless but, as soon as they are brought out into the UV of sunlight, there is a photochemical reaction and the mineral turns a dark blue as the result of vacancy defect sites. These imperfections add value.

Diamond – the ultimate gemstone

Diamond is mechanically even harder than sapphire and so its main industrial application is in cutting and grinding structures, such as drill bits for mining. In this usage it does not matter if the stones are coloured or flawed, rather, the need is merely to have large quantities at low price. To synthesize diamond is possible, but it is still a much more difficult challenge than for sapphire. The aim is to turn any form of carbon (such as graphite) into the compact diamond crystalline structure. Not only do we need a high melting temperature, but any simple attempt to cool a carbon melt leads not to the formation of diamond, but instead to the production of soft black graphite. Graphite is certainly not a hard material, nor is it of value for gemstones. The underlying problem is the same as I mentioned for quartz, where several different structural phases can exist as we cool from the melt down to room temperature. The familiar crystal structure of natural diamonds is formed because, in addition to high temperature, the growth takes place at extremely high pressure within the mantle of the earth. Efforts at synthetic growth are forced to duplicate this high-pressure situation.

Scientists are a fairly determined breed (obstinate might be a better word) and so are attracted to the problem because it is challenging. Consequently, growth of synthetic gem-size diamonds is now feasible, albeit expensive. Production of large gem quality stones is not yet economic, but for smaller discoloured diamonds, synthesis is a realistic and commercial option, and indeed it has a very large industrial market.

Diamonds may be a girl's best friend or a man's road to ruin, but the market value is artificial as the price is set, not by rarity, but by deliberately limiting the number of stones released for sale. Their attraction is that they have a very high refractive index so can be cut to scatter the light. They are hard and chemically stable, so in romantic terms they are a symbol of pure enduring love. Unfortunately, they are not the conventional symbol of purity. Their mode of production, deep inside the earth, means that not only do they have the structure of highly well-ordered atoms of carbon, but they always include traces of all the other rubbish which was present when they were molten in the mantle. I remember a friend from a diamond company saying that he had never analysed a diamond which did not show traces of at least forty impurity elements. As for the ruby example, the synthetic diamond material may actually be purer than the natural examples. The one exception may be the presence of traces of a chemical catalyst that helps in the crystal growth, which in some cases has been nickel.

More recently there have been deliberate attempts to control nitrogen impurities in the synthetic diamond, as this can be used as a high temperature semiconductor in applications where silicon, or other standard semiconductors, cannot function.

The market for lower price imitations of diamond is considerable and, in principle, one only needs to make a material of very high refractive index (to give similar optical properties) and that is transparent throughout the visible spectrum (as for diamond). This stimulus has proved to be a powerful incentive and resulted in great progress in crystal manufacture, and led not only to high refractive index diamond substitutes (such as the so called cubic zircon and strontium titanate) but also many other compounds with an interesting range of properties that are now used in modern optical systems and lasers.

Giants of natural crystals

The price of gemstones tends to rise with size, as larger stones are found less frequently than small ones. Natural examples of large, uniformly coloured crystals (e.g. many centimetres across) are quite rare for most materials, but there are some classic examples of Brazilian quartz which weigh in at around a ton. Such examples are never totally clear and perfect, as they always contain regions of discolouration. One common quartz impurity is aluminium, which produces a brown, smoky quartz but, relative to most other minerals, quartz crystals (SiO_2) tend to grow as fairly clean and clear specimens. There are few impurities which are soluble in the quartz matrix. Exceptions are the classic examples of doped quartz in the formation of amethyst, (Figure 9.2) which variously includes iron, manganese or titanium in the crystalline structure. To a field geologist the amethyst may be hidden in the interior of rather drab looking geodes. These geodes are the pressure vessels in which crystal growth took place, and so the colours and gradations of material formed on the internal walls depend on the concentrations of starting materials, plus the high temperature of the chemistry and crystal growth. There is preferential uptake of some salts, so the fingers of crystal which form within large geodes change colour along their length. Growth is always on the inside walls of the rock. Figure 9.3 shows different types of geodes, including not only familiar amethyst examples (with specimens over 2 metres long), but also some extremely large gypsum geodes. These amethyst geodes are possibly the largest that have been discovered. My two substantial Spanish geologist friends are sitting comfortably inside an immense gypsum geode that they found on the Brazilian coast (see Figure 9.3). These crystals of gypsum are not the largest that have been discovered. Instead, supersize crystals include selenite specimens from the Naica mine in Mexico, where single crystals as large as 11 metres long and 4 metres across have been found. For non-geologists this is a confusing literature as selenite is just a different name for a variant of gypsum. Both are calcium sulphate dihydrate ($CaSO_4.2H_2O$). Despite the name, selenite should not normally contain any selenium. The name derives from the Greek for the moon (selene)

(a)   (b)

(c)

Figure 9.3 Natural gypsum and amethyst crystals within Brazilian geodes. In (a) are Javier Garcia Guinea and Martin Fernandez inside a geode they found. In (b) is Roberto Almarza in front of some large amethyst-coloured geodes. In (c) is a view inside a gypsum geode. Photos were provided by Dr Javier Garcia Guinea.

because the appearance of the translucent mineral is associated with moonlight.

Can diamond crystals be equally large?

By contrast, even the largest known diamonds are on a totally different and much smaller size scale. For example, the original Koh-i-Nor gemstone weighed about 280 carats (57.5 gms). Because there was a region of discolouration it was cut down to a mere 109 carats (22.4 gms). Note quoted values may differ as the weight of a carat varied, but is now re-defined to be 0.2 gms. Several larger diamonds have been recorded, but invariably they had serious errors, such as a large black spot in the centre of the Cullinan stone, or were larger, but too poor for any good gem quality. The Cullinan weighed in at some 3,100 carats (621 gms) and may have been only part of a larger stone. It was cut into nine large gemstones, the greatest of which was called the Great Star of Africa (at 530 carats or 106 gms). No value is cited for the original stone but, currently, the Great Star is estimated to have a value well in excess of $500 million.

The units of carats do not give an obvious feel for the size of a stone, but originally a 621gm diamond would have been 177 cubic centimetres. If spherical, this represents a diameter of ~7 cm (i.e. roughly the size of a tennis ball).

Part II Defining the size and shape of natural minerals

The natural progression from admiring and collecting natural minerals, and selling gemstones, was to consider how the atoms of the structure are assembled. As a result of such thoughts over the last 130 years, there has been increasing understanding of both the possible visible shapes into which crystals can grow, and the details of how their atoms are chemically and physically bonded together. This is now standard crystallography, physics or chemistry, but to a large extent it is conceptually no more difficult than assembling children's building blocks.

Many minerals, and laboratory crystals, show a natural preference for a basic shape of the building block. Breaking pieces of calcite ($CaCO_3$) just gives smaller pieces of the same prismatic shape, with the same angles defined between the different main surfaces. Similarly, the grains of table salt (sodium chloride, NaCl) are cube shaped with right angle corners. All minerals display similar tendencies to form building blocks of a characteristic shape, unless they are particularly impure.

Tourist photos from Northern Ireland invariably include pictures of the very large blocks of basalt which have crystallized from a lava flow into the Giants Causeway in County Antrim. They are clearly impure, but still have grown in a pattern of similar sized blocks (in a pattern copied in many concrete street pavements). The basalt blocks are not identical. About 50% have six sides and another 35% have just five sides. In an early paper describing them, in 1694, Sir Richard Bulkeley said they were some 18 to 20 inches across (~50 cms) and the pattern of blocks has a mix of six- and five-fold symmetry. This is interesting as it is precisely the pattern used in both the leather patches for the design of a modern football; and in the design of geodesic domes by the architect Buckminster Fuller. The same five and six mixed pattern was noted in a form of carbon involving a sphere of sixty carbon atoms, and these were then named Bucky-balls. This form of carbon, discovered in the twentieth century, is part of a series of materials called 'fullerenes'.

Well before the modern views on atomic structure and chemical bonding, there had been a fascination with models of packing objects together in the most efficient way. Early references considered packing of metal cannonballs and, here, the solution is similar to that of the greengrocer for stacking apples (i.e. pre-supermarkets). The atoms (apples) of the first layer are placed in contact. The second layer, instead of being vertically above the first, is displaced sideways into the hollows. This is stable and is a fairly efficient packing route which leaves little free volume. It only requires that the atoms (or apples, or cannonballs) are carefully placed, one by one, in the correct location. Since we cannot build visible sized crystals atom by atom, a model closer to real crystal growth would be to put spherical sweets or marbles in a large jar to find the maximum density that can be reached. Merely pouring them into the jar will give a random glasslike network of some zones of tight packing with boundaries where the next block of material is differently aligned. Adding thermal energy to simulate the conditions near the melt temperature (i.e. gentle shaking of the jar) will allow the 'atoms' to minimize the free space and give an entire ordered structure in three dimensions.

Dense packing of spherical objects is not always good news as it can happen accidentally in ball mills, which grind metal spheres together to make uniform size ball bearings. If the balls go from a random liquid like pattern into a compact crystalline one, then the ball mill ceases up. To avoid this, some very large balls are included in the mill. They inhibit the perfect crystalline stacking from ever occurring.

Another very simple arrangement would be to have the first layer built as before, but with the second layer identically arranged with each sphere vertically above the ones in the lower level. This is called a simple cubic structure. It is energetically less favoured for arrangements of solid objects, such as apples or cannonballs, as it needs some forces at the edges to hold it together. Under gravity, such a structure would merely collapse sideways. With the addition of forces exerted by electronic bonds it can, however, be stable. For equal sized spheres there is far more unused space between the atoms/apples etc., than for the first model. Patterns are shown in Figure 9.4.

The right-hand model could stack more compactly with the next plane centred above the gaps in the lower level. Similarly, for the third plane there could again be a displacement into the valleys, or it could duplicate the base plane. In this model there could be an ABCABC arrangement, or an ABABAB structure. If a grocer is stacking apples where there are no sideways bonding forces and the only force is from gravity, the apple stacking used is ABCABC. Similar views have been sketched in ancient naval manuals for stacking cannonballs on a ship.

In metals, there is an attractive force between all the atoms as they share the electrons involved in the chemical bonds. By contrast, for insulating materials, such as salt (sketched in Figure 6.1) the ions of the lattice are held together by electrostatic attractive forces between positively charged ions of sodium and the negatively charged chlorine. This gives a three-dimensional chessboard type layout of the two types of ion. Effectively there are two interlaced lattices of sodium and chlorine. A chlorine ion is always in the middle of each face of a sodium cube (and vice versa). Hence these are called face-centred cubic structures.

(a)   (b)

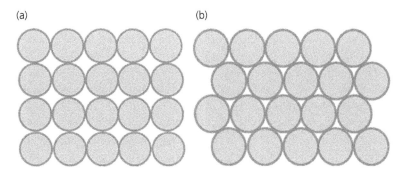

Figure 9.4 Simple packing arrangements of atoms. There are several viewpoints to consider in stacking atoms in a lattice. The sketches above are on a flat plane and the spheres could be in a regular chessboard type array, or packed in more tightly, as shown on the right side. Consider this as plane A. If we add higher levels, B, C etc., then there are more options. The first would be that the next plane (B) is balanced directly above the lower ones, and where there are forces bonding the atoms together this can be stable (called a simple cubic structure).

The basics of atomic structure and the resulting chemistry

For some reason scientists are not always brilliant at communicating their knowledge to the general public. The result is that the type of materials science I am discussing here may seem difficult and complex to the non-scientific public. This is really unfortunate as the ideas are extremely simple and all we need are a few key facts to understand basic chemistry and physics. The bonus of looking at the basic ideas is that they give useful insights into simple chemistry, crystal growth and the structure of imperfections in solids.

Atomic models differ from the models of cannonballs and apples because we just have to remember that atoms are made of a very small dense nucleus, which has a positive charge, and a large volume of negative electrons, which are circulating around the nucleus. A century ago it was assumed that there is similarity between atomic structure and the solar system. The massive sun was considered as the equivalent to the positively charged nucleus, and the electrons are like the planets, comets and asteroids circulating in fixed orbits about the sun, as summarized in Table 9.1. The gravitational force between the sun and the planets etc. decreases rapidly with distance from the sun (i.e. it falls as the square of the distance).

An early, simple model of the atom assumed electrons orbited the nucleus in much the same way that planets orbit the sun. For example,

Table 9.1 Planetary pattern of planets circulating the sun.

	Mass	Distance to Sun	Annual period (years)	Daily rotation (days)
Sun	333,000			~27
Mercury	0.05	0.4	0.24	59
Venus	0.89	0.7	0.62	244
Earth	1	1	1	1
Mars	0.11	1.5	1.88	1
Jupiter	320	5.2	11.9	0.4
Saturn	95	9.5	29.5	0.43
Uranus	17	19	84	0.75
Neptune	17	30	165	0.8
Pluto	0.002	40	248	0.3

in the solar system 99.85% of the mass is contained in the sun and the planets add only a very tiny fraction. Similarly, the nucleus defines the mass of an atom. In the solar system, the spacing of the planets does not follow an obvious pattern, except that the planetary years increase with distance. No pattern emerges for the planetary masses, or even the periods of their 'daily' rotation. Examples of solar system data are indicated here relative to earth values.

Very similarly, the electrostatic strength of the bond towards the positive atomic nucleus and circulating electrons also falls as we move to orbits that are further away. For fixed orbits this means, in terms of energy, that we have a staircase of energies and it requires a fixed amount of energy to jump from an inner orbit to an outer one. Eventually there is the possibility of escape from the system. Chemists call this escape energy an 'ionization potential' (therefore the energy needed for escape decreases for the outer orbits).

We also need to consider how many negative charges, the electrons, will be circulating around in an atom. In total, this will merely be the same number as the number of positive protons in the nucleus. So, since carbon has 6 positive charges within the nucleus, it must have 6 associated electrons to make it neutral (nitrogen has 7, oxygen has 8 etc.). As a first guess at atomic structure it was assumed the electrons circulated in fixed orbits. The main weakness of the early model of electron orbits as matching planetary motion is that real electrons require quantum mechanics to describe their motion. Effectively they have wavelike properties in some of their responses, so they are not hard, solid objects following clearly defined planetary type orbits. Instead we must think in terms of a probability, or an average value, of where they are likely to be. The problem is no different from defining the positions of players on a football pitch. They may start in fixed positions but can cover a large area of the pitch. Those on the left side may have an average position which is to the left of the field. The defenders will stay further back than the forwards, but citing average values disguises how much ground they cover. An indication of the electron orbit probabilities is shown in Figure 9.5.

How do electrons determine chemical reactions?

The very simple model is quite good, as it says that there are electrons centred around distinct orbital distances but it does not say how many

Orbital patterns

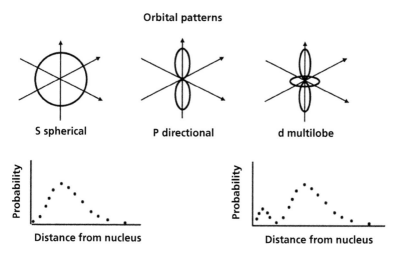

Figure 9.5 Probability of electronic orbits around a nucleus. Even the simplest model of electrons moving around a nucleus is complicated as there is no fixed orbital distance, only a most probable distance, even for those electrons which take nominally circular orbits. Many electrons associated with a particular orbit do not move in circles but, instead, track back and forth. They still have an average position which is at a particular distance from the nucleus, but for some of the time they could be very close to the nucleus. So instead of nice, neat planetary orbital circles (or ellipses) at constant distances from the sun, we have to think in terms of the probability that the electrons are at any particular distance. A probability diagram for a planet would be a vertical line of 100% probability at a fixed distance value (ignoring our slightly elliptical orbits). This is not the case for electrons, as sketched above. If planetary motion were similar then, for some of our time, we would be close to the sun but, at other moments, far further away. The labels of s, p, d etc., are conventionally used to indicate which type of orbit is relevant.

electrons can be accommodated in each orbit. This is essential, as it is the number of electrons in the outermost orbit which defines the chemistry of the atom, and the way in which atoms stack together into molecules and crystals. I will propose a very simplistic view of different atoms where they are like a set of hotels. There are a number of floors (the orbital levels) with different types of room on each floor (the s, p, d etc. type orbits). For economy the hotel managers prefer to operate with completely filled rooms and filled floors. The manager therefore puts guests into the lowest room on the bottom floor. When this is full, the next customers move to fill the next floor. If there is a stray guest, or just one or two vacancies on a floor, they do a trade with a neighbouring hotel and transfer guests between them. Hopefully both hotels then finish with completely filled floors and happy hotel managers.

The oddities of quantum mechanics say that the lowest floor has space for only two electrons, there are eight spaces on both the second and third floors with two types of room, eighteen on the next etc., and higher floors with more room types. Also they can have mezzanine type sections. These would be funny shaped hotels like an inverted cone, and the room heights (i.e. energy distance between floors) decreases as we move upwards. If the floors are all filled then there is no need to exchange guests with neighbours, and so these match the inert gas materials of helium, neon, argon etc. One too many guests are the alkali metals (Li, Na, K . . .) and so they give a guest to a neighbour. At the other end of the scale, the floors missing just one guest are fluorine, chlorine, bromine . . . Hence, electron exchange results in Na^+ and Cl^- ions and a pair of linked hotels called NaCl (as sketched earlier in Figure 6.1).

Building a salt crystal

Once having realized electron transfer produces ions and there are attractive forces between the positively and negatively charged ions, we can design crystals. As mentioned in Chapter 6, for the NaCl salt crystal there is an ionic arrangement like a chessboard with black and white squares, as this will surround every positive sodium ion by negative chlorine ions. Moving to three dimensions means we must maintain the chessboard pattern vertically. Hence, we have designed a cubic structure for sodium chlorides (and sixteen of the other possible alkali halide materials). The radii of Na^+ and Cl^- are about 0.102 and 0.181 nm respectively (e.g. relatively, apples and small melons). This means the chessboard pattern, and repetition size, is dominated by the large chlorines. The smaller sodium ions are happy sitting in the gaps between the halogens. The relative volumes are ~5.5 to 1, so there is little free space left over as the sodium ions fill the gaps. Our model from this ionic binding was excellent. It produced a stable structure, chemical bonding with full outer electron shells and minimized the free space.

The method is in trouble and will break down once we reach very large ions, especially when they are of a similar size. This happens with the alkali halides such as caesium chloride, where the caesium positive ion radius, at 0.170 nm is very similar to the Cl⁻ ion radius of 0.181 nm. Such large spheres of nearly equal volume do not fit well together in the chessboard pattern as there would be too much free space. They reshuffle into a new structure (called a body-centred cubic structure).

Such models of packing spheres, either into a square structure (simple cubic) or in the triangles of the cannonballs (face-centred cubic), do not require any modern physics, and were discussed by Harriot of Oxford in a letter to the astronomer Kepler in 1611. Rather than consider packing spheres, an alternative task is to stack cells which will contain other materials. This is also a standard problem for any marketing company, with examples from milk cartons to biscuit tins to patterns in brick walls. In many ways it also mirrors the design of natural honeycombs for storing honey and larvae.

Addition of complexity in the structures

For elements and alkali halides we only worried about spherical type building blocks but, in most materials, there is more complexity. A material roughly similar in structure to sodium chloride is sodium azide (NaN_3). It grows in a similar chessboard type pattern, but with a group of three nitrogen ions sitting in a row at each of the equivalent halogen sites. However, we now need to consider how to tighten the packing, so instead of a line of nitrogen ions along the line of the chessboard, the triplets sit at an angle. The next question is to decide if the angle is always the same way at each site. Azides may not be a familiar set of compounds but lead azide, $Pb(N_3)_2$, is a popular primary explosive in terms of power delivered when it goes unstable (bang!). Azides are not ideal to handle as they can be very temperamental. The equivalent barium azide was, by contrast with lead azide, very stable and one could drop a sample without any danger. Nevertheless, I remember some colleagues improved the growth process and removed impurities (primarily traces of water) and produced nice large crystals, which were then as dangerous and unpredictable as the lead azide. Depending on one's viewpoint the water impurity had stabilized the lattice, or removed the explosive potential.

In our ball model for crystallography, we need to consider chemical groups (e.g. nitrogen azide chains, carbonate, nitrate, sulphate radicals etc.) and many groups of ions into more complex structures. Spherical balls are no longer a sufficient model and one needs to include other shapes. Our building blocks have moved from football spheres to include rugby ball shapes, rods and other variants of the building units. Such factors reduce the possible way we can assemble a crystalline structure. It was realized that there are only some fourteen basic structures which can be formed into repetitive three dimensional lattices (called Bravais lattices). The logic for this was developed by Bravais back in 1848, and his nomenclature of simple cubic, tetragonal, orthorhombic etc., is still widely used by mineralogists and physicists. The detailed proof of his models came with establishment of X-ray crystallography and the alternative later structural methods of analysis.

Bees and honeycombs

Packing together empty cells with the minimum of material and the highest possible density is clearly a problem in beehive technology, as wax needs to be produced and heat is better retained in a small structure. The honeycomb solution of six sides, therefore, appears to be from an intelligent minimization of the amount of wax needed to build the structure and maintain rigidity and stability. The solution is pretty impressive and is fine for the two-dimensional array of the honeycomb. It would not necessarily be the best design if the bees were planning a three-dimensional crystal of comb cells. Free ranging bees, as distinct from those conditioned by the design of honeycomb supports introduced into the hive by apiarists, may not go for three dimensions, but they do build two combs back to back, accessed from opposite sides. The cells are not in line, as this would require a complete wall of thick wax to give strength between them. Instead the two sets of hexagons are offset so one pattern centre is in the middle of the other pattern. This is a solution that has fascinated scientists for many years and it appears that mathematicians may be slightly more intelligent than the bees, as a Hungarian, Fejes Tóth, offered an alternative packing which would use 0.4% less wax. He displaced the rear cell unit sideways by half the length of the cell edge. I thought about this solution for a time but now have doubts that, overall, the bees may still not have the best design for a small honeycomb. It depends on how he considered what happens at the final boundary of a complete, but small, honeycomb. His model appears to me to have a slightly larger perimeter for one side of the double comb, and would need additional structural strength, so perhaps the bees know best.

My crystallographic models are simplistic but they emphasize that a non-physicist, (and even a baby playing with toy blocks) will have more than enough intuition to see the principles that can define the ways ions stack together into crystalline structures, and for my purposes, the types of building error that can occur. Further, a comforting thought for a non-scientist is that even professional biologists, chemists, and physicists often have very real trouble in visualizing anything more complex than my simple examples.

10

Valuable imperfections in crystal lattices

Part I Building crystal lattices

Building with bricks and mortar is far simpler than growing crystals but, despite our hopes, only the most naive expect perfection in the design, construction and performance of a building. It is not surprising, therefore, that if we fail with a few thousand bricks, and rooms that can be corrected, we should expect some problems when growing crystals involving billions of atoms. Crystal growth is often treated by academic scientists as a minor technical problem and not mathematically or philosophically challenging. There is a clear class divide along the lines of academics and technicians, which is totally unjustified. In reality crystal growth requires very considerable skill, imagination and perseverance, but is rarely rewarded with fame and recognition. Industrial companies may be better, since the quality of materials is linked to financial success. Indeed, without crystal growth skills we would have no modern sophisticated technology, such as electronics, computers or high-power single crystal jet engines.

The range of possible growth problems is quite daunting. Firstly, we normally want to control the exact composition of the components, and then consider adding in precise amounts of other elements. Rather than a few bricks we need to do this on a scale with billions of atoms, whilst keeping undesirable impurities down at the level of parts per billion, and dopants at controlled levels as low as parts per million in specific locations. This precision is outside our normal experience. Rather than building brick by brick, we cannot see or control the local growth conditions and must rely on self alignment of the atoms and hope that they do not arrange in variations that we do not want.

If we use any natural minerals as a reference then they will show all too clearly that, whilst the natural samples may look attractive, in

The Power of Imperfections. Peter Townsend, Oxford University Press.
© Peter Townsend (2022). DOI: 10.1093/oso/9780192857477.003.0010

terms of technology and crystallography they can be disastrously poor and irreproducible. The second problem is that, even with a perfect mixture, many crystals can grow in rather different structures. The builder's equivalent is that the same number of bricks can be used to make a tall tower or a set of garages. For technology, the example with carbon atoms is producing either diamond or graphite.

Crystallography and Murphy's Law

The popular law attributed to Mr Murphy is that if things can go wrong, they will do so. (In fact, this was only the popular version of the law coined by Edward Aloysius Murphy, and he did not like this interpretation!) In crystallography terms, building mistakes are immediately evident, even without the addition of impurities. We can quickly guess at the simpler types of error which could occur, and I have listed some of the most familiar errors in crystal growth in Table 10.1 and Figure 10.1. I think of this as a beginner's starter kit of standard imperfections at individual lattice sites. By contrast, Figures 10.2 and 10.3 sketch growth problems of larger scale features. From the house builder's view, these are errors in the layout of the streets.

We should not be too ashamed if we fail to grow large, perfect blocks of crystal, as the harsh view of reality follows immediately from an understanding of thermodynamics, which unequivocally says it is impossible to grow perfect material. Imperfections are an inevitable consequence of the physical laws which determine our lives. If we see this as a feature for crystalline imperfections, then maybe we should be more tolerant with ourselves and assume that personal faults (e.g. such as the seven deadly sins) may also be genetically programmed into us as part of the gene equivalent of thermodynamic factors. What a relief – goodbye guilt!

In a really simple material composed of a single chemical element, such as copper, tin, silicon or carbon, we can start our list of errors in the crystallography by asking if all the lattice sites are occupied. Inevitably the answer is no. Unoccupied sites are called vacancies and, at room temperature, they will typically occur at some tens of parts per million sites. This is not bad in engineering terms. For example, if we consider how many errors we make doing familiar repetitive tasks, such as typing, then an error rate of one per thousand key strokes (i.e. about three

Table 10.1 Typical basic errors in building a crystal lattice.
There are two basic types of standard defect. Firstly, there are defects which mainly influence a small region of the lattice. Initially it was assumed if only one lattice site was imperfect it should be called a 'point defect'. In reality even these alter many other nearby lattice sites. Secondly, there are defects influencing very long rows, or blocks, of atoms termed 'extended defects'.

Point Defects come in a variety of forms such as:

Vacancy an atom missing from a lattice site
Interstitial an atom sitting between normal lattice sites
Anti-site atoms of a compound AB where A is on a B site, or B on an A site
Impurities the wrong atoms are included in the lattice
Non-stoichiometry the overall composition is not as predicted by the chemistry
Pairs and clusters groups of defects which minimize strain or charge problems
Precipitates inclusions of blocks of different compounds
Nanoparticles typically these are clusters of atoms that are not soluble in the host lattice

---

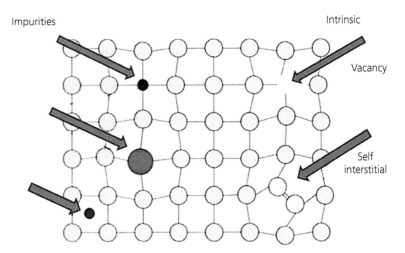

Figure 10.1 Simple local defects in building a lattice. The sketch shows a simple lattice of atoms with impurities that are replacements or interstitials, either smaller or larger than the host sites. Intrinsic defects shown here include a vacancy and an interstitial pair of host atoms.

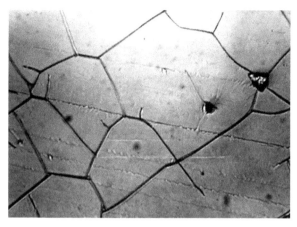

Figure 10.2 Extended defects such as grain boundaries. Grain boundaries (i.e. analogous to the joints at the edges of paving stones) can appear as shown below. At such places the chemical bonding is distorted and compositions may vary. They are ideal sites to both attract rubbish (as shown at one of the boundary junctions on the right side of the picture), or preferentially allow entry of solvents. The analogy with paving slabs is apt. These boundaries have significant problems, or benefits, of lines of weakness and altered chemical activity, and easy mechanical fracture. On the positive side, they are ideal as sites for catalysis or the fracture as needed to make stone tools.

errors per A4 page) is actually very good. After detection and correction there will still be maybe a hundred per million retained. Bank transactions aim for a level of 'one in a million' error rate, as a minimum quality standard. Since most of us can recall at least one bank statement error this is probably substantiated. Nevertheless, the one in a million error rate is still very significant if we consider how many bank transactions are being made each hour.

Along with vacant lattice sites, atoms can finish up away from their correct locations, to sit somewhere between the 'official' sites, and we call these interstitials. Since even for diamond or silicon we cannot start with perfectly pure material, there will always be impurity ions in all materials. Impurities can settle on to lattice sites to replace host ions, drop into vacancies, bond together to make nanoparticles, make little separate inclusions of different compounds etc. These features are all consequences of the laws of thermodynamics which try to minimize the energy of the overall assembly of atoms. In house building terms

(a)   (b)

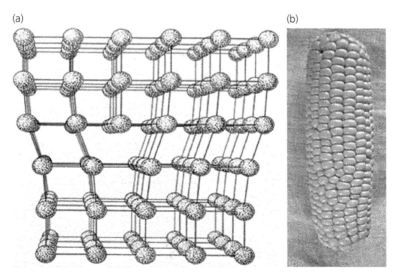

Figure 10.3 Examples of dislocation type extended defects. Dislocations occur where part of the crystal has more planes of atoms than the neighbouring region. In the figure on the left, imagine there are 6 atoms across the upper horizontal rows, but only 5 in the lower layers. I have drawn the mismatch at the third column. The outer boundaries fit together nicely but the rows around the 3rd column have a misfit distortion extending back into the crystal. This dislocation changes the bonding, and is a line of weakness with modified chemistry. Rolf Hummel, of the University of Florida, noted the situation resembles the structure on the surface of sweetcorn cobs. My photo of a corn cob shows disorder of the arrangement of the seeds where the top half, with more rows, meets the lower end of the cob.

we are trying to minimize costs, and often this means shortcuts and errors.

On a small scale of a few million atoms it may not be that difficult to arrive at a crystallite which is reasonably well ordered, especially if we can start growth from just a single seed point. In most situations, growth will commence randomly all through the melt volume. This means we will finish up with blocks of crystallites that are slightly skew relative to one another and there are boundaries between the various grains. The boundary zones will not just be weaker than the grains, but will also have differences in the stability of the chemical bonds. They therefore are lines where material will fracture. Excellent, as without them we could not have reached the Stone Age. However, the changes in strength and chemical reactivity can variously be useful or problematic.

One common type of long-range error which occurs even within a single grain is called a dislocation. This appears if the number of atoms along one row in a block of crystal differs from the number in the adjacent row. The bonding will distort to make as many atoms as possible tie up across the mismatch plane, but inevitable distortions alone cannot satisfy the mismatch. A line of error appears, which is the dislocation. There are many variants on this theme and Figure 10.3 shows a classical physics type view of such an error for lattice atoms, together with a very familiar one which appears on many corn cobs.

Dislocation chemistry

I have included an image, in Figure 10.4, of a crystal surface where a dislocation had emerged at the surface, as it hints at the complexities which can result. In this example, the altered chemistry of the dislocation has attracted impurity atoms. This produces *three* different etch rates against chemical or mechanical attack. These rates may also change with the type of chemical solvent that is used. A normal value for the 'perfect' material is very much slower than for the impurity-rich dislocation core, but it is a much faster rate for the weakened bond structure around the dislocation. In this example, because the dislocation has attracted some impurities, the etch rate is different and is much slower than for the original material. The consequence is that we see the tip of the spike has been protected and remains at the original surface level. These three consequences are a protected zone (the tip of the spike), a deep pit from the fast etching and an intermediate rate of the normal material. In many ways this is just like the problems of soft and hard rocks which are etched away by wind-blown sand. Classic examples are seen in the pictures of the pillar structures in Bryce Canyon (some even still have hard rocks balanced on top of the sandstone underlay).

Impurities and strain in growing a ruby lattice

In ruby we gained the red colour by replacing some of the aluminium ions of aluminium oxide with chromium. Chemically this is fine, as both are trivalent so bond well with three oxygen ions. In terms of

(a)   (b)

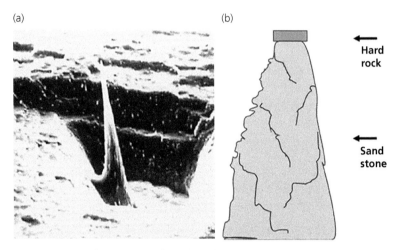

Figure 10.4 Dislocations and surface chemistry. The changes in chemistry at a dislocation line can add together into complex surface patterns. In the image on the left there was initially a dislocation line that reached the surface, which attracted impurity ions. When the surface was attacked three etch rates developed. Firstly, the bulk rate of dissolution gave a background plateau. Secondly, the distorted zone around the dislocation, the lattice, is weakened and etches faster, but where the line was decorated with impurities it has scarcely been attacked at all. This gives a spike, with the top still at the original surface level, sitting in a pit which is deeper than the etched level of the bulk material. The sketch on the right shows how a layer of hard rock sitting on soft sandstone can be wind sand-blasted to leave pillars. Classic examples are seen in Bryce Canyon, where the features can be 100 metres high.

ion size there is some problem, as Al^{3+} and Cr^{3+} ion radii are 0.054 and 0.062 nm (nanometres) respectively. This sounds like a small mismatch (~15% in radius), but in terms of ion volumes ($4/3\pi r^3$), the chromium ions are 50% too large for the aluminium site. Significant strain energy is needed to grow the lattice. However, if we insert pairs of chromium ions at adjacent sites, then the lattice can relax over a slightly larger volume and the total strain energy of the pair is spread and reduced. Critically, it is less than the strain energy of two separate impurity ions. Strained lattices also influence the ease of motion of the ions when the material is growing, or if it is subsequently heated to a high temperature.

In a social analogy, we can consider a densely packed commuter underground train into which we introduce individuals of different size, or carrying baggage. From simply ionic size effects, consider the presence of a suitcase (an impurity in a vacancy), a parent carrying a baby (i.e. oc-cupying a small space and therefore a sort of small impurity, but not at the vacancy centre) or a supersize bodybuilder (either as an interstitial or a large on-site impurity). These will not only distort the location of the immediate neighbours but also disturb the packing density of quite a few of the more distant passengers. Such disturbances in the packing will influence movement along the train, or exiting from it. These are features exactly analogous to the measurements of ionic diffusion and mobility in a crystal lattice. Pairing the non-standard size impurities can ease the problem in the train, as well as in a crystal.

Perhaps surprisingly, if we put very small impurity ions into a vacant lattice site, the ions will not sit in the middle of the site but will prefer to move off centre to a side position. Again, this is matched by the behaviour of people. A person entering a large empty space with many chairs will rarely go to the middle of it, but will prefer somewhere nearer a side or corner. Physicists make this sound more impressive by citing the mathematical justification and call it a Jahn-Teller distortion of the site. (If you want fame, always publish your simple equation or idea with a co-author with an exotic name. A double-barrelled name at-tached to a new scientific law is remembered more often than a single name one – unless you are really famous.)

Finally, in the range of simple errors there are many crystals, such as gallium arsenide, as used in light emitting diodes (LED) and laser light sources, in which some of the Ga ions can sit on the As site, or vice versa. These are termed anti-site imperfections.

Errors in growth and non-exact composition

Whilst I initially presented silica as a respectable and simple stable oxide of silicon with one silicon ion coupled to a pair of oxygen ions, such ide-alized perfection need not exist. With silicon oxides it is possible to grow, or modify, material which varies continuously between the composi-tion of just silicon (Si), all the way through the oxide concentrations (SiO_{2-x} with $0 < x < 2$), up to the silica case of SiO_2. Such oxygen de-ficient variations are called non-stoichiometric. This pattern is found with many oxides. A possible example of this happening occurs with white paint, where the whiteness in part comes from the presence of titanium dioxide in the paint. Changes in the oxygen concentration by reactions with the atmosphere destroy the correct composition and, as the material moves to be deficient in oxygen, it starts to absorb blue light. Hence the white paint takes on a yellow, faded tinge. In principle, adding impurities might inhibit the chemical reactions which degrade the paint. This sounds interesting from the viewpoint of the user, but is a less attractive improvement for the paint manufacturer as it would reduce repeat sales.

My model of the titanium dioxide yellowing is certainly feasible, but it is not a universal explanation as there are alkyd type paints which yellow in the absence of light. In general, moisture and oxygen reactions can induce colour changes in some resins. Changes in the imperfections and surface chemistry are complicated.

Part II   Lithium niobate – a key industrial material

One of the most important synthetic crystals used in the electro-optic industry is called lithium niobate ($LiNbO_3$). The layout of the lattice is similar to that of sapphire (Al_2O_3) but the presence of alternate Li and Nb ions, rather than two Al ions, adds some wobble to the crystal axis. It does not occur as a natural material but was synthesized and is now incredibly widely used, for example in TVs, mobile phones and optical fibre communication systems. In essence it is a mixture of lithium oxide and niobium oxide. Unfortunately, when the two materials are melted together the melt does not sit nicely at the 50/50 composition. Instead, it is lithium deficient. The properties of the crystals which then grow are significantly inferior to the possible lithium niobate properties, and so there has been intense industrial effort to overcome the resulting difficulties. This is a minefield of problems, as for the optical applications (e.g. for making optical waveguides) we need to make regions of high refractive index to confine the beam of light. Commercially this was achieved by treatments with impurities such as titanium (Ti) ions or protons (H). The doping and substitution reactions introduce even more phases of new compounds and/or distorted lithium niobate structures. Because there is major commercial interest in controlling the refractive index of lithium niobate, it generated thousands of research years of development. The only obvious reasons for such effort and financial investment are that the commercial return for the photonics industry is immense.

Unfortunately, these chemical style routes to changing the refractive index of lithium niobate often degrade the electro-optic performance, and fail with the structurally identical crystals of, say, lithium tantalate. I have a personal interest in this topic as I saw the opportunity to use a totally imperfection-based route to overcome these difficulties. Not only did it succeed with $LiNbO_3$ (and $LiTaO_3$) but it enabled the fabrication of waveguides, waveguide lasers, and a host of other opto-electronic devices in many hundred types of material where no chemical routes to waveguide fabrication have yet been found. The method is now widely employed.

Optical waveguides formed by imperfections from implantation

The key step in making an optical waveguide is to trap the light within a pathway of higher refractive index than the surroundings. For $LiNbO_3$ the methods are very specific and achieved by chemical dopants of titanium or hydrogen to *enhance* the index, but, as just mentioned, the techniques reduce the performance of the $LiNbO_3$ electro-optic properties and are not applicable in other materials. My alternative approach, using imperfections, was not to write the waveguide, but to *reduce* the index in layers adjacent to the guide to define the low index boundaries. This retains the desirable features of the guide material and, if the boundaries are still transparent, then it does not degrade the guide in any way. The concept is trivial, but I chose a technique that is applicable to virtually all optical materials used in photonics, based on damage formation and destruction of the host lattice from ion beam implantation.

During the implantation the energetic particles start to slow down as they deposit energy. Initially, for high energies, this starts with lattice ionization (with minor damage to the lattice that can subsequently be removed by a modest heat treatment). However, near the end of their track the energy transfer is more effective and is called nuclear collision damage, which destroys the target lattice. For a transparent crystal, such as quartz, the damage zone becomes amorphous silica and the refractive index falls by some 5% (an enormous amount in waveguide technology). Hence, implanting ions (e.g. helium) and generating lattice damage is a simple route to leave a near surface layer of unaltered high index material sitting on a substrate of low index that isolates it from the bulk (Figure 10.5). To define the waveguide boundaries just needs additional lower energy implants.

An example of guide formation in quartz is shown in Figure 10.6, helium ions with an energy of 2.2 MeV (Million electron volts). There is minor electronic damage in the guide region but nuclear collision damage from helium ions convert the quartz into amorphous silica. The refractive index profile maps the damage and penetration when using a small ion dose, but higher doses develop a simple broad region of low index silica confinement for the waveguide. After a modest heat treatment, any electronic damage is removed and it is an excellent, very low loss, waveguide. In Figure 10.6 I have chosen to plot the vertical axis as decreasing refractive index, as this conveys the presence of an optical well that confines the light. The effect is dramatic and especially impressive when one realizes that, in the damage zone, the helium (or other) ions do not contaminate the waveguide. Further, they involve only 0.1%, or less, of the total number of ions in the entire waveguide structure. Even smaller concentrations are required when using heavier ions.

My original aim and trials using lithium niobate were successful. The behaviour is basically the same as for quartz but, with a more complex and non-stoichiometric target composition, the electronic energy transfer introduces some permanent small changes that differ for the refractive indices of light aligned along different crystallographic axes,

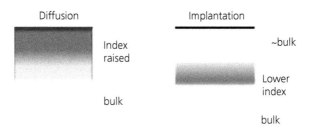

Figure 10.5 Treatments for waveguide formation. Dopant diffusion increases index in the guide region, implantation 'damage' lowers the index below the guide. In $LiNbO_3$ titanium diffusion requires many hours at high temperature and degrades electro-optic properties. By contrast, implantation is rapid and can even be made below room temperature without altering the crystal performance.

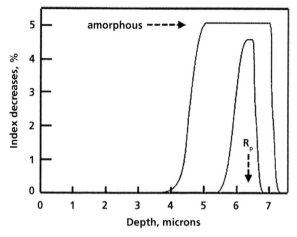

Figure 10.6 Ion implanted quartz waveguide. The crystalline quartz waveguide extends down to the nuclear damage region which has converted into amorphous silica. The implant was with helium ions of 2.2 MeV energy. The projected ion range (Rp) is apparent with the smaller dose of 10^{16} ions per square centimetre. The extended low index volume developed by a dose of 8×10^{16} ions/cm^2.

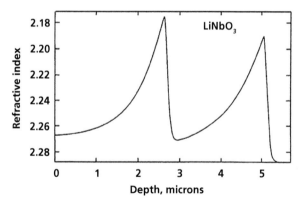

Figure 10.7 A pair of buried waveguides in lithium niobate. Two different helium ion energies were used to make a pair of buried waveguides. The central barrier is controllable to enable coupling between the guiding in these two optical wells.

but does not degrade the material. Figure 10.7 shows it is equally easy to build waveguides at different depths below the surface, which is not feasible with chemical diffusion guide formation. Further, many key photonic materials have phase transitions that would negate the use of high temperature diffusion.

The truly key value of the method has followed with waveguide fabrication in several hundred materials in which no alternative route had been considered or developed. Using the materials has generated a wide range of photonic devices, such as waveguide lasers, systems for doubling the frequency of a laser (second harmonic generation), optical switching and routing, plus many more sophisticated examples of photonics. Overall, one sees the basic idea was trivial, but controlling imperfections is extremely useful.

Precisely the same strategy is now used in ion beam cancer treatments, where an energetic ion deposits most of the destructive energy at the end of its track, at the depth of a buried tumour. Ionization of tissue on the way in is unfortunately undesirable, but this is taken as a trade-off against the destruction of a buried tumour.

11

Impurities and the growth of semiconductors

Part I Impurities that have changed the world

In a book that is trying to highlight the opportunities that controlling imperfections can have on the world, semiconductors look to be the most spectacular example for our current generation. By the 1960s, semiconductors had moved from being an interesting academic topic in physics, to items that could be commercially fabricated. The growth of a dominant industry then steadily transformed virtually every aspect of our lives. Without semiconductor electronics we would not have computers, smart phones, satellites, the web, internet, and all the small gimmicks built into a host of equipment from toys to cars and aeroplanes. The list is endless. All this has become possible because we learnt how to accurately control and locate impurities on the scale of parts per million into the surface layers of elements such as germanium and silicon. The methods were extended into a wider range of semiconducting materials for electronics and optical applications, such as light emitting diodes and lasers, etc.

Is this all good news?

These are fantastic technologies which not merely improve many aspects of our lives, but now totally control them. Without them, civilization as we now know it could not exist. Our enthusiasm is immense, but we are equally blinkered to the many ways in which all the electronics have undermined other key factors of our lives. Some of these problems will be discussed in a later chapter, but here I shall briefly mention a few extremely negative examples. The first is that a fixation on continuous electronic links has seriously altered our ability to view the world around us and communicate with real people. This frequently

The Power of Imperfections. Peter Townsend, Oxford University Press.
© Peter Townsend (2022). DOI: 10.1093/oso/9780192857477.003.0011

destroys intimate social contacts, empathy and real-world personal interactions. Electronic music, broadcasting downloads and CDs are part of this panoply of electronics defining and changing our lives. Even concert halls are not immune, as it is claimed that modern audiences want the sounds to be more like that of a CD, rather than a live performance. Acoustic engineers are now redesigning electronics to distort live classical orchestral sound within concert halls to have the inferior quality of broadcasts and recordings. As a musician I find this unbelievable.

A second feature is that instant access to an immense database of information has dulled our memory skills, and hence its use in creative imagination. This is combined with an increasing problem of separating facts from fake and/or prejudiced items. My recognition of this dumbing down of humanity is partly summarized in my book, entitled *The Dark Side of Technology*, which explored how our initial eagerness to accept new technology blocks us from considering any negative consequences. Writing it totally transformed my own perspectives. One very simple example based on electronics was our total dependence on it for items as diverse as satellite navigation, data storage on the 'cloud', every aspect of fibre optic signal transmission, and other long-range communication. The crunch realization was that all of these could be lost in a single naturally occurring event. Back in 1859, Carrington described the effects of a solar emission which hit the Earth and, instead of just producing the normal lovely Aurora at high latitudes, it was visible as far south as Cuba. This ionization of our atmosphere generated enough electrical disturbance to induce voltages in the very basic telegraph systems that were then in use. Later examples of smaller solar events have caused fires in power stations, interrupted satellite links and blacked out airplane navigation. Fortuitously, since we moved to an electronic age, no solar emission has yet matched the scale of that of 1859. Were it to do so, or actually be more intense, we face the prospect of instantly, and permanently, losing satellites, ground communications, and/or destruction of the power grid network (i.e. the grid cables provide superbly long antenna systems that would pick up electromagnetic signals from the Aurora sufficient to destroy major electrical grid components). Having recognized the obvious weakness, I am also concerned that similar damage could easily be duplicated by terrorist activity. Extremely large and powerful solar events are numerous on the surface of the sun, but fortunately we are a small target, and so most of the emissions are not directed towards the Earth. The current continuous Aurora signals are just the sum of more randomly directed minor emissions. A repeat of the 1859 event (or a larger one) is inevitable. We can observe the frequency of solar flares and their angular spread, therefore one can easily predict that there should typically be one major strike on the Earth every 100 or so years. Prediction of when is impossible, but there is 100 per cent probability that they will occur.

Equally clear is that, until the late twentieth century, the only electrical consequence would have been major radio interference and fires at localized power stations. Indeed, these happened. Power grid losses would have the most significant impact in terms of deaths in northern latitudes that include cities and continents such as Washington, Beijing and Europe. However, communication loss might be total. Governments have admitted that perhaps as many as 80 per cent or more of the population in these areas would die, especially if the event was in the winter time. Even the much simpler problem of loss of GPS (Global Positioning System) has not seriously been protected. It is vulnerable to deliberate interference and some 10,000 incidents have been reported. President Bush, in 2004, directed the design of a back-up system and better defence for military purposes, but as of 2019, the money has never been spent. No consideration has been made for loss in terms of civilian consequences. The proposed aim was to protect against terrorist activity interfering with the signals, not loss of the satellites and ground stations.

Our dependence on electronics means we may no longer keep written records but use our own computers or the so called 'cloud' storage. For many applications this is useful, but if one wishes to have an historic example of the dangers of keeping all the information in one single place, then an obvious example was the library of Alexandria. At the height of the Mediterranean culture, 2000 years ago, a vast library and storehouse was constructed in Alexandria, to which all the documents and greatest writings were assembled in one site for ease of communication. It attracted scholars from all over the region until a great fire destroyed everything. Interestingly, various sources say this was in 48 BC or 641 AD! I assume there were probably two separate events.

The explosive impact of electronics

Having offered a warning that technological progress can, itself, have imperfections, I will return to the more positive aspects of the growth of technologies, although they inevitably include many that were funded for development of weapons of war and nuclear bombs. Metallurgy and glass science had gently matured over several thousand years but electronics totally changed the world. In the first half of the twentieth century, electronic circuitry was made possible by the development of thermionic electron valves (or tubes). These were simply a glass bottle that had been evacuated so that electrons could be sent between two sections. In the earliest version, a heated filament emitted electrons which were attracted by a positive voltage on a collector. The first demonstration of this was made by the English inventor John Ambrose Fleming in ~1904. There were just two parts, the heated negative cathode (-) that emitted electrons (−), and a positive anode (+) to collect them. With two parts it was called a diode. Whilst apparently a very simple idea, it meant electrical current could be controlled to flow in just one direction.

An elegant variation on this was introduced in the USA by Lee de Forest (~ 1907) who added a wire grid in the path of the electrons, close to the filament, to make a triode. As the grid was closer to the cathode than the anode, just a small change in voltage on the grid controlled the flow of electrons from the cathode towards the anode. Small voltage changes could then influence the electrical current emerging from the anode. The current was sent through a resistor to produce a large change in voltage, hence the triode geometry meant it acted as an amplifier. Figure 11.1 sketches the basic triode design.

Size of electronic devices

The ideas are simple to understand, and very effective. The triode could take a weak microphone signal to boost it to have sufficient power to drive a loudspeaker. Later developments added more control elements and better performance, but basically such devices established and advanced electronics for the next sixty years. The downside of electron tubes was that they generated a lot of heat from the filament, consumed considerable power, were bulky and had a relatively limited working lifetime. This was not too serious for radios, amplifiers or gramophones,

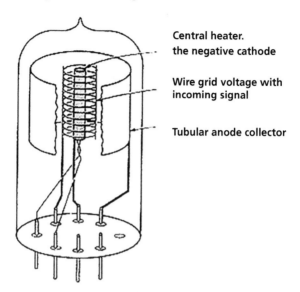

Central heater.
the negative cathode

Wire grid voltage with
incoming signal

Tubular anode collector

Figure 11.1 Simple triode vacuum tubes. A heater (the cathode) emits electrons and a metal plate (the anode) collects them. There is a voltage difference between them and the cathode is negative, this repels the electrons and the anode is positive to collect them. This is the key feature as it is a valve, so electrons can only flow in one direction. (A vacuum is needed to stop electrons from being scattered by air.) The triode valve included a wire grid close to the cathode, so only a small voltage is needed to alter the electron emission. A changing signal voltage applied to this grid controls the current flow, small changes of the voltage on the grid voltage have a large effect on the current. This means a weak voltage signal, as from a radio signal, can produce a big change in a current, hence it is an amplifier.

but it restricted the complexity of the type of signals that could be used. Valve lifetime limitations were initially very evident in the first programmable computer, ENIAC, which had 17,500 vacuum tubes and a failure rate of at least one per day. It was then noticed that, as with light bulbs, most failures appeared when the machine was turned on or off, so continuous running helped, but still with limited reliability.

To a modern generation with continuous use of mobile phones, it is worth trying to put into perspective the size and weight problems associated with valve electronics. During the Second World War, the British Intelligence agents dropped into France tried to communicate back to Britain with 'portable' radio transmitters. State of the art, in

1942, was a radio with a range of nearly 500 miles which neatly fitted into a suitcase, and weighed a mere 30 pounds (~15kg), plus of course one needed a 6 volt car battery to power it. It was not mobile phone performance as it was necessary to set up many metres of aerial and then carefully tune the transmitter frequency. We must also remember that, in this case, radio meant a Morse code transmitter, not direct speech. To change frequency so as to avoid detection during a long transmission, one physically changed some of the components. Not too surprisingly, detection of an early secret agent with a transmitter system was not very difficult. The average life expectancy before detection (and execution) of the agents was a few weeks. The latter fact was not emphasized when recruiting volunteers. A 'miniature' state of the art version, at 20 pounds, was developed in 1943. Whilst such instruments steadily became better, even by the 1950s a clandestine radio transmitter was of a large briefcase size, and still weighed about 20 pounds (~9kg). It ran off the electrical mains, but had a speech channel as well as a Morse code transmitter. A major upgrade was that the range had increased to maybe 2000 miles.

The portable radio transmitter example emphasizes one reason why there was a very large market opportunity to find a way to reduce size, and power needs, as well as increasing performance and reliability. This was achieved via the introduction of semiconductor equivalents of the diode and triode devices. Conceptually, the semiconductor diode or simple transistor perform the equivalent functions of allowing current flow in only one direction, or by adding a voltage to the central section, to control a large current flow via a small bias voltage from the signal. A rather simplistic view is that once we understand these initial processes in a semiconductor device, then we have the key knowledge to appreciate the ideas and operation in a modern computer chip.

The surprising feature is that every aspect of the modern version of electronic technology is totally based on understanding and control of imperfections in semiconductor materials. Early examples were with a single chemical element system designed around germanium semiconductor properties. Devices based on germanium unfortunately were prone to overheat and burn out, so the industry moved to the related element of silicon. With a wider range of needs and applications, the devices expanded with dozens of alternative compound semiconductors. Many semiconductors are constructed with two or more elements (e.g. gallium with arsenic, or gallium with aluminium, arsenic and phosphorus, GaAs, GaAlAsP etc.). The other key feature is that instead of having individual units, as in the vacuum valve, millions are now packaged together in the surface layer of a single electronic chip.

Electrical conduction

For insulators, we saw that atoms preferred to bond together by electronic transfer so that the outermost electron shells of energy levels are full (i.e. Na^+ and Cl^-). By contrast, other materials such as copper or aluminium have metallic conduction. The reason for this is that they only have a few electrons in their outer electron shells, plus many unfilled energy levels in that shell. If they interact with other identical atoms it is impossible for there to be identical electron energies for each bonding electron (this is a basic fact called the Pauli Exclusion Principle). The levels therefore split very slightly. In a crystal there are billions of interacting metal atoms, so this gives billions of extremely close energy levels which then blur into a band of energy levels, as in Figure 11.2. For such metals there were more energy levels than electrons in the outermost shell so the same situation continues in the band of levels for the crystal. This is good for conduction because, if electrons are excited, then there are empty levels where they can move across the metallic sample. Conduction needs both electrons and empty states for movement.

A textbook analogy for this is traffic flow on a motorway. A densely packed motorway has lots of cars but no space to move (a traffic jam). This is like the electrons at the top of a filled valence band, and so the system is an insulator with no conduction. No cars in the next band means it is just an empty motorway and also means no traffic flow (i.e. the insulator situation). By contrast, a metal such as sodium is an excellent high flow rate example as there are both a high density of electrons and a high density of unfilled energy levels. The third option is if there are only a few empty spaces on the motorway, or very few cars, then these are alternatives which result in a small current flow, and this is the situation for semiconductors.

Conduction in semiconductors

In addition to ionic or metallic bonding there is a third option, called covalent bonding. This happens for materials such as carbon, silicon or

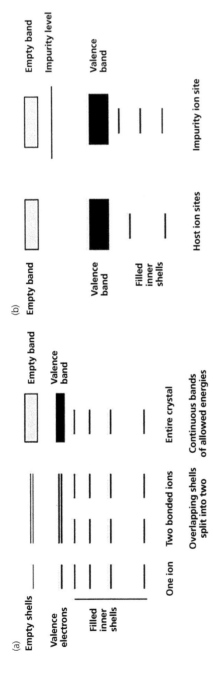

Figure 11.2 Energy levels of atoms and crystals. (a) Electrons orbiting an atom have a staircase of energy levels so that no two electrons have identical energies. When two atoms join together, the outer shell of electrons (the valence electrons) interact so the outer electron levels split into two closely spaced new levels. With three atoms the split is into three levels etc., so in a crystal there are millions of extremely finely spaced levels which appear as a continuous band of possible energy states. In this sketch for an insulator, the valence band is full and the next set of levels is empty.

(b) If the valence band is full and the next higher set of levels is empty then electrons cannot travel through the crystal. Electrons can use thermal energy to make the transition to the empty upper levels if the energy gap is small enough, but for insulators such as quartz or glass, the gaps are large and the jump is not possible. At low temperature even silicon is an insulator. To control the electrical properties and make a semiconductor, impurities are added which have a filled electron level that sits in the energy gap near to the upper band. The distance is now small so heat will allow jumps to the upper levels and this gives a controllable electrical conduction.

germanium. These atoms have four electrons in their outermost shell and four empty levels. Transferring four electrons to make ions is unrealistic, so the compromise is that they share electrons with neighbours. This pairing of electrons in chemical bonding means the tetravalent elements have the impression that the outer electron shells are fully satisfied. It is called covalent bonding. The net effect is that the shell performs as though it is completely filled and there is no conduction within this energy band.

This need not mean covalent materials are insulators, as not only are there energy bands which contain electrons, there are also higher energy bands which are empty. If we can move electrons from the filled to the empty band then electrical conduction will occur. The processes are not efficient so the overall effect is to have a poor conductor, called a semiconductor.

In totally pure and perfect crystals, the electrons might move from the filled bonding band to the higher (but empty) conduction band by using thermal energy. The prediction of this happening depends on the spacing between the bands and the temperature. Carbon, in the form of diamond, has a very wide separation between the bands, silicon less, and the germanium energy gap is even smaller, as listed in Table 11.1, with examples spanning an immense range of 10^{20} in terms of conductivity. Pure diamond is always an insulator, silicon at room temperature is a very poor conductor, but germanium can switch into being a poor conductor if it is heated. This was a weakness of early transistors where the germanium power transistor became hot, started to conduct for purely thermal reasons, and then heated more and burnt out. Silicon does not have this problem in normal electronics, and so became the semiconductor of choice for equipment near room temperature.

Table 11.1 Energy gaps and conductivity of C, Si, Ge and Cu.

Element	Energy gap, eV	Conductivity at room temperature Units of Siemens/metre
Diamond, C	5.5	~10-13
Silicon, Si	1.11	~1.56 x 10−3
Germanium, Ge	0.67	~2
Copper, Cu	N/A	~6 x 107

Conductivity values range enormously from an insulator (diamond) to semiconductors (Si and Ge) to a metal (copper). The electrical conductivity is proportional to the number of electrons per second which manage to jump to the empty conduction band. The jump rate per second (dn/dt) depends on the number of available electrons (n), the frequency at which they are trying to make a jump (f), and the probability of success (P) that they make it upwards. Hence mathematically the overall success rate is $dn/dt = nfP$. Where n is the number of electrons in the lower level and f is the natural vibrational frequency of the atoms in the lattice.

We can guess at the form of P. The key factor is the ratio of the energy needed compared with the energy available. Throughout chemistry, physics and biology the general rule is that if we have a small amount of energy, for example from thermal energy $E_{thermal}$, and we need a jump or reaction energy $E_{reaction}$, then the likelihood of the event occurring falls off logarithmically with their ratio (i.e. the probability $P = \exp(-E_{reaction}/E_{thermal})$. For example, if the ratio is 2 to 1 then the probability is $\exp(-2)$ which is 13.5 per cent, at 4 to 1 it falls to 1.8 per cent, and by 8 to 1 is down to 0.034 per cent. This is equally familiar in changing speeds of chemical reactions, or plant growth with changing temperature. The value of thermal energy for an electron in a solid depends on the temperature in degrees Kelvin (e.g. 27 °C is 27 + 273 = 300 K). This gives the energy for $E_{thermal}$ as kT (where the constant k is called the Boltzmann constant). For electronic transitions events will occur if $E_{reaction}/E_{thermal}$ is less than about 25 times.

Adding a dopant to silicon can provide an energy state containing an electron within the energy gap, which thermally can move up into the empty upper band and give conductivity. At first sight, thermal energy may seem valueless for this as the packets of thermal vibrational energy at room temperature are just ~0.027 eV. Our energy units are electron volts and Table 11.1 is telling us that for silicon the energy gap is ~1.1 eV, so for an impurity level within 10 per cent of the conduction band the barrier is still 0.11 eV. Our simple intuition will assume that, if we have the energy and muscle to jump a particular height, then that is our limit, even if we are in a crowd of people. My analogy for the impurity electron in silicon is that it is trying to jump an impossible height of 1.1 metres when we can, in fact, only jump a miserable 27 cm.

However, for the electron, several packets of random vibrational energy can arrive simultaneously, and just occasionally the electron will gain the necessary energy. It is a highly unlikely event but, basi-cally, even a very unlikely event attempted many times can sometimes succeed. We see how the odds change once we realize that packets of vibrational energy are arriving some 10^{13} times per second. And there are billions of starting electrons. (Such a high input of attempts would make winning the lottery a certainty.) A very simple analogy of help from friends is seen in a rugby line-out, where the player jumping to catch the ball is helped up by other players, so he goes higher and stays there for longer than he could do on his own.

We might assume that we cannot input energy from many peo-ple, but in fact there are ways to add energy inputs. For example, this showed up in the problem with people walking across the newly opened Millennium Bridge over the Thames. Each step input a tiny amount of energy into the bridge. When more and more people walked in step, the bridge began to oscillate. Worse was that, as the bridge started to sway, people changed their walking pattern to stay in syn-chronism with the swaying. Fortunately, the problem was not as se-rious as for the Tacoma Narrows Bridge, where wind-driven swaying hit a resonance that gave a spectacular and catastrophic collapse of the structure which, at the time, was at the forefront of suspension bridge design. Both bridge examples show how small individual sources of energy can combine together. This is precisely the situation for the vi-brational energy of the lattice enabling a crucial energetic electron level change.

Part II Imperfections that control electron flow in semiconductors

Ideal impurities to consider are phosphorous and boron. They are close to silicon in the periodic table and similar in size, but phosphorous has a fifth valence electron so it is readily moved up into the silicon conduc-tion band of energy levels. Alternatively, boron has only three bonding electrons so will act as an electron trap to make the doped region ap-pear somewhat positive. The semiconductor equivalent of a diode is to have adjacent zones doped with P and B ions, so electrons will only ef-ficiently flow in one direction if we apply an electric field. The notation is to call the different areas p or n type, where n (negative) means there is an excess of electrons and p (positive) means it is electron deficient.

The triode valve equivalent (i.e. a transistor) just means we need, say, an n-p-n structure.

Impurity dopants enable the production of not just these very simple semiconductor devices, but also all the complex circuitry which has been developed. There is no more basic science, just an immense amount of fabrication technology. My message is that all electrical conduction and processing in semiconductors is controlled entirely by a combination of two things: locally varying the concentration and types of impurity, and controlled variations of voltages which give energy to move the electron and holes. These tools are equivalent to the letters, pen and paper of writing. Transistors were words, modern chips are books, computers are libraries.

For the electronic motion in the p doped region, the convention is to talk in terms of a positive 'hole' rather than electron movement. This is sensible as there are thousands of silicon ions per impurity. We could think in terms of electron movements rearranging into the empty space and moving, say, from left to right. Alternatively, we can focus on the simpler movement of an empty space, which for practical purposes is moving in the opposite direction, from right to left. We prefer this description and the 'space' is called a positive hole. The two types of charge transport fit the analogy with cars moving on a motorway. In a nearly empty motorway, the cars (our electrons) move easily and rapidly and are the objects in motion. By contrast, in a nearly static traffic-bound situation each car advances only when the one in front has moved forward to make a space. Viewed from a distance one does not see individual car motion, only the backward progression of the space (i.e. the positively charged slowly moving hole).

The ease of electron motion in the almost empty conduction band is greater than for the hole (i.e. lots of electrons have to reshuffle to make space for the hole movement) so the mobility of the charge is different in the two bands of levels. Experimentally this is true, so our little model is offering good intuition. In real semiconductors both types of process are exploited. There is no more physics but, in technological terms, the problems have scarcely begun. The dopant concentrations must be exactly defined in each region of the device. Unwanted impurities, as well as defects of the silicon lattice, must be excluded if they are likely to participate either in the number of electrons available, or act to impair the free transport of electrons. The current levels of control have a target aim for impurity concentrations as low as parts per billion compared with the silicon. Technologically this is an amazing achievement. However, the huge effort to remove unwanted imperfections should not cloud our view that semiconductor physics is still totally a function of controlling desirable imperfections.

Growth and development of electronics – Moore's 'Law'

The phenomenal growth of electronics and associated technologies within the last seventy years is unprecedented. One measure is to cite how many transistors have been built on to a single semiconductor chip, often cited as Moore's Law. Moore noted that the number of transistors produced on a chip was doubling per two years. With improved chip design this rose to a doubling every eighteen months. It is not a 'Law', but a roughly logarithmic growth trend. Nevertheless, the semiconductor industry used it as a target milestone for the speed of their development. The pattern is sketched in Figure 11.3 for the first seventy years of transistors and semiconductor chips from the initial invention in ~1947 in the Bell laboratories.

Such logarithmic patterns of improvement occur in many technologies, but I do not find this pattern surprising. There are competitive

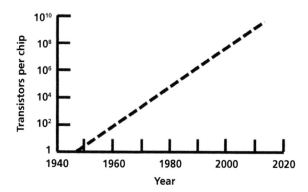

Figure 11.3 The increases in transistors per chip with time. The number of transistors per device initially increased logarithmically with time since the invention of the transistor in 1947 to a modern chip with more than a billion components. This plot shows a doubling rate of about eighteen to twenty-four months.

market forces between companies, and new ideas which need to be gradually tested and introduced, so realistically we expect better performance products to emerge year by year. Any company that decided to move from this pattern whilst it developed a major step innovation would be in severe danger of going out of business before they could see a public that liked the new product. Hence gradual improvements are the natural pattern. Despite this I find it quite disconcerting that the trends are so sharply defined for many products. Are we making the changes voluntarily? If this is not true, then who, or what, is actually in control of our destiny?

Similar logarithmic growth patterns with time for earlier electronic devices can be viewed over a much longer historical period. Ray Kurzweil showed a similar logarithmic rate of development occurring through the twentieth century period from the onset of switches and electromechanical relay devices to vacuum tubes, simple transistors and early computer chips onwards, used in making calculations and switching devices. This pattern has been sustained over more than a century with an approximately logarithmic rate of increase, as in Figure 11.4, which increased on moving from mechanical to electronic devices. Kurzweil has taken a different and more commercial approach by factoring in a cost term as operations per dollar. This is a crucial

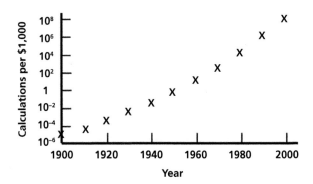

Figure 11.4 Costs of making calculations through the twentieth century. Here is an overview of the speed of numerical computations per second per $1,000, with techniques from controls with paper cards, to electrical switching devices, mechanical relays, vacuum valves and the semiconductor variants. Logarithmic trends apply but the slope increases with the semiconductor versions.

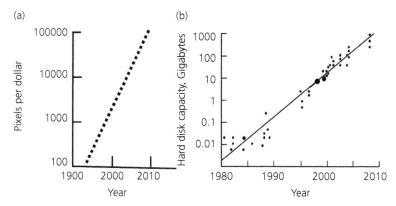

Figure 11.5 (a) **Early developments in CCD cameras and,** (b) **hard disc storage.** The figure on the left looks at the time dependence of CCD imaging in pixels per dollar, and the figure on the right summarizes the million-fold increase in the capacity of hard discs during their first thirty years. In both cases there were logarithmic trends with time, but considerable changes in the technologies being used.

feature to include, as an improvement of a million times in processing speed without a corresponding reduction in unit cost would not sell.

Many similar trends have been noted. Figure 11.5a by Hendy describes pixel resolution per dollar on a digital camera. Equally dramatic is the rise in hard disc capacity (Figure 11.5b). Companies use such progress for advertising, but developments are not always sustainable and advertising is altered to hide the limitations. For the CCD camera chips, modern marketing focusses on the size of the pixel, rather than an increasing number per chip. In computers the shift is to the number of processors, since they became limited in terms of handling connections and wiring capacity. Basically, it means the Moore type patterns have hit an upper limit.

Can growth rates be predicted and maintained?

Since the growth rates are trends, not laws, prediction is uncertain except that with major industrial markets there will always be attempts to outperform competitors. There are often physical limitations set by the size of components. In silicon chips, electrical interconnects could not be reduced further once they start to overheat from the current flow. Many of the 'advances' with new chips and marketing of electronics are minor, and are merely a tool to use with advertising to trigger replacement of existing equipment with new. Commercially this is essential as electronic reliability is frequently excellent and radios and CD players etc. can run smoothly for more than twenty years. Only the packaging may look dated, until it comes back into fashion. Finding reasons for obsolescence are fine for the manufacturer but less desirable for the consumer. A prime example is in the marketing of smart mobile phones, where there is still enthusiasm to buy the latest version, often as a fashion accessory. In terms of performance very few users use the inherent power of their systems and updates are irrelevant.

Less obvious with genuine progress is that advances frequently require totally new technological solutions that are totally incompatible with earlier processes. Figure 11.6 indicates how signal transmission speeds have logarithmically steadily increased over 200 years by switching between technologies. Figure 11.7 then indicates how improvements in types of optical fibre and/or transmission wavelength

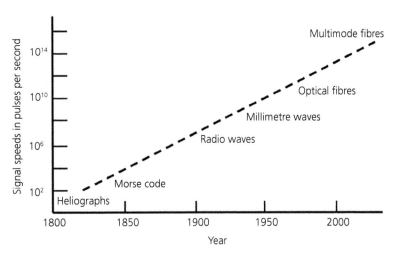

Figure 11.6 Evolution of optical communication rates. The plot looks at speed of communication over 200 years, from techniques as diverse as a heliograph and Morse code, to modern optical fibres. Nevertheless the pattern fits well to a logarithmic plot. Most impressively it spans a speed range of almost a hundred million million times.

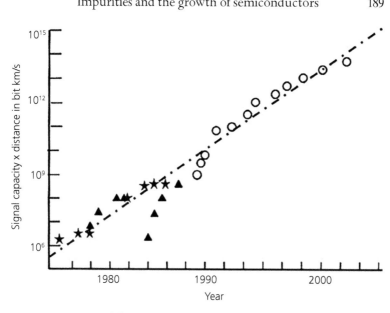

Figure 11.7 Optical fibre communication. Data rates times distance performance only advanced by changing to totally new technologies (indicated by different symbols). Details are unimportant, but the improvements have been nearly a billion times in forty years.

became limited and had to be seamlessly replaced by a completely new processing technology, without disruption to the users.

Good predictive estimates of what will be needed and will be marketable can make or break an industry, and particularly for innovative items it is extremely difficult to be accurate. Leading technologists, and otherwise perceptive people, have often totally misjudged the future market for technologies. I have already mentioned the example of ignoring optical fibres in preference to microwave systems, but to indicate how easy it is to be wrong I will cite a couple of other examples.

> . . . to transmit the voice over wires . . . would be of no practical value (1865)
> One day there will be a telephone in every major city in the USA (1880, AG Bell!)
> There is a world market for as many as five computers (1943, head of IBM)
> There is no need for an individual to have a computer in their home (1977)

Similarly, the use of automobiles was initially disparaged, and it was said it could never displace the horse. Modern predictions which I suspect may be equally wrong are that we will cheaply and efficiently generate power from fusion reactors. Scientific progress has been made and the aim is fine, but for more than half a century the prediction of a system was at least twenty-five years into the future, and the size and costs have escalated considerably. If they are to economically succeed, a totally new method is required.

A far more difficult guess at the future is in the role of computers based on qubits (quantum bit). Laboratory scale devices indicate they have a real potential to function in calculations, but at this stage their complexity and reliability is a decade away from any general usage. (Maybe my comment will be added to the preceding list of failed predictions.) Nevertheless, very positive hype is essential to maintain funding for their development, and the underlying science is interesting.

12

Small anomalies and long-range consequences

Part I Intuition and defect structures

For semiconductor chips, it was obvious that packing higher and higher densities of transistors in the surface layers hit a barrier once the interconnects started to approach atomic dimensions. Equally, their reliability reached the limits of masking and fabrication technologies. Nevertheless, for electronic devices the dopant effects seem to be remarkably well localized, hence the ability to tightly package so much on to a chip. This is not the case with insulator responses, where the defects that control optical properties appear to have much longer-range interactions, and therefore are potentially more troublesome. Having spent many years considering how to understand imperfections in insulators, I will try to summarize these features. Fortunately, they are rarely as extreme as those mentioned earlier, such as pheromones, where mating signals and other natural human chemical reactions to people can be sensed from a great distance, even if in a background of parts per billion.

The basic principles of the construction of an atom which has electron shells around a nucleus are relatively simple, and the idea leads smoothly into an initial understanding of chemistry and crystallography. Real crystals are of course never perfect, but even here it requires little imagination to guess at the more familiar simple types of imperfection, such as how impurities in minerals and gemstones define colours, or how they respond to heat treatments. Understanding defect behaviour has been a slow process even when we understand the concepts. A key reason is that influences can be extensive, and it has taken decades to develop the experimental and theoretical modelling techniques that unlock their behaviour.

The Power of Imperfections. Peter Townsend, Oxford University Press.
© Peter Townsend (2022). DOI: 10.1093/oso/9780192857477.003.0012

How localized is a 'point' defect?

I will take a historical and well documented example of a very basic defect, a vacancy, in the alkali halide lattice of sodium chloride. Science textbooks make life simple for students and discuss such defects as though a vacancy is totally localized at a single lattice site in the crystal and just interacts with its immediate neighbours. This is inadequate, but there is a historic reason, linked to quantum mechanics, why this view has persisted. In the 1930s it was found that, if sodium chloride crystals were irradiated with X-rays, they changed colour. Alternatively, with a little chemistry one could make salt with slightly too much sodium, and this produced the same colour that absorbed blue light (near ~450 nm). Both experiments could be understood if we had removed a halogen (chlorine) ion from its site on the NaCl lattice (Figure 12.1), and left an electron in the little empty box of adjacent sodium ions. The electron maintains the charge balance in the crystal. This was just an experimental result so attracted only minor fame for the observation.

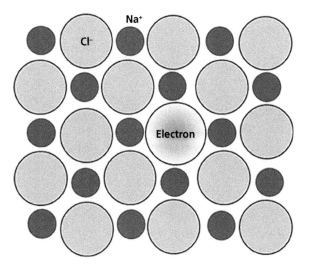

Figure 12.1 A vacancy in NaCl. This model of the F centre in NaCl shows that charge balance is maintained if one removes a chlorine ion and then leaves an electron trapped in the space. In reality, the electron wave function spreads rather further than the original empty site, but to a first approximation it behaves as though it is an electron trapped in a box.

In 1937, a scientist called Robert Pohl realized that the electron in a box was a perfect test model for the newly proposed quantum mechanics. An electron moving like a wave can be reflected by the walls of the box. Just as with a violin string, several related wavelengths may be trapped. For the electron in the box, there is not a continuous range of options, but only special energy levels which match the size of the box (similar, but not identical to the number of notes and harmonics which come from a violin string, which has fixed ends).

The energy needed for the electron to jump up from the lower energy level to the next level on the staircase of energies defines the energy of light that would be absorbed. The absorption of a particular energy (e.g. blue light) causes the colour of a salt crystal which contains such 'electron in a box' imperfections. Considering the simplicity of the model, the result was excellent and Pohl predicted, within a few per cent, the photon energy (and so the absorption wavelength). He could not do better as the wavelike motion of the electron means it penetrates slightly into the region of the immediate neighbours and the box dimensions taken from the crystallography of a perfect salt lattice do not allow for this, and so the crystallographic box is a little too small. The model was so good that it scaled correctly (well almost) with the other sixteen alkali halides with the same crystal structure. In 1937 quantum mechanics was the exciting new face of physics, and the results attracted considerable attention. The feature was called a 'point' defect. Errors caused by interactions with more distant neighbouring ions were glossed over and initially forgotten. Neither the quantum mechanics nor experimental techniques were sufficiently advanced to do otherwise.

Perpetuation of errors by age and repetition

The model and 'point defect' phrase became embedded in our textbooks and minds as facts. This has influenced and hindered progress with the subject ever since. There have been steady improvements in experimental methods and equipment that reveal longer range interactions. They show effects of the defect, not just at the nearest, or next nearest neighbour sites, but in extreme situations one can quantifiably measure the responses out to fifty or sixty shells of neighbours (i.e. more than 100,000 nearby ions). Quantum mechanics has improved and much of this experimentation is supported by theory. We can now appreciate how many atoms are sensitive to the presence of some impurity or imperfection in the structure. I made a plot of such data which suggests that, with time, there is a smooth logarithmic trend in how many neighbours we believed were influenced by the imperfection core. A version of my plot is shown in Figure 12.2. It appears we have gradually been convinced how many neighbours should be considered, and the number doubled roughly every five years. This logarithmic plot means that some eighty years later we must accept that one really simple little defect can be having measurable long-range influence on more than 10,000 ions in the lattice. Experimentally, we need a lot of skill to detect their influence and the changes, but it is possible. Such distant interactions imply that no part of a crystal is free of defect interactions, impurities, vacancies and dislocations. Crystalline perfection cannot exist!

An interesting philosophical observation is that the steady advance has precisely the same logarithmic pattern as was evident in Moore's law and all the other technological examples in Chapter 11. I thought this trend was fascinating and tried to construct similar plots for results

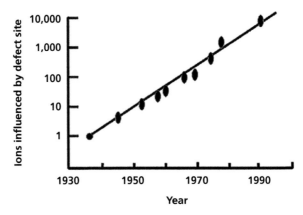

Figure 12.2 The figure shows how improvements in experimental techniques have allowed us to recognize the number of neighbouring ions in a crystal lattice that are sensitive to the presence of an imperfection in the structure. This example is for data on imperfections in alkali halides, but similar patterns emerge with progress on other materials. If there are impurities or imperfections at levels of say 10 or 100 parts per million (ppm) then we need to recognize that every site in the crystal is influenced (i.e. there is no region which is perfect!).

from people working with equivalent imperfections in both the quartz lattice and silica glass. Perhaps unsurprisingly, the literature offers very similar patterns with the scale of defect interactions doubling every five years. However, the quartz literature only started around 1960 and the silica glass literature even later. Neither community of workers seemed to have recognized that they could have leapt forward twenty or thirty years by considering even longer range effects than were currently fashionable. Equally, the data also show there can never be a 'perfect' crystal. Theoretical models built on perfect crystals are greater simplifications than most of us ever consider. The presence of defects need not destroy the local lattice structure but only distort it. A tempting misquotation might be that the Bonds are shaken, not stirred.

Useful defect properties of optical materials

For insulators the optical properties of absorption and luminescence feature most strongly, not just in the visible spectrum but across the entire transmission range of the materials being used. In crystals, absorption occurs at high energies where electrons are excited from the valence band up into the conduction band. Equally there can be transitions within localized defect sites, or from defects into the conduction band. Similar patterns occur in molecules. On decay the energy states light is emitted (termed luminescence or phosphorescence). The absorption and luminescence processes are sketched in Figure 12.3. For atomic transitions, both absorption and emission are at sharply defined energies. However, in molecules and solids, thermal energy can vibrate ions and modulate atomic spacings. Instead of the sharply defined energy staircase of electrons orbiting an isolated atom, the levels have some flexibility within parabolic well shaped buckets of allowed energies as the result of additional thermal energy. The formal title of such a sketch is a 'configurational co-ordinate diagram'.

The diagram shows that the absorption energy that leads to luminescence decay is greater than the energy returned by the light emission because, in the upper levels, the base of the well is displaced. This is the familiar example of energetic UV disco light being absorbed in impurities added in washing powder. The molecule relaxes in the upper state, followed by a decay with lower energy blue emission. The intentionally added imperfections produce blue light, and our expectation is that more blue light only reflects from clean garments, such as white

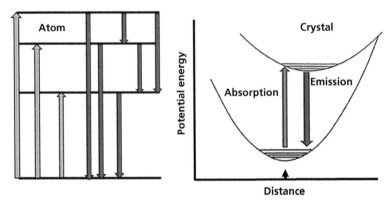

Figure 12.3 Absorption and luminescence. The line diagram indicates transitions for isolated atoms. In a molecule or solid, thermal energy adds a broadening of the energy levels within a potential well, and the energies vary with position. Higher levels are less strongly bound, centred further from the core position and have shallower wells. Thermal energy offers a wider spread in position for excited electrons. Hence one expects absorption bands to be at higher energies than the luminescence, and have a smaller spread in energy across the band. The sketch looks the same for an electron based on a defect site in a crystal, or the vibration of a molecule which relaxes a little when it is excited.

shirts. Such luminescence is an example of 'what goes up must come down'. Pumping energy into a system, such as a transparent mineral or a biological cell, means electrons are excited to higher energy levels. The higher state for the electrons is invariably a shallower parabola and, therefore, there is a small range of accessible energy levels. When the electrons return to the lower set of levels their energy emitted is less, and it has a greater spread than during the upward absorption process. Figure 12.4 sketches data for potassium bromide.

Part II Luminescence in radiation dosimetry

There is a plethora of practical applications of luminescence examples, but here I will select just one, namely luminescence signals used in radiation dosimetry. Equipment to monitor luminescence emission is extremely sensitive and we can often detect impurities and defect states that exist well below parts per million of the normal lattice sites. This

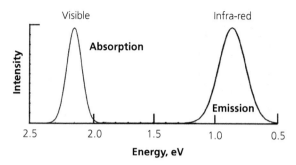

Figure 12.4 Absorption and emission patterns for potassium bromide. Note the emission is at lower energy than the absorption, and spreads across a greater spectral range, as predicted in figure 12.3. The plot has high energy on the left as this matches the pattern we think of with colours from ultra-violet to visible and infra-red light.

offers very high dosimetry sensitivity. One example is to monitor accumulated radiation exposure (i.e. from X-rays, gamma or cosmic rays, and from natural or manmade sources). Radiation is an inevitable background to our daily lives. Radiation exposure is not confined to medical use of X-rays, nuclear reactors or fallout from bombs. A zero background is of course impossible. Instead, there are huge variations in background radiation levels from rocks such as granite, crockery, building materials, cosmic rays, flying at high altitude or even potassium etc., in our bones. There can even be radiation associated with smoking (e.g. from radioactive alpha particulates on the tobacco which can reach surface lung tissue) and impurities in the atmosphere. Flying and mining both raise our exposure to natural radiation.

To quantify the radiation received we need dosimeters. An early commercial search for a truly sensitive dosimeter, with density and atomic mass characteristics similar to the human body, focussed on lithium fluoride (LiF). Back in the 1960s the brightness of the LiF signal attracted commercial interest and the manufacturer tried to improve the quality of the product by purification. Unfortunately, they found that the efficiency arose from the presence of impurities. After this great leap backwards, they realized the impurities were fortuitously excellent and, instead of removing them, spent the next year trying to discover what they were and how to optimize both their concentrations, and the way they were incorporated into the lithium fluoride crystal structure. Key impurities were magnesium (Mg) and titanium (Ti), plus harder to analyse quantities of oxygen.

Experimental methods and guesses at impurity sites

The basic idea was to use the radiation that was being monitored to excite electrons from their normal sites in the material, and allow them to move and become trapped at an impurity site. When we wish to read out the accumulated radiation dose, the material is heated to liberate the trapped electrons. During their return to the lower normal energy levels they dumped the excess energy as photons (light). Since heat is needed to release the light energy, the process is called thermoluminescence (TL). The Mg ions helped with the trapping and the Ti with the subsequent light production. The choice of impurities is critically linked to their ionic sizes relative to the original Li and F ions. Ions of similar size are acceptable in the lattice. A straight substitution of a Mg^+ ion on a Li^+ lattice site is not possible as the Mg^+ is far too large. However, ionic size varies with the charge state of the ions, so a divalent magnesium can replace a monovalent lithium ion. In size terms this is possible as the original Li^+ radius is 0.76 nm and the Mg^{2+} is 0.72 nm.

Having solved the size problem Mg^{2+} creates a charge problem, so the trick is to remove two adjacent lithium ions, and insert only one magnesium ion, Figure 12.5. Charge and size effects are then satisfied. This imperfection produces some minor distortion in the lattice which acts as a weak electron trap. Nevertheless, there is still strain energy involved and this can be minimized by heat treatments to allow clustering of the units into a ring of three.

The problem is similar, but messier, for the titanium. Titanium is tetravalent and the Ti^{4+} state is a suitable size for replacement on a lithium site, and substituting oxygen ions on adjacent fluorine sites helps to restore the charge equilibrium without straining the lattice (O^{2-} at 0.140 nm is only slightly larger than the fluorine ion at 0.133 nm). The package is effective and sensitive, and over the subsequent seventy years such dosimeters have been empirically improved by using additional impurities and heat treatments. Realistically we are still uncertain as to the precise details of the sites involved, and there are very many alternative materials in routine usage. In the application the

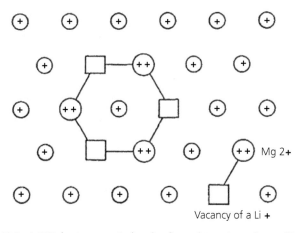

Figure 12.5 A LiF dosimeter. A sketch of metal ions in a plane of LiF. Some of the Li^+ ions are replaced by Mg^{2+} impurities. In size terms they fit into the Li^+ vacancy sites and maintain the charge balance when Mg^{2+} ions are added. For radiation dosimetry we need a more stable and deeper trap than the right-hand example. This is achieved with the ring of three units that form when heating the crystal to ~150 °C for a few hours in the sample preparation stage.

dosimeter is exposed to the radiation that is to be monitored. The sample is subsequently heated and during charge release light is emitted. The luminescence signal intensity is directly related to the radiation dosage.

How old is ancient pottery?

A major problem facing many archaeologists and museum curators, as well as those marketing antiques, is to decide the age of a particular piece of pottery. Key questions range from it being an original, to a later copy or an intentional fake. Items such as Chinese ceramics occur in each category and the market value is immensely different. At the more academic level, many examples of Mediterranean pottery have used the same types of clay, design, and colouring pigments for many hundreds of years and solely relying on an expert on style can easily lead to error. Indeed, when one renowned museum moved from 'expert opinion' to quantitative TL methods of age dating they found more than 75% of certain types of ceramic were misclassified. Wishful thinking had aged some of the later items by 2,000 years.

One such dating route uses thermoluminescence where the radiationx dosage occurs because of radioactive inclusions in the starting clays. Fortuitously, firing of pottery to a high temperature removes all the formerly trapped electrons, and so firing initiates a zero on the 'clock' type signal. There are several variations of the method and, despite the complexity of using natural materials, it works. In the hands of skilled exponents, the method is often able to give age determinations to within better than 10%. For an object from around 1000 AD one can usually bracket an age range from, say, 900 to 1100 AD. Therefore, to decide if a piece of Greek pottery is 2,000 years old, a later replica, or a nineteenth century copy, is relatively easy.

Original and fake artefacts differ so much in value that, of course, there are sustained efforts to produce fakes which pass dating tests such as thermoluminescence. If there is no cause for doubt, or the object is of low value, some of these attempts will succeed. However, with high value objects more sophisticated variants of the tests are added which effectively reveal any faked pottery. This is not the end of the game as it is still possible to fake the validation certificate – *Caveat emptor*.

Look to your garden gnome

A classic example of the change in value between an assumed later copy and an original, which was justified on the basis of thermoluminescence dating, was given by a Sotheby's sale of a garden statue called the Dancing Faun. This is a small bronze statue (32 inches high) that is now attributed to Adriaen de Vries. It initially was thought to be just a nineteenth century copy, rather than an original from near 1600. Fortunately, the bronze still contained the terracotta mould and it was possible to date this internal ceramic material of the mould by using thermoluminescence. Dates are never precise but the TL clearly bracketed the terracotta production in the period between 1450 and 1630. This fitted well with the view of the bronze as being from near 1600, and totally rejected the option of a nineteenth century copy. Thermoluminescence therefore justified the de Vries attribution. The event was headline news at the time as the Faun recorded the highest ever sale price for such a bronze. Instead of around £1500 as initially predicted, the statue sold for £6.82 million! It had been bought in the 1950s for seven guineas (i.e. around £7.50).

Radiation exposure from atomic bombs or nuclear accidents

In general terms, the ability to detect extremely weak light signals implies that studies of luminescence generated by irradiation sources can form the basis of sensitive radiation dosimeters. The preceding pottery, terracotta and ceramics examples showed that an accumulated radiation dose could be estimated by subsequent thermoluminescence of many non-conductive materials. The choice of the insulator for personnel dosimetry has, therefore, been actively pursued as there is a major market for such dosimeters. In uncontrolled radiation events subsequent dosimetry has been made with materials that were available, rather than those optimized for such measurements. Two classic examples have been in estimating the radiation pattern and intensity produced by the atomic bombing of Hiroshima and Nagasaki, or from radiation accidents such as in Chernobyl. Thermoluminescence is a very common effect of most insulating materials, such as minerals and ceramics, and the only dosimetry requirement is that, prior to the radiation exposure, the pottery had been heated to a high temperature to set a background zero in terms of signal. Pottery is ideal as this signal zeroing occurs during firing. In these three nuclear examples it was possible to assess radiation doses by studies of thermoluminescence from bathroom ceramics, in the form of basins or toilets.

Crystal colours from imperfections

I mentioned that adding impurities to glass induces colour, and the same impurities produce different effects depending on their charge state, chemistry and bonding of the material in which they are incorporated. In aluminium oxide, classic examples were chromium and titanium to make rubies and blue sapphire. Note the colour is not produced by the crystal, but represents the residual white light that has not been absorbed. Chromium causes absorption of blue (and ultra-violet) light so the gem has a red colour and we called this chromium-doped aluminium oxide, ruby. Less obvious is that absorption of UV and blue light in ruby results in a red luminescence from the chromium as relatively narrow lines. This was exploited to make the first ruby laser.

For diamond, producing colours is much more difficult. Some natural diamonds come in coloured versions, but generally they are not nice and uniform, which would increase their value. Instead the impurities are localized and so the patchy coloured parts degrade the value of the stone (as for the Koh-i-Nor gem). Heat treatments of gemstones to change the colour can be very effective, but in the case of diamond this is a very risky process as any significant heating can result in the collapse of the diamond lattice into black specks of graphite. This may happen locally in the material and ruins the gem qualities. Back in the 1920s, it was discovered that some diamonds appeared of a higher quality after being irradiated with X-rays. This is a little unexpected as certainly my guess would have been that they would have looked darker or changed colour. The improved gems went up in value and all was well unless they were worn in bright sunlight, when they faded back to their original state. Recent attempts to change the colour of diamonds have been more permanent and these have variously involved irradiation with high energy electrons, or neutrons (in a nuclear reactor), to displace individual carbon atoms from their lattice sites. If we can maintain a low concentration of displaced atoms then the formation of graphite is not a problem. Based on the irradiations, coupled with some further heat treatments, there have been successful results of both green and blue diamond formation.

Radiation, including the UV from sunlight, can excite electrons in diamond and move them from one impurity or defect site to another. Greater exposure to radiation means more electrons have been shifted to new sites. If the material is subsequently heated, the electrons leave their unstable sites and return to the lowest possible energy levels and, in the process, can emit their excess energy in the form of light. The intensity of the light is related to the amount of radiation it received. This is just thermoluminescence, and diamond is interesting in this respect as it was the first recorded example of such an event. Back in 1663, Robert Boyle reported to the Royal Society that, whilst in bed, he had placed a diamond that had been exposed to sunlight against a warm part of his naked body and the diamond gave a 'glimmering light'.

Diamonds are normally thought of as insulating materials but, just as with silicon or germanium, additions of impurities can turn these materials into semiconductors and, because diamond has a wide band gap, it could provide UV, blue or visible semiconductor lasers. Semiconducting diamonds exist naturally but are relatively rare, and probably result from nitrogen impurities in the gemstones.

13

Photonics in the twenty-first century

Part I Photonic concepts

The twentieth century saw the processing of information, data storage and communication advance phenomenally as the result of improved electronics. In parallel there were examples of handling light signals contained in optical fibres and waveguides. The term 'photonics' encompasses the marriage of these two technologies and includes attempts to make devices that centre on the optical properties. A future item could be the commercial achievement of optical computation. Optics in diagnostic medicine and treatment is already successful, but less than routine. Unfortunately, there is considerable inertia in the medical profession to accept new approaches, particularly where they involve new areas of expertise, or unfamiliarity with knowledge of successful treatments. Nevertheless, future photonic technologies will emerge once their commercial and medical benefits are recognized.

A simplistic overview of progress and opportunities so far are sketched in Table 13.1. The advent of optical fibre, radio communication from satellites, and high densities of local transmitters have transformed our use of telecommunication from email, to the web, TV, films, interactive games and surveillance. Mostly these are seen as beneficial advances, but the social impacts, and the potential for control of our lives as these intrude into them, must not be ignored. Once established, it will be impossible to remove them. They already have destroyed many aspects of privacy, and are the enabling technologies for escalation of computer and cyber crime, as well as being actively developed to be instruments of sabotage and war.

Technological blunders are not new, and a current (2020) argument about the frequencies used for new G5 communication is typical of commercial interests taking precedence over international benefits. In this case, the frequency channels may overlap the ones used for weather forecasting and detection of water vapour etc. This seems nonsensical,

The Power of Imperfections. Peter Townsend, Oxford University Press.
© Peter Townsend (2022). DOI: 10.1093/oso/9780192857477.003.0013

Table 13.1 Technologies and structures that make up photonics.

Application	Main idea and uses
Optical fibres	Communication by pulses of light Very high data rates with different wavelengths. Struggling to match current expanding demands on signal capacity.
Laser light sources	From low continuous power from semiconductors to high power pulses, to gas lasers available for welding metals and ship building
Photon detectors	Mostly semiconductor based, including CCD imaging Fairly efficient but plenty of opportunity to improve; Imaging at low level photon counting also exists
Waveguide switching and signal routing	Waveguides via doping and treatments of electro-optic crystals Key items
In line fibre amplifiers	Mostly with erbium impurities for telecoms at 1.54 micron Used for multi-wavelength optical fibres to increase capacity a hundredfold with closely spaced wavelengths.
Solar energy conversion	Semiconductors with photonic energy converters Efficiency improving and enormous potential, even for bodywork of future electric cars.
Optical remote sensing	Everything from eyesight to mineral prospecting An immensely important and expanding subject area
Fibre sensors and control circuitry	Measurements of pressure, strain, temperature, humidity etc., over large regions inaccessible to humans Sensors and clothing for local and remote medical monitoring
Materials analysis	From archaeology, paintings, glassware, wine, to banknotes, radiation dosimetry etc. A vast expanding topic
Security coding	From devices to home surveillance, but equally an opportunity for criminal activities
Photon entanglement	Complex to understand, but with the potential for extremely fast computation
Medical diagnosis, Optical biopsy etc.	Potentially extremely effective and a major growth area from analysis to treatment, but inhibited by traditional medicine
Military usage, Optical and cyber weapons	Battlefield safe communications and weapons Decryption and cyber warfare Heavily funded but with spin-offs

as the forecasting community say it will set back their predictive ability to the level of around 1970 and the G5 system is mostly just faster transmission of social media trivia.

Advances in these twenty-first century ideas may be controversial but it is certain they have the power to change our lives even further and, without exception, all the techniques considered in the table use imperfections in some form or other in enabling the components. Precise predictions are impossible and many are still cloaked in commercial secrecy.

Photonic components needed for optical fibres

Fibre optic communication is probably the most familiar example of photonics and here it is obvious that we need not only the technology of fibre growth and mass production, but also a range of compact, intense and directional light sources, as well as detectors that function at extremely high speed. Data rates are measured in billions of pulses per second, but fortunately semiconductor lasers that are run continuously with good stability can have their output modulated for such data rates, and detectors can cope with such signals. Switching, routing and mixing signals is challenging as we need to send our emails, internet, TV channels and other items along the optical fibres. To do this we need ways to mix, to separate, to encode and decode all this overlapping information. Signal strengths fall with distance along the fibre because of absorption and scatter, so we must amplify the signals when they have faded. There are many components which are likely to improve, but I will just outline the operation of a simple optical switch, a source of clean rapid pulses and an optical amplifier. The concepts are simple but fabrication is challenging because of imperfections in materials.

Optical switching and routing

Mixing and routing optical fibre signals is, in principle, no different from the early telephone exchanges – where the operator took a wire with a plug and joined the incoming caller's telephone line to the one needed to reach the other phone. The modern equivalent is fully automated, has no moving parts, and is made with optical waveguides to route the light signals. It is many millions of times faster, and more secure. The routing improvements are built into switching structures on a material, such as lithium niobate, in which the light is trapped in pathways of optical waveguides near the surface region of the crystal. As discussed earlier, the guide paths have a higher refractive index than the surrounding material. This is usually achieved by adding impurity ions to increase the refractive index with dopants, or to use structural damage and disorder to lower the density and index of the material around the guiding zone.

Optical behaviour is not always identical with our normal experience of how objects behave, because light can either be viewed as little packets of energy (photons) or as waves. If we drive a car on one of two parallel tunnels with a thin divider, we can only change between them when the barrier is removed. By contrast, sending photons (light) along one waveguide gives a rather different behaviour. The light has wavelike properties, so it is not perfectly confined and it has a sideways leakage tail of the light energy extending beyond the apparent physical boundary. If this optical energy can penetrate into the second waveguide then signals will tunnel through. Photons will jump between the two guides and resonate back and forth between them if the guides are parallel. However, if they only meet briefly, and then separate, the coupling section can be set to operate as a switch from one path to the other. The resonance of oscillating energy transfer between the two is, effectively, the same as that which we can see for energy switching between two coupled pendulums. This is not a new concept, as the pendulum pair effects were discussed by Christiaan Huygens in 1673.

For controlled switching we use a material where the refractive index can be changed by an electric field. Many optical crystals can be deformed by an electric field, or pressure, and are termed electro-optic, or piezo-optic, etc., such as lithium niobate ($LiNbO_3$). The main feature of such materials is that they have a directional arrangement, so the end faces are different. Crystal structures which have sequences of planes ABCABC . . . are potentially ideal. We might initially guess that for $LiNbO_3$ there would be two types of plane with either Li or Nb (i.e. ABABAB . . .) but the structure includes a third option which has a set of metal sites that are unoccupied. This gives the sequence in lithium niobate as ABC sets of planes with lithium, niobium and vacant sites. Each plane is separated by layers of oxygen ions, and their spacing is not identical. The asymmetries in the structure mean the crystal has directionality and one end can be charged positively compared with the

other. Applying an electric field distorts the crystal and changes the refractive index. Thus, it can act as a switch to change the barrier between
two waveguide regions. The effect can be used in the opposite sense by
squeezing to generate a piezoelectric voltage for a spark, as a gas lighter.
The material is sufficiently key to photonics that the $LiNbO_3$ structure is
sketched in Figure 13.1. As stated in Chapter 10, the structure is almost
identical in size and ion arrangement to sapphire (Al_2O_3), but sapphire
has Al ions on each metal site and the sapphire axis has all the metal ions
neatly aligned one above the other, where, by contrast, the $LiNbO_3$ has

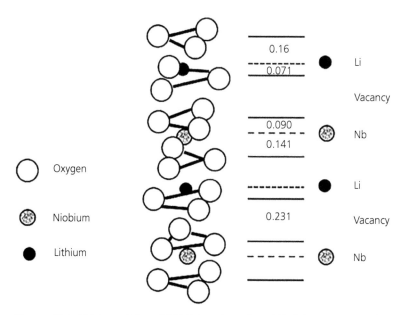

Figure 13.1 Lithium niobate ($LiNbO_3$) and sapphire (Al_2O_3) are quite similar
in that the building block has triangles of oxygen separated by planes containing metal ions or vacancies. Distances are in nanometres. The oxygen defines
the cell size, both have a threefold symmetry axis when viewed from the top.
Sometimes obvious in crystals cut as star sapphire. For $LiNbO_3$ the ABC sequence of lithium, niobium and vacancy means there is a top and a bottom
with one face slightly positively charged relative to the other end. These distortions of unequal spacing between the oxygen to metal planes, and the unfilled
vacant set of sites, all contribute to give a ripple in ionic alignment along the
vertical axis, and this makes it easy to distort the lattice structure, which alters
the refractive index.

a slight wiggle along the axis which makes it far easier to distort and crumple under pressure, or an electric field.

Having found a material that is strongly responsive to electric fields, it is possible to design switching and routing on waveguides within the lithium niobate by controlling the refractive indices with an electric field. The plan is to build two waveguides of high index in the surface region of the niobate. The waveguides are far enough apart that light does not efficiently tunnel between them. If the index difference between guide and surrounding material is large then this spacing can be small, as it poses a bigger barrier for the optical tunnelling. Having carefully separated the two guides we can then use an electric field on the region between them to distort the lattice and lower the barrier. When this happens light will tunnel across from one guide to the next. Over a long length of material, it will swing backwards and forwards between the two. The electric field makes a switch for the optical signal. Lithium niobate is far from unique, but it is transparent, quite responsive, and it has been carefully developed so is the photonics equivalent of silicon as the favoured electronics material. The layout of a switch design in an electro-optic crystal with waveguides defined by impurities is sketched in Figure 13.2, together with the refractive index profiles with and without an applied electric field. I have chosen to plot the refractive index axis vertically, but decreasing, as this defines the optical well that traps the light. One can thus envisage the equivalent pattern of water flowing in channels.

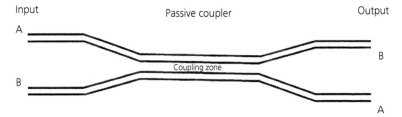

Figure 13.2a Passive waveguide coupling. Two parallel optical waveguides built into the surface region of the material. Light tunnels from guide A across to guide B and then oscillates back and forth. If we choose the correct length then it acts as a coupler to transfer power from one guide to the other.

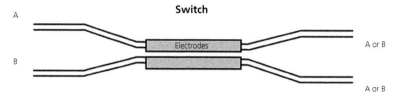

Switch

A

B

Electrodes

A or B

A or B

Figure 13.2b Optical switch. An electric field can alter the refractive index of the central barrier and couple the guides to switch power between them.

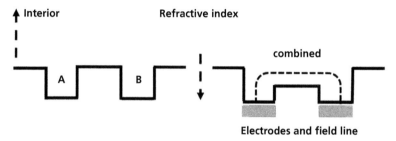

↑ Interior

Refractive index

combined

A   B

Electrodes and field line

Figure 13.2c Guide coupling. The left figure shows the refractive index profiles of two isolated optical wells. Adding electrodes and an electric field controls the refractive index of the barrier region and the energy flow across the wider guiding zone. In this example the guides have a higher index than the bulk material.

Making stable optical pulses

If we start with a laser light source coupled into an optical waveguide, the signal will flow off to the detector. For digital communications we need to turn the signal on and off extremely rapidly. Unfortunately, turning the laser on and off gives pulses which are not totally stable in intensity and frequency (i.e. wavelength). This would limit both speed and the use of several closely spaced carrier wavelengths being sent si-multaneously down the same fibre. Therefore, it is necessary to run the various lasers continuously and chop the signal. A device which can do this is an interferometer. As usual, the idea is based on familiar ex-amples, in this case looking at ripple patterns on the surface of a pond where there is an island in the pond. We split the laser beam down a pair of identical waveguides and, when we recombine them, the two parts are still exactly in step so they just add back together and con-tinue as a signal. However, if we can change the speed (i.e. refractive

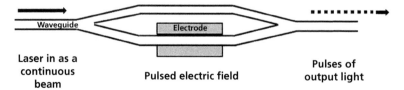

Figure 13.3 An electro-optic interferometer. The sketch above shows optical waveguides built into the surface of a material such as lithium niobate. The aim is to turn the continuous laser signal into on and off pulses. The pulsed electrodes alter the refractive index in one arm so that the two signals are anti-phase when they recombine, and the two waves cancel.

index) in one arm, then the two waves can be exactly out of step and cancel one another out. The two conditions define a zero and one. Pulsing the electric field on the modulator arm means we cleanly control the optical signal pulses with no frequency shifts or intensity changes. The arrangement is sketched in Figure 13.3. This is identical with the island in a pond example, where the wave patterns on water add or cancel at certain positions.

Wave interference patterns are not limited to small ponds but exist in oceans as well. In the Pacific Ocean, because of the large scale, the initial wave patterns can be a series of straight lines (a plane wave). Any islands in the ocean set up a standing wave pattern with intensity and crossing directions fixed by the positions of the islands, which diffract the original plane waves. In Polynesia, people have made stick patterns to represent the intensity of the various waves and their pattern of crossings to use in navigation when they are out of sight of the islands. It is the same mathematics of plane waves hitting objects, but devised many hundred years earlier than for the optics.

In-line fibre optic laser amplifier

As signals travel along the fibres there are tiny losses from optical absorption and scattering. Additionally, the pulse even from a laser source has a very small spectral spread in wavelength (and therefore speed). This is visibly familiar from a prism or rainbow, where blue light travels more slowly than red light and separates into a spectrum. Spreading blurs the pulses into one another. Both problems limit the useable length of a fibre. Early devices for amplification and regeneration of clean pulses within the fibres attempted to convert the optical data into electrical signals. This allowed the use of electronics for mixing and re-timing them. It then meant reconverting everything back to optical pulses. It was feasible and certainly was a useful way of starting up the fibre technology for low data rates. The obvious disadvantages are that all such components add cost, need maintenance and, with more complexity, are likely to have shorter working lifetimes and the occasional failure. Failure is not totally desperate, but clearly incredibly undesirable if it occurs in the middle of a trans-Atlantic or Pacific fibre link. A less obvious problem associated with technological advances is that fibres now transmit maybe 100 or more different signal channels, which are colour coded with extremely closely spaced wavelengths in the near infra-red. If there were separate amplifiers for each colour coded signal, the first step would have to be to separate the channels into their tightly spaced colour (wavelength) channels, followed by conversion back to electrical signals, amplification and laser regeneration with precisely the same wavelength. An immense challenge, unrealistic with 100 channels, and it would be required at every amplifier stage. This is definitely unrealistic. The problem was avoided by a very successful and inspired development of all-optical amplifiers built into the optical fibres. The modern germanosilicate optical fibres are particularly transparent in the near infra-red region around 1.54 microns (as shown in Figure 8.3) and use the transparency window with up to ~100 signal wavelengths with photon energy separation as small as ~20 nm (equivalent to ~0.01 eV in the spectral region near 1.54 microns).

By doping the fibres with ions of erbium, we fortuitously are in precisely the spectral region where a broad luminescence band of erbium matches the fibre transmission window around 1.54 microns. Exciting a section of erbium-doped fibre with a simple local laser power source means the excited erbium ions can relax and emit any wavelength within the range of interest. Erbium is one of the so called 'rare earth' ions (lanthanides). This group of fifteen elements have an unusual property that the energy levels of the three outer electrons, which take place in chemical reactions, are at a higher level than some of the unoccupied inner electron levels. Hence chemically they are similar, but they differ in terms of spacing of their energy levels, and thus differ in the wavelengths of their optical transitions. For a different transmission region, we could have used luminescence from a different rare earth ion.

It is also fortunate that there is a fundamental difference between ordinary light and laser light. For ordinary light sources, the photons are emitted in all directions and every little photon event is independent. In a laser, a few photons heading in one direction stimulate the production of other photons moving in precisely the same direction, and all of them are in phase and have the same wavelength. It needs a high light intensity to get the process started, but once it begins all the power heads in a single direction with every photon united in the way they produce effects. Non-scientific examples would be a crowd walking at random in different directions and then a few key events force them to march in step in a single direction. A Mexican wave at a football match is similar. The end result is a light source which can achieve things that are not feasible from the sum of random individuals.

A simple, continuous, local optical pumping laser excites the erbium in-line fibre amplifier. Without any external stimulus, electrons would eventually relax to give any of the wavelengths in their broad erbium emission band near 1.54 microns. However, even a feeble laser pulse passing through this region will trigger the erbium decay at exactly the same input wavelength and phase as the signal pulse. This rebuilds the data pulse and reshapes it. Even better is that every such signal amplification can run independently, and operate in parallel, for all the carrier wavelengths. Figure 13.4 offers a simple sketch of this design.

Waveguides in semiconductors

Waveguide definition is equally important in semiconductors to make LED and waveguide lasers. The semiconductor approach is always via control of impurities and imperfections to vary the composition, by differently doping the various regions of the crystal. The laser input energy is from electrical current into the junction region between the electron-rich and electron-deficient zones of the material. By contrast, in insulating materials the input energy to pump the output wavelength can be another laser (normally of a shorter wavelength). No matter which method is used, the pump energy stimulates some luminescence in the target (semiconductor, crystal, or glass). The switch from luminescence in random directions to a tightly directed laser beam in a single direction can only occur if the excited intensity is sufficiently high. Confining the luminescence in a waveguide concentrates

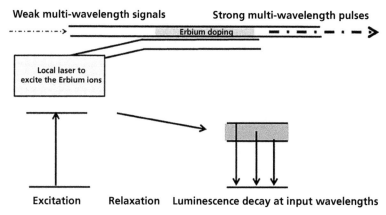

Figure 13.4 An optical fibre amplifier. Erbium is excited by a local continuous laser. Natural decay and emission would be between 1.4 to 1.6 microns (i.e. a perfect match to the most transparent region of the fibre). Feeble incoming signal pulses trigger their same wavelength decay and rebuild the intensity of the outgoing laser pulse. Each signal channel acts independently, so just one simple section of erbium-doped fibre selectively amplifies all the 100 signal channels, as needed.

the power in one direction and allows signals from the entire length of the guide to contribute. Waveguide structures greatly raise the efficiency of laser action, and so reduce the need for high intensity pump power. Waveguides are therefore included in all compact laser sources, whether they are based on semiconductor or photonic mechanisms. Once again, we have a classic example of modern technology enabled by control of imperfections.

Part II Waveguide sensors and imperfections

Photonic devices include many types of sensor and, for optical fibres, these may offer behavioural changes from different points along the length of the fibre. Fibre sensors can not only be used as sensors at a fixed location, but also offer information from different sections of the fibre within a continuous length of a sensing device. They are variously engineered by inclusion of suitable impurities or other defects to have properties that change as a result of heat, pressure, radiation, strain, chemical changes, or humidity, etc. If a local region of the fibre is modified (e.g. by heat) then the refractive index will be altered and so, when we send a pulse of light down the fibre, most of it will go to the far end, but a small fraction will be reflected from the boundary with the region of different index. Merely by timing the gap between sending and seeing the reflected pulse we know the distance to the modified section of fibre. The intensity of the reflected signal gives the scale of the event, and by changing wavelength we can probe other properties. Conceptually it is like bat sonar where the echo, frequency response, and intensity all contribute to give distance, size and type of reflector. Engineers wish to emphasize they are making totally new and radical advances, so they have invented a new name for the fibre equivalent of bat sonar, and called it 'time domain reflectometry', termed TDR.

It is now standard practice to build optical fibre sensors into buildings and bridges. These offer fire detection options as well as monitoring effects of earthquakes, wind distortion on concrete or other component flexing and failure, damage from road works, etc. Major failures break the fibres so the TDR indicates this position within a few millimetres. Not only glass fibres can break with stress, but also the steel cables of suspension bridges. In recent years the worrying scale of this problem has been realized and optical fibre detectors, as well as microphones, are being used to record each fracture event and try to locate it. Many older suspension bridges have lost as many as 10% of their continuous steel cables and the concern is that heavier traffic loads and a weakened structure will set up a downward spiral of collapse.

Distributed radiation sensors employing optical fibres are ideal for monitoring regions near, say, the core of a nuclear reactor, as these are areas where humans are not able to operate and many electronic systems are seriously damaged by the background radiation. More recent usage includes medical examples, as fibre sensors can be used in detection of blood flow and oxygen content. Fibres can be built into clothing, and with associated transmitters these fibres give continuous signals without cables and tubes etc. being attached to the patient.

Fibre sensors offer different types of data recording, but this is not knowledge, unless we know what we are looking for, and the types of signal signature that exist. Examples of oversights are many. One classic case was the failure to recognize the development of ozone depletion over the Antarctic, even though the airborne recorded data were available long before the problem was detected by ground measurements.

Photonic structures

Photonics has many future and exotic possibilities, including topics such as invisibility cloaking of an object. There are also many unexpected ways to guide light or be selective in rejecting or transmitting a particular colour. Structures which can do such tricks can be uniform in composition but have a solid central core with a set of parallel holes in the surrounding material. If the holes and the structure are on the scale of the wavelength of the light, then the wave just senses the average glass density. Glass with holes has lower average density, so a lower index. This is one route to making a waveguide. Figure 13.5 offers an example. Indeed, there are natural examples of such materials. Two familiar ones are butterfly wing colours and opal. The light transmission and reflection are determined by fairly large-scale interference between the waves, which is often on the same scale as the wavelength. Such absorption and colouration effects are on a quite different scale from the nanometre atomic scale of simple 'point' crystal defects and absorption sites. For example, butterfly wing type interference patterns

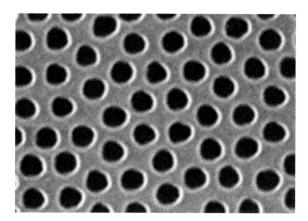

Figure 13.5 A photonic structure. Control of a pattern of holes on the scale of the wavelength means light can be transported, bent around corners, selected or blocked, or different wavelengths can all move at the same velocity.

maybe around 500 nm. Other oddities in terms of properties are that the structures can block, or transmit, selected wavelengths and/or be made so all colours travel at the same speed (i.e. a constant refractive index). The maths is understood, but the concept of designing and making these photonic materials at optical wavelengths has only been around for about thirty years. I am absolutely sure new ideas and uses will become common place, and the photonic crystal materials will lead to ever more varied components of photonics.

Invisibility cloaks

The science fiction examples of an invisibility cloak may not be realistic, but to coat an object so that it is less obvious is definitely possible. Multi-layer antireflection coatings have been applied to optics for many years, so a pseudo version of invisibility has existed for transparent objects. The results of this are valuable for many types of optics from antiglare glass and non-reflective TV screens, to telescopes on snipers' rifles which do not give away position. Low reflectivity has a down side as it is possible to not notice a window or other object, and many bird impacts on clean windows clearly demonstrate this.

An invisibility cloak of a solid non-transparent object is far more challenging, but not totally impossible over a limited spectral range. In part the early examples have come from engineering of materials which have the opposite dispersion from normal glass etc. So instead of the refractive index increasing with shorter wavelengths (as for a glass prism), the new materials can show the reverse. This is as surprising as seeing a primary rainbow with the sequence of colours reversed. Small objects have been coated with such materials and the light made to bend around them and continue as though the object were not there. The present examples are mostly laboratory toys although I suspect this is a topic area which will grow, not least because it attracts large sums of military funding.

To emphasize that invisibility cloaking is not limited to tiny objects we can note that, in 2009, there was a minor collision between French and British nuclear submarines in the Atlantic. Both nations have very advanced and successful sonar invisibility cloaks. The sonar wavelengths are longer than for light, but the incident indicates what may be feasible for optics, and also that not all invisibility is desirable.

Part III Photonics in medicine

I have mentioned examples of optical sensors for medical uses, from in situ surface cancer detection by luminescence spectra to optical probes without invasive surgery. One can equally use light to trigger repair mechanisms, either of the body, or via chemical additives. Progress is slow as the expertise of the medical profession and those in photonics do not often overlap and so there is a lack of acceptance of ideas even where there is success. One example of an underused, very successful optical treatment is photodynamic therapy (PDT). This is not some recent invention but was first demonstrated around 1905.

This is a classic exploitation of imperfections. In this case the cancer cells have slightly different chemistry and optical properties compared with healthy tissue. The principle of the method is intrinsically very simple and approximately as follows. A cream is applied to the area (or added by injection, or orally). This contains a chemical, such as porphyrin IX, which preferentially attaches to cancer cells. After a short time when attachment has taken place, the area is illuminated with high intensity LED or laser light, typically with a red wavelength. Light excites the porphyrin IX molecules, and when excited they activate a chemically reactive state of oxygen. This kills the cell to which it is attached. It is a direct and cell selective treatment. It is particularly valuable for, say, facial skin cancer as there is normally minimal scar tissue after treatment. Such optical methods are likely to be more generally used in the future. The method can be combined with surgery to minimize metastasis in tissue surrounding the cancer removal. Cancer problems frequently spread beyond the original site so treatments that can destroy the problem, without large area surgery, are very attractive.

The imperfections of established medical practice are complex. In a personal example with total success, a large-area skin cancer was treated with the cream application and light exposure performed by a nurse. Total treatment time was under half an hour, no anaesthetics were needed, and there was no long-term scarring of the area. An alternative consultant wanted to perform surgery, which involved a large skin graft (which was not guaranteed to succeed). This would have involved several hours of skilled practitioners and a high cost, but kudos for a surgeon. PDT has been very successfully administered where surgery or radiation therapy are problematic, as in neck and brain cell tumours. With relatively minor imagination one can easily see how to extend it, for example to lung cancer surfaces, bowel surfaces, or other areas where optical fibre probes can be introduced. A key aspect in each case is that these often commence as surface cancers, and so could benefit from the photonic route, without surgery.

The Star Trek version of a photonic optical non-contact body scanner is still in the future as a universal diagnostic probe, but equivalents already exist for specific medical conditions. LED and laser diode units can measure blood flow, read the thickness and degree of burns without contacting the surface, or shine through tissue. In Chapter 7 I mentioned the example of optical biopsy, of say breast cancer. By changing the wavelength, it is possible to distinguish between healthy tissue, cancerous lumps and cysts. The responses differ both by measuring absorption of the light, and by recording the way they scatter patterns. Fluorescence from the tissue differs between healthy and cancerous cells. We should remember tissue is relatively translucent in the near infra-red region, so quite deep penetration is feasible. I believe photonic medicine has a great future. I am particularly hopeful that it will occur since it can be a cheaper option, plus bonuses for the patient that it is faster than surgery, and reduces exposure to hospital infections.

Light absorption, detectors and improvements in vision

Absorption of light can be detected, recorded and transmitted via mobile phones to offer remote medical advice and/or monitoring of a patient. Success merely means that we need a probe photon that is absorbed, and/or emits luminescence, that is specific to the molecules involved with the medical application. This is already a key feature of remote sensing as used in areas as diverse as photography, forensic images, medicine, studies of art works, agriculture and surveillance. These are not new topics as, even thirty years ago, I travelled in a car which included infrared sensors, imaging, and a head-up display so that one could drive with excellent vision at night time, without visible headlights. There are now many more advanced existing photonic examples. With them we have been able to sense thermal images (as seen by snakes), ultra-violet ones (as for insects and birds), and such views literally offer greater insights into their behaviour. More adventurous items under trial are pixelated sensors that include simple imaging when incorporated into a damaged retina. Basically, the ideas are simple, the implementation is feasible, and the desire to explore and help others means such devices will proliferate. I nearly forgot; their manufacture can be extremely profitable.

14

Chemistry and catalysis

Part I Impurities and imperfections in chemistry

It is only a short step from physics to chemistry and, at the boundaries of physics of materials, these subjects overlap. From the viewpoint of trying to extol the value of imperfections, chemistry is a remarkably fruitful topic. The earlier examples of imperfections in technology used aspects of physics, with materials ranging from metals to glass and semiconductors, but beneficial roles of impurities and imperfections are equally crucial in the chemical industry. If I were a chemist, the following chapter would have been my opening to reveal how to exploit impurities, trace elements and imperfections. Perhaps the most striking and obviously important chemical example of useful, carefully selected impurities and imperfections, is catalysis. Catalysis is the term used to describe how the presence of impurities or special lattice sites can totally transform chemical reaction rates. It is absolutely central to very many areas of commercial chemical manufacturing. Even in processes where catalysis is currently not in use, it is actively researched in attempts to try to introduce catalytic methods to reduce cost, lower energy consumption and gain processing speed. Scientists will often start 'in depth' research into an unfamiliar area by reading reviews etc. via scientific web sites. From such a starting point the importance of catalysis is rather easily apparent, as web searches for 'catalysis' offer over five million returns.

Catalysis has been part of our technological history since long before anyone realized that such processes existed. Examples range from fermentation of wine, to making vinegar, soap manufacture and leavened bread. Industries rely on catalytic processes for large scale production of compounds such as ammonia, or nitric and sulphuric acids.

The Power of Imperfections. Peter Townsend, Oxford University Press.
© Peter Townsend (2022). DOI: 10.1093/oso/9780192857477.003.0014

In biological systems the active sites are equally important, but are not normally described as catalysts, instead they are called enzymes. They frequently function via traces of metals in their structures. Crucially, we could not survive without them.

Many chemically based products need some type of imperfection during their manufacture, and one can find a very diverse range of products. In attempts to produce glues the products have spanned types of adhesives that were too weak, to glues that were too strong for their intended use. I used the term imperfection as both were failures in terms of their original design. An example of a 'too weak' case was the glue that apparently failed its original objective for bonding paper to other surfaces, but which was then a major commercial success as it offered a sticky edge which could be easily peeled from a surface. This is absolutely ideal for making temporary notes, and we can leave them in a visible place without destroying the background surface. We now use such notes by the thousand.

At the other extreme an example of glue 'failure' was the production of superglues which are incredibly strong, and react and harden within seconds. This again was initially thought of as a problem, as the setting time was too rapid. In hindsight, superglues are fantastic as they bond many materials and are ideal for making repairs, as well as in their role in construction. The apparent downside of superglue, that they adhere powerfully to skin, can also be viewed as a classic case of a property which is a nuisance in some situations to being absolutely perfect in others. Modern surgery makes considerable use of superglues as they set rapidly, are apparently non-toxic, but eventually dissolve into the body tissues. They are an excellent replacement for stitches and can be used in situations where skills with a needle and thread are challenging, or would be too slow. Their second benefit is that there are no stitches that need to be removed. Superglue is used in some countries by the police in riot control, by removing troublesome public by temporarily gluing their hands together.

If I am willing to count chemical developments such as these, which originate from initial errors in intended application, or the even more diverse definition of successful technology which came as the result of accidental factors, then the range of products and applications from serendipitous chemical imperfection-based examples is enormous.

Catalysis

Catalysis is the term used to describe a method which can speed up chemical reactions. Even without any background knowledge of chemistry, we can make an intelligent guess at the broad outline of a catalytic process. If we assume that a chemical reaction between materials X and Y results in a new material, Z, then this will take place at a rate which is partly determined by the energy needed to start the reaction. This is no more than we expect by cooking something in the kitchen. In cooking, as in much of chemistry, this energy is provided by heat. Catalysis becomes interesting if the normal reaction process only operates at high temperature, is difficult to control or if it is slow and costly. In commercial terms these factors mean it is an expensive process, and/or runs too slowly at lower temperatures. Now consider the role of a catalyst, C, which typically enters the reaction to offer an alternative lower energy route. The reaction costs fall, it works at a lower temperature and it may be faster. The key feature is that there are energy barriers that need to be crossed for a chemical reaction. We can easily jump a one-foot fence, but not a ten foot one. A ladder with one-foot spacings means we can reach the top. In essence this offers a catalytic route, which can be reused many times.

A typical example is sketched in Figure 14.1. A particularly valuable feature is that during the reactions the catalyst C is released after it has done its work. This means it has not been used up by the reaction and it can be reused over and over again. Consequently, catalysts may only form a very small fraction of the chemicals being processed, but nevertheless speed up the reaction by tens to thousands of times. Plus, even expensive catalytic materials are economically viable. Examples include platinum and rhodium, that are used in car exhaust systems to clean up the emissions. Catalysts drive the process and survive without being chemically consumed, so they offer both chemical and economic advantages. (Note some events which just initiate a reaction, or event, are often incorrectly called catalysts.)

I like easy analogies and see catalysis in the same way that one views a problem of many schoolchildren trying to cross a busy road outside a school. The crossing rate is very slow as there are few gaps in the traffic, or the children would need a lot of energy to run through the gaps when they appeared. A school traffic warden is the equivalent of

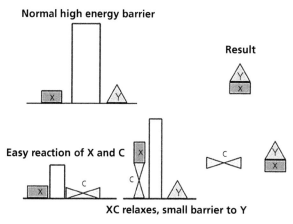

Figure 14.1 Simple catalysis. Imagine chemicals X and Y need to react to form a new compound, XY, but the process needs a large amount of energy, say 1,000 units. The reaction rates may be negligible at low temperatures. However, if a catalyst, C, can initiate a low temperature reaction, such as X + C to form XC (e.g. using merely 100 units), and XC can then easily react with Y, the overall efficiency has increased. Ideally XC + Y produces XY (the product we wanted) and liberates C for reuse. The second reaction will also require energy, even if this were 200 units it would define the overall reaction rate, but still be much faster than without the catalyst.

a catalyst. The warden walks out with a sign and stops the traffic, allowing the children to freely cross the road at low speed (i.e. less energy). The warden returns to the side and can repeat the process. The overall crossing rate has increased even though the energy needed is low.

The model actually has another intuitive feature which applies to certain types of chemical catalysis. In this version, once children start crossing the road, even without a warden, the traffic slows down or stops. The children who start the crossing catalyse, and make easier, further crossings (i.e. they themselves reduce the barrier). A familiar example of this is ripening of apples or tomatoes by putting them in a drawer with one that is already ripe. The effect is not psychological. The ripe fruit emit ethylene which stimulates ripening, so they all ripen faster.

Note that a one-step catalytic route is just the simplest to consider. Many examples of catalysis are more complex with several intermediary compounds formed between the start and final product. Indeed, each step may need different catalytic inputs. Quite often catalysis was first suspected and linked to trace impurities. This frequently has required very good intuition, coupled with highly skilled analytical methods, to find which were key players. The scale of detection sensitivity which may be required is astonishing, as in one example (to make aryl boronic acids) the presence of palladium at 0.24 parts per billion (ppb) had an effect, and routine production uses 50 ppb levels. If we were counting humans, 0.24 ppb is two people relative to the entire world population.

Types of catalysis

Chemists tend to differentiate between two types of catalyst: as those which are in a different phase from the materials that are being activated, and those which are totally mixed together with them. The former is a 'heterogeneous' reaction and a classic example is the use of particles of solid platinum and other metals in a car exhaust system (heterogeneous because there are both solid and gas phases involved). The platinum helps to break down the hot gases of the car engine. By contrast, catalysts which are in the same phase, such as a liquid in a liquid, cause a 'homogeneous' reaction. The heterogeneous catalysts operate with chemistry where all the excitement occurs on the surface. It is highly desirable to make the catalysts with very small particle sizes so there is relatively more surface area. The reactant (e.g. the car exhaust) has molecules which temporarily adsorb onto the surface and the chemistry begins. In reality the process is not so simple, as a perfectly plane and smooth surface might be very ineffective. Instead, surfaces which have steps, corners, vacancies and dislocations, etc. (all the stuff of imperfection crystallography) offer lower energy sites to temporarily bond with the reactants. Both impurities and intrinsic lattice imperfections are required in many good catalysts. I could ask why are some metals good catalysts and others not. The answer is tricky, but part of their reason for success is that the surface bond energies are strong enough to trap the materials that are going to react, but weak enough that the product materials can escape. Like Goldilocks and the three bears, some porridge is too hot, some too cold and some just right.

Margarine

One of the familiar early examples of catalysis is the production of margarine in the presence of porous nickel (Ni) particles. This is usually described as hydrogenation of vegetable oils. Vegetable oils contain chains of carbon atoms, typically between ten and twenty atoms long. If hydrogen gas is mixed with vegetable oils there is a very minor rate of chemical reaction at the temperatures suitable for food processing. Hydrogen exists as a gas in the form of a molecule containing two hydrogen atoms (H_2). This is relatively strongly bonded and so does not react. When hydrogen lands on the surface of the nickel, the gas molecule relaxes, and it falls apart into a pair of separated hydrogen atoms. At the same time the vegetable oil contains chains in which the carbon atoms are bonded via a link in which two carbons share pairs of electrons. This is called a double bond. The nickel surface weakens and breaks this carbon double bond and allows the pair of hydrogen atoms to become inserted into the vegetable oil by attaching to carbons ions. This new hydrogenated structure breaks free of the nickel surface and we have a new material. The nickel surface is now available to repeat the process. Figure 14.2 shows examples of before and after hydrogenation of the double carbon bond.

Historically, the original starting materials for margarine were a number of fatty acids in vegetable oils, but the number of margarine variants has increased to include raw materials such as sunflower, rape seed, palm oil and soya beans. Because they have the double-bonded carbon structures, they are termed unsaturated (i.e. not every possible

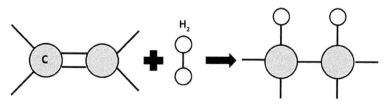

Carbon double bond   **Carbon–carbon single bond**

Figure 14.2 Hydrogenation of a double carbon bond. The left hand pair of carbons are joined with two electron bonds, whereas after hydrogenation this double bond type is removed to leave just a single two-electron bond between the carbons, plus two new carbon-hydrogen bonds.

bond goes to a different atom). The double bonds put a 'kink' in the chain of carbon atoms making it hard for them to pack into a crystalline solid. The overall effect is that they have very low melting points unsuitable for food processing. Hydrogenation with the nickel catalyst at about 150°C removes the double bonds and straightens the chain or carbon atoms, which increases the melting point. Catalytic hydrogenation has allowed many more materials to be used, and some two million tonnes of margarine are produced each year as the result of catalysis.

The margarine example of nickel catalysts was first made in 1897 by Sabatier, but this was not the only catalytic advance in that year. Another example was the discovery that mercury could be used to lower the production temperature of indigo dyes. Apparently, this was not a very successful process and many catalysts had been tried without any useful advantage. However, the anecdotal version is that a chemist was stirring and measuring the temperature of a dye mixture with a thermometer. He broke it. The release of the mercury then showed excellent catalytic action. Mercury catalysts are also used in many other examples, including fluorescent dye production, manufacture of PVC plastics and vitamin B2.

Catalysis in the oil and automotive industries

Catalysis is a key component of the oil industry, from every stage of oil processing in the refinery to attempting to remove harmful residue from vehicle exhaust gases. The original product, crude oil, is a complicated mixture of at least one hundred chemicals. The majority of them are called hydrocarbons as they are compounds which include only carbon and hydrogen. The number of carbon atoms runs up to tens of carbons per molecule. In principle, the different molecules can be separated by distillation as the heavier molecules boil at higher temperatures. In automotive terms if there are, say, 15 to 25 carbon atoms per molecule the material is similar to diesel, whereas kerosene (paraffin) would be in the 10 to 15 range and gasoline (petrol) is lighter at, say, 5 to 10. Compounds with even fewer carbons per molecule are quite likely to be gaseous at room temperature.

Separation of these components from the original oil is called cracking. In order to lower the energy required, and therefore lowering the costs, the refineries use catalysts. The objective is to break the larger components down to the molecular size suitable for petrol and/or to make reactions which generate the material with 8 carbons, called octane. The actual conditions and details of the process are too complicated for our purposes, and many chemical reactions are taking place. Details are irrelevant, but the example clearly emphasizes this is a major industrial process benefitting from impurity-controlled catalysis.

One of the catalysts used in the petrochemical process is a fine powder (i.e. to give a large reactive surface area) of a material called zeolite. This is a compound with a very loose and open structure made from hydrated aluminium silicates, plus ions such as sodium, potassium, calcium and barium (i.e. a complex and variable mixture). The loose open structure is mechanically an imperfection but, much like water entering a sponge, it allows oils to penetrate to the interior of the powder. The effective catalytic surface area is increased. People have estimated one gram (say a teaspoonful) of zeolite material has an effective reactive surface area of an acre (40 per cent of a hectare).

The role of catalysts to minimize the poisonous exhaust gases (e.g. from a car) typically needs rather expensive metals, such as platinum and rhodium. A particular villain for humans is carbon monoxide (CO). It is clear and tasteless but very toxic and can prove lethal. The catalyst in this case induces a reaction between the CO and a second pollutant, nitrous oxide (NO). Pairs of the pollutant molecules result in the formation of carbon dioxide (CO_2) and nitrogen (N_2). In chemical notation this is:

$$2CO + 2NO \rightarrow 2CO_2 + N_2.$$

The nitrogen production is not a problem as it is the major constituent of the atmosphere and locally, at ground level, the carbon dioxide is not a problem either. Indeed, it is useful for plant growth. The panic and focus on reducing the carbon dioxide emission is because, in the upper atmosphere, it becomes a serious threat as it acts as an absorber and reflector of long wavelength light (i.e. heat). This greenhouse ceiling drives an inevitable warming of the Earth (as discussed in Chapter 5). This has similarities with catalysis as a minor temperature change can produce significant differences in reaction rates. A rise of mid-Atlantic water temperature of, say, ~2°C (~3°F) gives a logarithmic rise in water vapour pressure that, in turn, drives the scale of cloud formation, tornadoes, flooding and changes in wind patterns, or desertification.

Not all catalysts are beneficial and one obvious problem has followed as the development of materials called chlorofluorocarbons (CFCs), with compositions such as CF_2Cl_2. CFCs have excellent industrial properties and were designed as refrigerator fluids, and used later as aerosol propellants. They were initially prized as the basic ingredient for the refrigeration industry. Destruction of old equipment, or other routes which release the CFCs, allows them to drift to the upper atmosphere where they very effectively drive catalytic reactions with ozone (O_3). Their effect is to destroy the ozone layer. Since it is merely a catalyst, the major problem is that a single CFC molecule does not just attach to one ozone molecule, but instead can destroy thousands. Normally the ozone concentration is quite low, and it is only slowly formed by sunlight exciting oxygen in the upper atmosphere. Ozone plays a major role in absorbing energetic ultraviolet light. Loss of ozone allows in the UV light which can have significant effects not just on our eyesight, but also eyesight of other animals as well as interfering with, or destroying, agricultural and plant growth.

Catalysts for plastics

More than 300 million tons of plastic are produced worldwide each year, of which half are either polyethylene or polypropylene. Their commercial importance is outstanding. Yet their discovery and development have a history of accidents, impurities and defects. Polyethylene was first developed in ICI in the late 1930s using very high pressures on chemical reactions. Sometimes heating ethylene ($H_2C=CH_2$) under pressure produced a waxy solid, polyethylene. This was attributed to traces of oxygen impurity in the ethylene to catalyse polymer formation and polyethylene became a highly profitable product. Not only were imperfections crucial (in the form of catalysts), but they define the stiffness and toughness. Around half the solid plastic is made up of tiny (ordered) crystals, whilst the rest is (disordered) amorphous material. The individual chains pass through multiple crystalline and amorphous regions and lock together to form a 'nanocomposite' of linked hard and soft regions. The polyethylene chain is made up of -CH_2- units which are relatively 'smooth'. They can crystallize easily, but a mixture of short- and long-chain branches (because of imperfections) define the degree of crystallization and control mechanical properties.

Side products from oil refineries include both ethylene and higher olefins, particularly propylene ($H_2C=CH.CH_3$). The polymer from propylene has a methyl (CH_3) group attached to every second carbon atom. Attempts to produce polypropylene using ethylene polymerization catalysts were largely unsuccessful as the arrangement in space of the methyl groups around the chain is critical to crystallization. Imagine the -CH_2- chain of polyethylene pulled out to its full extension, then one sees the angle between carbon atoms produces a flat zig-zag. Attaching -CH_3 groups to alternate carbon atoms produces different arrangements. They can be all on one side (isotactic), alternate (syndiotactic) or random (atactic). Only polypropylene chains with better than 99 per cent iso- or syndiotactic structures can crystallize. The rest currently lack commercial value.

In 1953, the German chemist Karl Ziegler was experimenting on the reaction of ethylene with aluminium alkyl compounds to make low molecular weight products which were of interest to the chemical industry. He perceptively recognized that traces of nickel contamination in the reactor, from previous experiments, acted as a catalyst for the coupling of ethylene molecules. He then found that a combination of titanium chloride and his aluminium compounds could polymerize ethylene very easily to make polymers with small amounts of branching. These discoveries were the basis for a range of polyethylenes, from soft, flexible plastic to harder, more rigid materials, right up to the ultra-high molecular weight polymer used for conveyor belts and hip joint replacements.

Optimizing the quantities of catalysts is important both for efficiency, cost and whether or not they remain in the product. The first attempts were successful but only generated around 5kg per gram of titanium, as the crystallites of titanium chloride only worked by reactions at the edges and other imperfection sites on the surfaces. Modern catalysts are more homogeneous and generate around 1000kg per gram of metal catalyst. An Italian chemist, Giulio Natta, showed that some of Ziegler's catalysts could polymerize propylene to give a material with isotactic arrangement of the methyl groups around the chain, and crystallize (so-called 'stereoregular' polymers). Their work led to the 1963 Nobel Prize for Ziegler and Natta, together with rapid development of industrial processes for making stereoregular polypropylene. The isotactic polymer is stiffer and higher melting than polyethylene, and does not creep under load. It is now widely used to make everything from car bumpers and other parts, to stadium seats, plastic crates and boxes.

Trace metals and enzymes

Enzymes appear to be the catalysts that make a very large number of our bodily processes operate, and enzymes have a subtle chemistry which usually includes some metallic content. Therefore, the intake of trace elements is absolutely essential for all aspects of our health. Probably foremost among these impurities is zinc. Some 200 of the major enzymes include this metal and a list of medical problems which occur from zinc deficiency, plus a list of processes which use zinc enzyme catalysis, could occupy a book. Zinc is needed in aspects as diverse as DNA and RNA production, in defence against viruses, fungal infections and cancer, as well as in growth and reproductive hormones. There is a particularly large need for zinc intake during pregnancy and the US estimate of daily need is around 15 milligrams for normal life, but as high as 20 to 25 milligrams per day during pregnancy. Vegetarian diets are often particularly weak in adding adequate zinc intake. To place these quantities in perspective, 15 milligrams is roughly 15,000 times lighter than the weight of meat in a half pound steak, or less than 100,000th the total weight of food and drink many people consume each day.

Despite the nominally small quantities of material needed, a large percentage of the population do not manage an adequate zinc intake and, among the many resultant problems, it has been suggested that this deficiency can lead to paranoia and aggression. One Nobel Prize winner has suggested that zinc-rich diets should be distributed in a certain area of the world where there is an inherent shortage of zinc in the diet, but an excess of ongoing conflict. This may be a very perceptive suggestion, and I certainly think it should be tried as an experiment, not least because under normal conditions zinc intake does not appear to lead to obvious problems, but in industrially zinc-contaminated areas the very high levels can be harmful.

Medical understanding of the role of trace impurities, even those as critical as zinc, has really only been developed over the last fifty years, so the possibility of dietary-controlled improvements in health are highly likely to appear. A second important trace metal is magnesium. This has a similar chemistry, compared with zinc, and accumulates to be the fourth most abundant element in the body as it is found not just in bones and red blood cells, but also plays roles in muscles, nerve and cardiovascular operation. Intake, in this case, similarly has a minimum essential dietary requirement, but for magnesium excessive intake has a number of serious side effects. The criticality of trace magnesium in the diet of horses is clearly apparent as, without it, they cannot produce foals. For humans, such clear-cut links to specific impurities and health or fertility are not always as clear, although pernicious anaemia is caused by a deficiency of vitamin B12, and this requires cobalt.

Another trace metal which is often cited, is lead. Lead poisoning is thought to have disastrous consequences ranging from infertility to madness. It has, therefore, appeared as a side effect of economic progress and civilization as we know it. It is claimed that it was a contributory factor in the fall of the Roman Empire as there was lead contamination from lead plumbing (lead is plumbum). In Rome, the soft water dissolved the lead. Such sanitary benefits (and side effects) were 'enjoyed' more by the ruling classes. The same problems reappeared during the Industrial revolution of the nineteenth century and, even now, lead plumbing exists in many older buildings. Bioavailability is crucial. For example, $BaSO_4$ is used in medical diagnostics because it is totally insoluble, whereas $BaCl_2$ is highly toxic. Lead is also an excellent additive in petrol, but the resulting contamination in the exhaust gases is a severe health risk, and so lead has been excluded from the petrol of many countries.

Heavy metals, such as lead or mercury, tend to have obvious deleterious side effects on the body and cause brain damage. Ingestion of mercury from mercury compounds (nitrates or chlorides) that were used in hat making had harmful health effects. It is claimed the phrase 'mad as a hatter' was linked to the fact that hatters could develop neurological problems from these chemicals.

There are many ways in which different minerals can enter into our diet. Some of the routes are unexpected. One such was a fashion at the end of the twentieth century for unglazed earthenware pottery, and this was used for dispensing health food drinks such as orange juice. Unfortunately, orange juice is quite an effective acid solvent and, if left in unglazed pottery, it can leach out heavy metal from the container. It also has the potential to release radioactive minerals which emit alpha particles that can cause surface cancers where they make contact.

I picked this short selection of examples of trace metal impurities to underline that, in many ways, they are essential for our well-being, but equally, adding too much, or the wrong trace metals, can be very detrimental.

One can sometimes sense the irony of great inventions as both the CFCs for refrigerators, and lead additives for petrol, were invented by the same excellent chemist, Thomas Midgley. Both were highly acclaimed and effective and, in neither case, were the negative aspects realized until after he died. Even that event was unusual. He had designed a hoist system because he became bed-ridden, but became entangled with it and died in 1944.

Part II Catalysis and cleaning windows

In the chemical examples of catalysis, I have mentioned they normally have a thermal energy input. In the case of window glass, an alternative energy input is sunlight. A problem with the outside of windows is that they become dirty, and particularly angled structures, such as greenhouses, conservatories or skyscraper cladding, accumulate various types of dirt on the surface, from dust to bird stains. Rather than rely on frequent manual cleaning there is an incentive to find a glass surface which is self-cleaning, albeit with the addition of some rain water and sunlight. A catalytic route to achieve this has been developed in which the surface of the glass has a coating of titania (an oxide of titanium). It is feasible, and details are offered in the caption to Figure 14.3 on the underlying science.

The energy for the chemistry has come from sunlight, and the cleaning layer from rain water. The titania is not consumed or worn by the cleaning action as it is firmly attached to the glass surface, but is effective via its role as a catalyst. The technology has also been incorporated into paint, although it was quite challenging to produce a material which did not attack its own binder. In experimental applications, coating walls of road tunnels with photocatalytic paint and illuminating the walls with low levels of UV light not only keeps the walls clean but also gives a 50 per cent reduction in NO_x pollution from vehicle exhausts.

Self-cleaning glass with structured surfaces

An alternative mechanism which can help to maintain a clean glass surface works on precisely the opposite approach. Instead of aiming for a surface that attracts water as a superhydrophilic surface, the chemistry is used to achieve the opposite approach and actively sets out to make a superhydrophobic one! As expected, there is a need for imperfections, but this time not via chemical catalytic reactions, but by moving from a

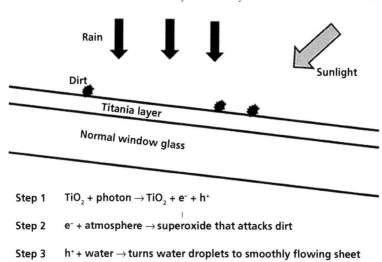

Step 1   TiO$_2$ + photon → TiO$_2$ + e$^-$ + h$^+$

Step 2   e$^-$ + atmosphere → superoxide that attacks dirt

Step 3   h$^+$ + water → turns water droplets to smoothly flowing sheet

Figure 14.3 Self-cleaning window glass. Titania (titanium oxide) includes a continuous range of oxides of TiO$_x$ with x values anywhere between 1 and 2. As a coating on window glass it will be close to the dioxide which is visibly transparent. UV from sunlight can excite electrons from the lower energy band to the higher empty band. Oxygen-deficient titania will additionally respond to blue light, as well as the UV. A blue tint may be desirable in a conservatory glass roof, as it gives the impression of a blue sky, even on a misty day. but I suspect one factor may actually be related to the inevitable blue tinge from oxygen-deficient titania (i.e. a good imperfection). The UV-excited electrons (negative charges) and the holes (positive) allow two types of chemical reaction to take place. Electrons react with oxygen in the atmosphere to form an excited and aggressive chemical version of normal oxygen, called a superoxide radical. The holes react with surface water to form a neutral hydroxide. The superoxide is highly reactive and will attack, destroy and break down microbes, bugs and other dirt on the glass surface. The neutral hydroxide changes the usual level of attractiveness between the glass and the water, and it spreads smoothly across the entire surface (termed superhydrophilic). The net effect of a smooth liquid surface with no droplets or points to impede liquid flow means all the surface dirt flows readily across it, and the glass is self-cleaning.

flat plane glass surface to one with a very fine surface structural pattern. The surface patterning requires bumps on the scale of a few microns in dimensions (i.e. a few per cent of the diameter of a human hair). The model of how this keeps water off the surface is thought to be that air is trapped in the spaces between the bumps. Water droplets that form on the surface are therefore pushed away from the surface. They have not bonded to the surface, but roll across the bumps and it stays super clean. If there is dirt as well as water on the glass then the dirt particles are carried along by the water.

The concept was the result of trying to discover why lotus blossoms always stay clean, via images of their patterned surfaces. The structural approach is not confined to lotus plants and glass. Instead, it can be built into a wide range of surfaces including fabrics. This opens up the future production of self-cleaning from clothing, to shop awnings and tents etc. The demonstrations of the effects are sufficiently new that there are far more market opportunities than have yet been proposed, but self-cleaning materials are now available.

Mining and recycling opportunities for trace elements

Whilst trace elements occur in the minerals that we find, the chemical separation of them is not always easy. Some elements have a very high intrinsic value per gram of material and it is worth contemplating how to find new economic ways to extract the valuable traces. To gain some perspective of the effort required, to extract platinum one needs to mine some 40 tons of ore, crush wash and process it in order to extract a single gram of pure metal. Other elements are even more troublesome, and/or exist in very few locations. Economically the end product from these impurities is highly desirable. High value materials such as platinum and rhodium are used in catalytic convertors. (One unfortunate consequence has resulted in many thefts of exhaust systems.) Catalytic convertors in car exhausts age, rust, and lose material during use and these contaminants appear as dust on the highways. From more major roads, such as the M25 motorway around London, it is economically worth vacuuming up the roadside dust to extract the expensive metal ions. It is cheaper than mining!

Detailed knowledge of the use of valuable materials in other products can be exploited. I was taught optics by a physics professor who, after the Second World War, bought very large quantities of disused klystrons, from radar systems. He then extracted the platinum electrodes. The profits were considerable and it funded the growth of his new physics department.

Low concentrations of valuable materials, such as gold or uranium, exist in the oceans, but only at concentrations of a few parts per billion. This does not sound like a great deal, but there are minimal mining costs, so novel ways to make the separation might be worth considering. Far more surprising is that there was a sewage treatment plant in Nagano, in Japan, where it was reported that ash from burnt sludge is returning almost 2kg of gold per ton of ash. This is far in excess of the 20–40 gms per ton found in the ore from a local gold mine. This is clearly a local anomaly resulting from the types of factory dealing in valuable metals that provide the input to the sewage plant.

People with a non-scientific or non-industrial background may realize that materials like gold and silver are expensive, but probably do not realize how volatile are the prices of other key metals. The price of some metals, such as aluminium or tin, tend to be fairly stable, but others, such as copper, silver or indium and rare earths have increased by 300 to 500 per cent, and more, over periods of just a few years. Indium (used in conductive smart phone screens) and rare earth elements are very vulnerable to price changes as the elements are limited to a few countries who, therefore, have a monopoly. Within the semiconductor devices the quantities are minute and so reprocessing dead devices is not effective, even though there are billions of obsolete pieces of electronics. Attempts to conserve and recycle such material is, therefore, an interesting, and potentially lucrative, challenge.

Part III Differences in properties of isotopes

In the simple model of an atom, there are outer electrons which define the chemistry and the way atoms bind together in molecules and solids. The number of electrons balances the positive charge of protons in the massive nucleus at the very centre of the atom. The design of the nucleus has a conflict because it has many positive charges which all repel one another and this should force it apart. It is stable because there are also a number of types of attractive force related to the mass, and other more subtle nuclear binding forces. The survival plan of the nucleus is to increase the mass neutrons. These add binding forces and dilute the charge repulsion. The neutron mass is approximately the same as for the positive protons.

Survival of stable nuclei is possible if there are about equal numbers of positive (protons) and neutral (neutrons) in the nucleus. Rather more neutrons are needed for heavier elements. Since the neutrons mostly play a stabilizing role, there can be several options as to the exact number in the atoms of a particular element. As discussed for radio carbon dating in Chapter 4, carbon has 6 protons, and normally 6 neutrons. This is nearly correct as 98.89 per cent have 6 neutrons and 6 protons. We write this as ^{12}C for carbon. However, ~1.11 per cent have a seventh neutron (^{13}C) which is slightly unstable, (plus a ^{14}C isotope). The chemistry is independent of the nuclear mass, so it is not immediately obvious what isotopic differences will appear and how they can be used in technology or analysis. In purely physical motion, as for molecules of carbon dioxide, or diffusion through a cell membrane, the weight matters. Mobility is set by the available thermal energy (i.e. heat). Simple kinetic energy depends on mass times velocity squared ($E = \frac{1}{2} mv^2$). Lighter objects move faster than heavy objects of the same energy. In a gas every molecule will have the same average energy, which is fixed by the temperature. In air most nitrogen molecules (N_2) have a mass of 28 units (i.e. a pair of ^{14}N atoms). There is a much rarer nitrogen isotope of mass 15 and a few so the molecules weigh 29 units (mass 30 is even less probable). Heavier nitrogen molecules have a 1.7 per cent lower velocity. This is noticeable. In a hundred metres race, a 1.7 per cent difference means you either win or lose the race.

In the atmosphere, lighter molecules are more likely to travel higher and the degree of separation of heavy and light molecules will depend on temperature. Similarly, water vapour diffusing inland from the sea means the heavier isotope of hydrogen (deuterium, mass 2) will move noticeably slower than normal water (the average velocity is less for deuterated water of the same energy). Normal water has two lightweight hydrogen atoms (of mass 1) and one oxygen ion (mass 16) to form H_2O. The molecular weight is 18. With one deuterium replacement it becomes 19, and the velocities differ by 2.6 per cent. This generates an imbalance of the isotopes with distance from the source of the water vapour, as the lighter ones travel faster and further. The ratio produces a signature of differences in plants (and humans), and how far they live from the sea. Analysis of the ratio is used in forensic identification, the likely origin of a person and archaeological studies of bones.

Photochemistry

Precisely similar effects of changes in isotopic ratios occur in chemical reactions which involve gas or liquid diffusion. Plant photosynthesis is crucial for our food and basically runs in two variants, called C3 or C4. The C3 route means three carbon atoms are involved in the first stage. Plants in Europe and North America are mainly growing with C3 photosynthesis, whereas in hotter, drier climates the C4 mechanism will dominate. Figure 14.4 gives a more formal definition of the two routes. C3 and C4 alternatives are very different in overall mass and so have an obvious effect on the ratio of normal to heavier carbon isotopes. The carbon isotopic ratio of the same end product reveals which chemical route was in use during the formation.

Isotopic effects appear from changes in diet and country of the food chain. In North America, many major crops which run with C4 photosynthesis are maize, sorghum, millet and sugar cane. These are the basis of the food chain for much of the animal feed. By contrast, different crops are used in Europe with a higher input from the C3 route. The same differences are revealed by the isotope ratios for people living in these different areas. Diet obviously influences the natural biological processes, and so the carbon ratios in our production of testosterone or other materials will relate to the food source, and the dominance of C3 or C4 type photosynthesis. For active, healthy athletes with a good healthy diet in, say, the USA, they will become typified by the C4 food chain.

Additives to the diet and many synthetic drugs may nominally produce 'natural' compounds, but the processing route is often based on

Isotopes and C3 and C4 types of photosynthesis

Carbon with only C12 e.g. 3 carbons in 3PGA 3-phosphoglycerate	Carbon with some C14 e.g. 4 carbons in Diacid oxaloacetate
C3 Calvin-Benson	**C4 Hatch-Slack**

Figure 14.4 C3 and C4 photosynthesis. Two types of photosynthesis that fix carbon dioxide into terrestrial plants are understood and are termed C3 and C4. C3, or Calvin-Benson, involves 3 carbons in a material called 3-phosphoglycerate (3PGA). An alternative is C4, or Hatch-Slack, which produces diacid oxaloacetate.

materials such as soya, which is a C3 type material. The implication here is that athletic performance which was achieved with the benefit of artificial drug sources will be evidenced by a careful examination of the isotopic content of carbon and other elements of the body, and the by-products such as blood or urine. Sports drug testing exploits these differences. The analyses have resulted in many athletic bans and returns of medals. Despite this ability to separate latent and drug enhanced performance, many drugs are still considered 'legal' up to various limits. I am disillusioned by super-fast sports heroes who were actually only winning through drugs, not training and natural ability.

The carbon isotope ratio has more general applications as one can immediately distinguish between plastic (and other carbon-containing chemicals) that were generated from ancient oil reserves, or from modern agricultural or other crop production methods.

Analysis and forensic possibilities of isotope variations

Any property which can show differences between materials from different sources can be exploited, not only for conventional analytical purposes, but also for forensics. Key inputs include imbalances in isotopic content, which help to identify the probable origin of murder or disaster victims. The equipment and skills needed to make accurate isotopic ratio measurements have been steadily developing and it is clear that they will be employed even more in the twenty-first century. The forensic aspects are particularly fascinating as it is obvious that there are far more possible opportunities than have so far been demonstrated. The analytical methods are destructive of the material, but this is irrelevant if one is tracking the history of a body by analysis of, say, a femur or teeth. This approach is less acceptable for a living patient, although hair or nails can offer records extending back several months, and blood or other samples reveal more immediate history of the food or drug intake.

In difficult autopsy (post mortem) cases, the problem may well be to decide the region of origin of the body. Isotopic ratio studies of bone etc. can offer some real clues in this respect, and some bones only replace atoms over, say, a time scale of ten to twenty years. Therefore, cross-sectional analysis of the isotopic ratios gives information very much like the dendrochronology of tree ring dating. The annual temperature variations vary the ratio. The country of origin and the climate have further input into the isotopic effects. There are also tell-tale signals with respect to the diet. Someone who, for example, has lived in Russia, then Central America and then the UK, will have all this information encoded in their isotopic bone pattern. The method is sophisticated, and it clearly has the potential to identify if someone is a recent immigrant, or had been in a country much longer than was claimed. Equally it will show if the person did not come from the region that was being declared.

The pattern of changes in isotopic ratios is potentially powerful and examples that have been discussed, or are in use, range from testing the origin of different food stuffs, checking the types of food that have been fed to cattle, checking the authenticity of particular supplies, and the possibility of deliberate contamination or dilution of high value products with lower value items. Materials in such categories range from olive oil, vanilla, honey and royal jelly, to wine, tequila, drugs and explosives. The techniques needed to undertake such analysis are neither simple nor cheap, but there is a sufficient demand for the results that routine and reliable data can now be an economic and certain option.

Atomic bomb making in the Second World War

It may not be obvious to consider the benefits of imperfections in the context of world history, but in fact there are several examples. One such was the attempt by German scientists to make an atomic bomb. They certainly had the intellectual ability and determination. Nevertheless, in their construction of a nuclear reactor they were remarkably unsuccessful. Their design was sensible, and it used graphite as a moderator to control the production of neutrons of a suitable energy to produce bomb-grade uranium. The trick is to slow the neutrons coming from natural fission of the uranium, but not to absorb the neutrons and remove them from the process. Unfortunately for the German scientists their graphite was quite impure and, in particular, it contained boron. In percentage terms it was only a small amount of material but, as an imperfection in a nuclear reactor, boron is a disaster. It is some 100,000 times more efficient at absorbing slow neutrons than is the carbon of the graphite. Hence the neutron flux was greatly reduced and remained sub-critical for a nuclear reactor. This simple impurity undermined the German efforts at atomic bomb development.

A summary of imperfections and trace elements in chemistry

My initial examples were really quite similar in that many emphasized that a great deal of chemistry, and biochemistry, are driven by reactions that are enabled by some type of catalysis. Catalytic processes are absolutely essential mechanisms for life of all creatures and plants, but the ideas and methods may not be immediately obvious to non-scientists. Catalysis in nature and technology are crucial for our existence and life styles. Other examples of chemistry linked to imperfections and impurities underline our widespread reliance on chemical technologies.

The inclusion of an introduction to isotopic analysis was intended to emphasize not just what has been done, but what it can do, and is likely to achieve in the future. The concepts are absolutely key for biological and medical developments.

Isotopic effects exist in all technologies, not just these chemical examples, and in most familiar materials we totally ignore them. Despite this, they influence items as simple as thermal conductivity as changes in isotopes contribute to the scattering of heat transport through materials. They will also be relevant considerations in the development of photonic computation, as isotopic differences will potentially impair the functioning of the technology.

15

Imperfections in music

Inanimate versus living systems

My objective is to present very positive examples of imperfections, and in the preceding chapters I have concentrated on inanimate examples that underpin a great deal of modern technology. The examples were potentially easy to recognize, from breaking flints, making mixtures for metal alloys, total dependence on tiny traces of impurities for both semiconductors and optical fibres, and even their role in gemstones and some solid-state lasers. Catalysis was less clear as inferring the mechanisms were often speculative. I could have stopped at that point but imperfections in living systems, including ourselves, are equally important. However, the difficulty is that they tend to be a mixture of features which in some situations are highly desirable, and in others where they are definitely problematic.

In this chapter I will consider examples of sound, language and music. These are so familiar that we may not recognize how key they are in our lives, nor the reasons why we enjoy them, or why precisely the same music has interest only for some. This diversity will become even more apparent in the closing chapters. There are no absolutes. For me a definition of music is 'sounds that we like'. We need no scientific background to enjoy it; it is incredibly diverse in content, but it forms a large fraction of our lives even when we do not consciously seek it.

Music is highly varied in style, not least as it has a foundation in speech and language. Both are individual and relate to our geographic and social backgrounds, but continue to develop and change throughout our lives. Diversity in the sounds of our speech are essential as they characterize the sentiments we are trying to express, reveal our personalities and allow us to distinguish one another. Voices may be similar, but no two are truly identical (a feature exploited in forensic science via voice prints). There are nearly 8 billion humans alive on the planet but, in terms of speech, ideas, backgrounds, likes and prejudices, we all differ.

The Power of Imperfections. Peter Townsend, Oxford University Press.
© Peter Townsend (2022). DOI: 10.1093/oso/9780192857477.003.0015

This is uniqueness, it is not an imperfection as we want to be individuals. Failure to achieve distinctiveness is currently obvious in robotic voices used in navigation systems in cars, where the same voice emerges from each model and we have no sense of personal communication. Computer 'assistants' are worse, and irritating when they fail to understand our requests.

Accents and cultural differences have their negative sides, as heard when trying to resolve problems via call centres where the people are thousands of miles away, and we mutually have accents and dialects that are difficult to follow. Overall, such limits on communication can be incredibly frustrating. It is an imperfection with no positive features. It is not a new problem as, in Britain, before broadcasting started to homogenize our accents, written words might have been understood, but verbal communication with local dialects were definitely not. Bernard Shaw's *Pygmalion* was a very valid comment on the problem. True communication is music to our ears and it can be via speech, song or instrumental performance.

Imperfections and the pleasure of music

An essential part of musical appreciation is defined by us, and how effectively we are concentrating on what is being presented. For music that is being played, via earphones, from our phone whilst dealing with emails etc., the concentration on the music is minimal. It may block disruptive background noises whilst travelling on a bus or train, but musical appreciation it is not. Once we push music into this wallpaper level of sound, we may never listen carefully, and thereby derive the pleasure that music can offer. Playing music – or attending live performances – are totally different, as in both examples we need to be giving 100% attention to the music. Once we do that, we can make value judgements on the event, excitement and how it moves us on that particular occasion. This will vary with our own state of mind as well as the environment and performance. Even hearing the same CD at home will equally have quite different impact in different circumstances.

The challenge in drafting this chapter is that no two performances will induce the same response, there are many genres of music, and we change our views and likes with experience, the listening environments and other distractions, plus an unavoidable factor that hearing degrades with age and/or exposure to loud sounds (including music). Therefore, there is no absolute standard by which we can measure musical appreciation. This is the crunch problem when moving from simple science into the world of human interactions and opinion. My solution is to fall back on the science underlying music where I can at least sense why we may prefer vinyl or CD, or understand the input of electronics and recording on what is available for us to hear. It is also interesting to recognize how music changes with developing supportive science, plus the conditioning forced on us by the selection of music that is continuously being broadcast, discussed and recorded. The one aspect where we have freedom of choice is in what we sing or play ourselves, but this has its own limitations of our ability, need for other musicians, and indeed the quality of our instruments and local acoustics.

It is a tiny step from discussing speech to move into the realm of song, and this is our entry into the world of music. Singing is not solely a human activity but is used by animals as diverse as whales, humans, birds and insects as part of courtship, as well as for pure pleasure. In languages such as Mandarin the same basic sound is used with different tonal patterns. Hence, a single 'word' can differ in the sound and will often have four different meanings. Failure to enunciate or listen carefully will result in confusion. For example, the inflexions in saying the sound 'nan' includes meanings of both 'male' and 'little girl', as well as 'south' and 'difficult'. Such diversity in usage implies that we expect everyone to have an equally superb ability to pronounce and hear these differences, and have brains able to process and interpret them. Many people will claim they have no musical ability but, in reality, our brains are continuously sensing, analysing, and interpreting ambient sounds and using this to control our actions and emotions. This is exactly the same as our response to music, even if we are not consciously aware of it. Virtually all TV and cinema programmes use background music to set our moods with everything from relaxation, anticipation, or excitement to fear and terror. Supermarkets use it in quite subtle ways to good financial effect to boost sales. Playing German music one day, and French on another, influences us and it has a measurable effect on the sales of German and French wines on those days. Stores and shopping malls also use slow music to reduce our pace so that we spend more time in the shops, and buy more. It is played in crowded areas to reduce tension and crime. We are not the only animals that respond to these musical signals, as in farming, playing calming music to animals is successful in raising both milk and egg production.

Our sensitivity to music is exploited with military music, patriotic anthems and tunes at football matches. Many thousands of love songs resonate deeply into our emotional lives. 'Music' is not confined to living creatures as we can enjoy the sounds of a breeze and the breaking of waves on the seashore. Even the sound of a motor car, or other engine, has its own personal 'voice' and we subconsciously hear changes when something is wrong.

Cults of vinyl, tapes, CDs and streaming

Since the majority of music is marketed electronically, I will briefly consider the history of how electronics and technology have evolved and shaped the way we listen. Attempts to record music commenced almost 150 years ago, and broadcasts around 100. These routes to listening have therefore become the dominant method. None offer perfect reproductions of the original, or replicate live performances, and will never be able to do so as room acoustics and sound production cannot be identical to the original. An extreme example is to contrast listening to organ music in a large cathedral, or via headphones, or a radio in a standard-size room with soft furnishings. The reverberation times (i.e. the time it takes for a powerful noise to die away) are radically different. Indeed, one of the key features for church organ music is the long persistence of notes that slowly fade away over, perhaps, two seconds. By contrast, room reverberation times will be well under a second and effectively dead. The other spectacular feature is the sheer power of organ bass notes that one can physically feel, not just by ear. These are both at lower frequencies than we may hear and well below the frequencies included in recording and broadcasting.

This is an extremely clear example of the original event being degraded because of the electronic pathways to us hearing it. To a lesser extent it applies to every other type of delivery system of musical sounds. The possible exception is electronically generated pop music, as here there is a strong possibility that a live concert (delivered via electronics) might be duplicated by broadcasts or recordings. This route allows pop music to include 'autotune' which corrects the pitch of notes being sung. This is shunned by good singers as bending the pitch can add real impact, and is used very obviously in jazz.

For non-electronic performances the content of the music, plus spatial information processed subconsciously by our brain, is lost and degraded in any type of electronic processing. Perhaps equally, or even more importantly, is that at a concert the music has our full attention, there is excitement of the crowd, and there are background and directional properties of the sound which stimulate our brain processing. It is also known that a combination of visual, auditory, and other senses enhances our sensitivity in ways that are not possible for just one of them in isolation. For example, a discussion programme on TV can hold our attention more than just a sound version.

Since most of the time electronics and recordings will be part of our musical input, we must accept that we have settled for an imperfect product, but benefit from access to a vast library from across the world of performances from top-grade people. There is no doubt that recording and broadcasting have steadily improved over the last century and this is reflected in the types of medium that are involved. Records of pre-1925 were short, poor quality, but long enough for dance music, and this established a market for record players and sales. Rapid rotational speeds of 78 rpm limited their duration to maybe five minutes. Juke boxes in the USA brought in 45 rpm small discs, these were followed by vinyl discs at 33 1/3 rpm. From all such structures the sounds were modified (i.e. distorted) so as not to have cross linking between different parts of the track. Small portable magnetic tapes then became fashionable, but both vinyl and tape were superseded by compact discs in the early 1980s, which could run long enough even for a major symphony, not be bothered by dust or scratches and were portable. They reigned dominant for nearly twenty years until the advent of downloads and streaming. Each of these technologies has quite different distortions and weaknesses, with a characteristic sound. None are perfect, and our preferences may just indicate the period when we first started to listen, rather than any real advantage of one system.

Within this century there has been a dramatic shift to using downloads and streaming. It has a heavy bias to pop music and/or music that is merely background. Hence the equipment quality and signal information are compressed to cope with the uncritical demand. Musically, this degraded the quality backwards by half a century. This is an unacceptable step for those who will truly listen, concentrate, and appreciate the music they are focussing on. For the rest (the majority) it is irrelevant. A final comment on the downside of streaming is that it is moving into the use of playlists, where someone else has chosen snippets of music which are strung together so that we have no freedom of choice. Having once listened, it is considered to be music that we like and we are trapped and curated in the playlist's style, to the exclusion of everything else. This trend runs through both pop and classical music.

Easy access is not all progress as we are overwhelmed by electronically transmitted music as a background rather than something we are intently listening to. This is a pity as it undermines one of the great pleasures of life. The failure is not in the quality of the music as, clearly, we do not care if we are not fully attentive, and are willing to listen via low grade systems or Bluetooth and other data compressive devices, whilst simultaneously reading emails or working. One can equally comment that the continuous use of smart phones to chat about trivia with friends, and/or share electronic images, is far inferior contact than genuine person to person meetings. They undermine real contacts when they occur. In cafes, I see groups of young people sitting around the table, all using their phones and none speaking to the others at the table. For me this is a great social loss and a direct consequence of the distortions driven by social pressures of electronics.

Variations between composition and performance

An inevitable imperfection in music is that the intentions of a composer may not be represented by subsequent performances. Rather than view this as a negative factor it needs to be reassessed, because music develops in style, changes in instruments and auditoria, as well as our exposure to other music, which, in turn, influences what we enjoy. At last I can cite a positive imperfection which applies to the last millennia of composition! This may also apply to recent examples of electronically generated sounds from synthesizers and modern technologies. For them, the sounds we hear may fairly accurately represent those intended by the composer, although electronic effects and sound quality evolve rapidly, but basically the results will be acceptable to the composer. We will like it if this is the musical sound that is fashionable. Musical styles alter rapidly in the field of pop music, and sound modernization is acceptable. Much of pop music is sold to the young and so there is an inherent bias for each narrow age group (maybe with a five year spread) as to what period and format they prefer.

As for music composed and played prior to good quality recording, we have absolutely no way of hearing the music as it would have been intended by the composer, or with the musical experience of the audiences at the time it was written. Playing baroque music on period instruments, with tuning that was possibly used at the time of composition, is highly fashionable. However, the players have to guess at the interpretations that would have been used, and in no way can they replicate the musical expectations and experience of either the original performers or the audience. It may be enjoyable for them, but instruments, and exposure to music from across the globe via broadcasts and recordings, have irrevocably altered what we hear and expect. We may like it, but must recognize that it is impossible to duplicate the same responses as when the music was written and first played. Emotionally, and musically, we have such a different background that there is no similarity.

As a personal comment on the oddities of different tuning with baroque performances, I find no difficulties with singers or lutes etc., which are using a lowered pitch. Whereas, for violin music moving from the tuning reference of an A at 440 Hz to something flatter, at say, 430 Hz, I find it unpleasant and out of tune. Maybe this is because I have spent years trying to play a violin in tune. The same music played at the modern pitch is absolutely fine. In reality the 440 Hz 'standard' was just an average of organ notes from churches in Vienna in the early 1800s. These varied over several *tones* as there was no official standard. The 'standard' has also varied in countries with time and/or conductor. In the late 1800s, the USA frequency for the A rose to 461 Hz (i.e. almost A sharp). Really bad news for top notes of singers. The underlying logic is that higher frequencies have better carrying power (as for a scream). A classic 'imperfection' that exploits this is in a crowded room with many conversations, where higher, squeaky voices are more easily heard. More subtle, but dangerous strategies can be used in, say, a violin concerto where the soloist plays _very_ slightly sharp compared with the orchestral violins. If well judged, the soloist emerges from the background, but it is incredibly easy to sound as though the notes are sharp and out of tune. An excellent, skilfully delivered imperfection is required.

Instrumental changes have taken place in virtually every instrument that is now played. Some are visually less obvious, as with a violin, but modifications have improved power, allowed use of higher notes, faster performance and better strings and bows. In the nineteenth century the gut strings (sheep not cat) were non-uniform and broke easily. Non-uniformity and an inherent taper meant it was essential to align the taper in the same direction, as this influenced the spacing of the notes. Modern bound strings avoid these faults. Trumpets gained keys, a full range of notes, and more flexibility with fingering; clarinets evolved over a century; new instruments such as saxophones were invented. Such evolution of instruments (and concert halls) dramatically altered what was composed and what audiences were able to appreciate. One should also remember that, at say the time of Mozart or Beethoven, many orchestras were random groups who came together for a performance, without any rehearsal (as they were not paid for rehearsals). They were, therefore, sight-reading in candle light, often from hand-written music, without benefit of good spectacles. The sound quality is unlikely to have matched that of the much larger modern orchestras, although small orchestras reveal inner parts more clearly. Haydn's employer had a small orchestra and so there were rehearsals.

Rossini is known to have written overtures on the day of the first performance, whilst copyists were waiting for the next page. Paganini carried the orchestral scores for his violin concerti to the concert and collected them afterwards. This certainly undermined performance of his work in his lifetime and blocked it after his death. For his fourth violin concerto in D minor, the orchestral score was only discovered by a ragman in paperwork thrown out by his descendants in 1936. It was later reunited with the solo part and then next performed by Arthur Grumiaux in 1954, a century after its last performance. Similar anecdotal stories for Beethoven include his violin sonatas, where at one first performance, the violinist was sight reading, and Beethoven had not yet written the piano part as a score. Audiences would not have appreciated how bad this was, as rarely would they be able to hear the same work many times, unless they lived in a major musical city such as Vienna.

Novelty and new styles are frequently rejected (equally true of new ideas) as this is human nature, and so many of the major groundbreaking innovations in music were criticized disastrously at their initial performances. With time they have become the operas, concerti, symphonies etc., that are mainstay musical literature. I mention this as originality is often considered an imperfection, but with repetition and understanding it can move to being mainstream, but an inhibitor of later styles.

Changes from technology

Composers are human and there are no fixed styles, nor boundaries set by different nations. Instead, they exploit opportunities that are offered by improved technologies with the manufacture of different or more powerful instruments, understanding of acoustics and architecture that offer better concert halls, and, of course, electronics that brought about recording and broadcasting. Indeed, attempts to record music were the driving force behind electronic amplification. Composers, however, would not have anticipated that their works would be not just recorded, but balanced and 'corrected' by sound engineers during the processing of recording and/or broadcasting.

Amplification offers more power and reaches larger audiences, which often outweighs the distortions of the early instruments. This is eminently true for keyboard music when, some two centuries ago, the works were frequently intended to be played on a harpsichord. Harpsichords have low power, cannot sustain notes and drift rapidly out of tune. Early pianos were an advance in each area and could sustain modest power. Inevitably this changed compositional objectives. What we may forget is that such early pianos differed greatly from modern ones, so playing those compositions on a modern piano offers a very different sound. For example, in the Beethoven Archduke trio (for Rudolph of Austria) written in 1811, he had three *equally* powerful, matched instruments in his partnership of a fortepiano, a violin and a cello. The power balance was fine. Modern performances with a current grand piano involve, even with the best pianists, one powerful piano with two much weaker partners. For piano concerti, the technological advances have increased piano power since then by over a thousand times! This is not quite as bad as it seems, as our brains work with logarithmic intensity systems so a thousand times in power is perceived as maybe a factor of ten in loudness.

The use of keyboard instruments added an unexpected difficulty that, once tuned, they are inflexible. By contrast, singers can, and do, vary the pitch of their notes in subtle but recognizable ways, and different cultures use scales with slightly different tuning. If we look at a modern piano keyboard, Figure 15.1, we can play a standard Western scale with eight notes (an octave) running up and down the white keys from C to C. The rest of the twelve available notes will be required if we start on different keys. This pattern is only one of around eighty

Figure 15.1 The piano keyboard. This layout of notes can conveniently be used to define a simple scale, if one plays the sequence of white notes from C up to the next C. The range uses eight notes and is called an octave. Black notes, as between C and D, are either termed C sharp or D flat, depending on the musical scale that is being used. If one plays just a note sequence using the black keys, then the pattern resembles a pentatonic scale (e.g. from F♯ up to the next F♯).

scales that are in use around the world, and the intervals between successive notes can be very varied. In our terminology, the octave (i.e. C to C) doubles in frequency, and this is true for all scales. There is an intuitive choice of notes that includes a fifth, where the ratio of frequencies is 3/2 (C to G). Historically most cultures started with scales of fewer notes, typically five different ones (hence called pentatonic). A rough guide might be to play a piano scale using only the five black keys. This could give a scale starting from, say, F sharp (♯) on up to a doubled frequency for the next F♯. The relative spacings of the notes is quite variable in different scales of, not just pentatonic, but also those with twelve nominal semitones. Note that, although Western music has drifted into eight notes per scale, out of a selection of twelve, some twentieth-century composers, as well as certain types of jazz, include even more finer divisions, often called quarter tones.

People have a range of voices, from bass through baritone, tenor, mezzo and soprano, so instead of singing in the key of C many find the song is better suited to their voice for a different key (e.g. F or A sharp, ♯). This was fine until instruments with a keyboard were used, as each song would require retuning of the instrument every time there was a change of key. For a keyboard tuned to, say, the key of C, moving to some other key meant many of the notes are very clearly clashing and discordant. For modern Western music we use a set of twelve semitones and, for a fixed keyboard, the frequency ratio of every adjacent pair must be the same if it is to be used with different starting notes (i.e. an intentional imperfection). Over an octave the frequency doubles, so mathematically this is possible if for our twelve-note octave the semitone ratio is the twelfth root of two ($^{12}\sqrt{2}$ which is 1.059). It is called 'equal temperament tuning'. Useful, but musically slightly imperfect and limiting in terms of the flexibility of singers etc. For moderately good normal hearing one can recognize the differences.

There are further problems for a piano, even with equal temperament tuning, that are compounded by having both multiple strings, and many notes played simultaneously. The higher components (called harmonics or partials) not only do not fit a tidy musical scale, but will include frequencies that are not in any musical scale, and often clash with those from other notes that are being played. A good piano tuner can reduce the distortions by slightly detuning. The typical pattern is to flatten lower notes and sharpen the higher strings. It gives a characteristic piano sound, but high 'notes' will clash with those from a voice or violin. The faults are recognized in electronic keyboards and many have an option of offering the 'piano' sounds which are not as distorted as those on a genuine piano. The alternatives are sometimes labelled as 'stretched' tuning (i.e. as for a mechanical piano), or 'flat' tuning (using a reduced selected set of pure electronic harmonics). This is feasible as the harmonic contents are defined electronically, not by vibrations of strings and sound boards. For singers and string players this is slightly preferable, but obviously it no longer has a high-grade piano sound. In principle, the electronic piano sound could be developed with more complex software to improve the quality, but since many keyboards and synthesizers are aimed at popular music, which is brief and less caring about tone quality, I doubt that the market will justify the development.

Play me a note

Hidden behind all the tidy mathematics of equal temperament tuning was the assumption that we play simple notes at one frequency. It is feasible with a tuning fork, but the sound is boring, and unrelated to any voice or instrument. We cannot sing or play at only a single frequency, although some electronics can offer an approximation. An almost ideal classical physics example is the set of notes from an isolated string fixed at each end, like a violin string. The violin bow has rough horse hair coated with sticky rosin that grips the string and pulls it steadily sideways. At some point the string escapes and rapidly flies back. This gives a sawtooth motion to the string, Figure 15.2, which mathematically can be analysed into a set of harmonics with higher frequencies that are

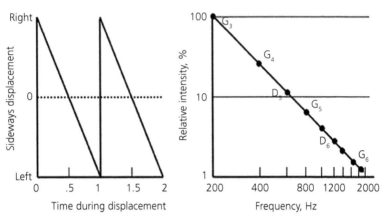

Figure 15.2 Frequencies that produce a sawtooth wave motion. The violin bow causes a sawtooth type displacement of the string which can be analysed into a set of equally spaced frequencies. Some are harmonics of the fundamental. An idealized pattern is seen on this logarithmic plot of intensity versus frequency. G_3 is the frequency of the violin G string.

exact multiples of the basic note (1f, 2f, 3f, 4f . . .). In isolation from just a string, the power of each harmonic decreases logarithmically (1, ¼, 1/9, 1/16 . . .). A real violin is designed to be an amplifier and the actual intensity pattern from a complete violin will differ considerably from that of the isolated string, and the amplification factor depends on both the note and the instrument.

What is produced, and what we think we hear, is a result of our brains trying to process a great deal of information and finding an acceptable rapid simplification. It includes not merely analysis of the component frequencies, but also cross checks to see if differences between frequencies also provide us with the fundamental. Clearly this should be fine with a violin as the frequencies are equally spaced (so 6f - 5f = 1f etc.). Figure 15.3 includes some measurements from a modest grade violin. For the A string (440 Hz) one sees a set of frequencies spaced by 440 Hz and steadily decreasing in intensity (roughly as predicted). The G string (at 196 Hz) is fascinating as the fundamental is incredibly weak but there are some strong harmonics and the brain extracts the fundamental from their differences (Figure 15.3). In these measurements, my students and I all thought we heard the note G being played, despite the fact it was missing the fundamental 1f component. Exactly the

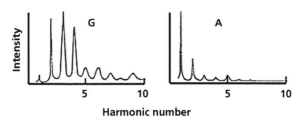

Figure 15.3 Power output from G and A violin strings. For this violin, we still imagine we hear the fundamental of the G as the frequency processing in the brain includes searching for harmonic difference frequencies. For the A string there is no problem.

same can happen with a double bass where both the power of the lowest notes is feeble, because to make an equivalent scaled up violin would mean a back section approaching 2 metres high! Additionally, our hearing is very poor at those low frequencies. What we think we hear is an imaginative data manipulation by the brain. Early telephones used exactly the same trick of distorting the earpiece sound to provide difference speech frequencies that were not sent over the limited bandwidth of a simple telephone lines.

Bells are even more confusing as the analysed set of component tones are anharmonic, which means they do not fit the tidy, equally spaced pattern, as for a violin. Instead they contain a wide, and apparently unrelated, set of frequencies. These may not include the note that is claimed to be the dominant tone of the bell, or of any obvious harmonics! We mentally analyse the bell frequencies and our brains make a decision (i.e. a guess) as to what note was intended. This is highly subjective, so different listeners can disagree on what was the 'fundamental'.

A good violin offers power at the lower notes. Broadcast and recorded systems do not transmit the fundamental of a double bass, and rely on us adding brain power to fill in what had been excluded. Beauty is in the eye of the beholder, but music is what the brain chooses to convince us what we heard. I enjoy my violin playing, but this is probably a minority opinion.

The sound analysis between string displacement with time and the intensity of the frequency components (called a Fourier transform) can be measured, but controlling it depends greatly on the skill of the violin

maker, who is constructing a wooden box amplifier. Electronic amplification is much easier, and can be adjusted to be flatter over a wider
frequency range, but since it does not have the imperfections of the
wooden box designs, the tone quality can be monotonous and boring.
Imperfections are the musical engineering of choice. In terms of cost,
the view may be different from a simple violin (e.g. around £130 as for
Figure 15.3), to a top-grade instrument by Stradivarius or Amati etc.,
which can be measured in millions. Electronics has a place in instrument manufacture from a modern luthier, as it can be used to analyse
the frequency responses of the various plates of the violin during the
shaping of the wood.

My second example is far more difficult to analyse and that is to look
at the harmonics and other stray notes generated by a classical singer.
Figure 15.4 offers a real-time view of the fundamental, together with a
vast array of partials that emerge from a professional singer. The figure
includes just two examples for the same note. In one the larynx is raised,
which gives more power in the higher registers – this is valuable for projection in a large concert hall. The other is with a low larynx that offers
a softer, deeper sound. We instantly claim we hear the intended 'note'
at the lowest frequency, and recognize there is a different tone quality. When we see the analysis, we find it hard to believe that there are
as many as fifty frequency components. They range from the intended
baritone note (a C at 130 Hz), plus a few of the immediate harmonics,
to a plethora of unrelated frequencies defined by our vocal system. We
have partial control over these by the position of the larynx, the tilt
of the head, tongue position, level of the soft palate, and shape of the
mouth and lips. Once we realize how many variables are involved, then
the number of extra frequencies become credible. Equally surprising is
that there are components (even from a baritone) over some five oc-
taves which extend to ~5000 Hz, which is in the frequency range to the
very top notes on a piano keyboard.

Rather than describing this as a mess of an unbelievable mixture
of selected frequencies, we praise it as being a lovely, powerful baritone sound. A clear victory for imperfections. The high larynx note has
good carrying power because of the high harmonics, whereas the lower
version has a smoother, deeper tone.

My section heading of 'play me a note' was intended to emphasize
that it is an impossible request and even the most basic music, from

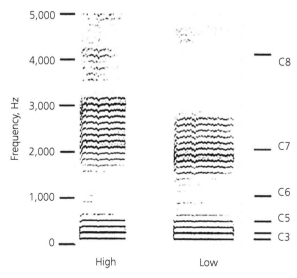

Figure 15.4 Tonal effects from the larynx. Data recorded during five seconds from a baritone singing a low C at 130 Hz. 'High' and 'Low' indicate the relative position of the larynx (see text).

singing, has a wealth of depth and tonal characterization, which is precisely what makes it exciting and attractive.

Directionality and location of sounds

Two ears offer displaced signals that give a sense of direction and distance so, in a concert, not only do we have information on intensity and balance between the performers, but also can identify the location of individual performers, plus signals from reflections. These features offer us a feeling of presence, both of the music and the environment of the hall. Trying to duplicate this electronically is incredibly difficult, not just because speakers or headphones and room acoustics do not match the responses of the concert space, but we hear a mixed reception from a variety of microphones which may be suspended far from any normal seating. This is equally obvious when contrasting speech in a room with a TV or radio.

Sound directionality is a serious factor, not only for recording music, but equally for choosing a concert seat. For a trumpet aficionado, the forward beam of high frequency sound in an orchestra may be roughly aimed towards the conductor so, if this is your favourite instrument, your seat should be along that axis. If you choose a seat at the same distance, but to the side (e.g. to view the performer), the power level can fall by up to 100 times, which in the way we perceive loudness is about as quarter as loud. In many concert halls, for equivalent price seats, this is a standard range of power that one hears for different instruments. Each one will be differently biased to favour, or suppress, different sections of the orchestra or choir. Such reality is lost, modified, or disguised in multi-microphone recording and broadcasting (even when it is nominally in stereo), and is one of the reasons why live music and recorded music are noticeably different. The balance of instrument power is equally 'adjusted' by recording engineers, so soloists stand out artificially from the rest of the music. Live music also benefits from the adrenalin of the players (and occasional mistakes), plus the reflective acoustics and sound polarized in different planes. We subconsciously process all this detailed information and it contributes to our enjoyment, but such subtleties cannot be delivered by the electronics.

I recently read that modern sound engineers are trying to add electronics to high grade concert halls to make a noise that is degraded to be closer to that of a CD, as this is what they believe audiences expect to hear. Unfortunately, my thoughts on this are unprintable.

Overview

The clear conclusions are that music is an essential part of life for humans and many other animals. We enjoy it because every performance seems unique, even for a CD, where the change is how we feel and concentrate at the time of listening. We definitely do not want perfection as the variations are the spice of our life and the ways we identify with the music, performers and ourselves. Even our musical taste can be highly diverse and alter with time. For the purpose of this book, the inherent imperfections in reproducibility of listening and playing are precisely what makes music enjoyable.

I cannot fail to add that there is a simple, but far wider ranging, consideration of music on how, and why, it has developed, in my book *The evolution of music through culture and science*.

16

Evolution achieved through imperfections

Part I Species variation and evolution

Over the last 150 years, there has been immense progress in understanding the workings of biological processes and the realization that species
are not constant, but can occur with many variations in each successive generation. This ability, and uniqueness of individuals, emphasizes
why we should appreciate the differences we once would have termed
imperfections. The genius of Darwin and Wallace was to recognize that
these variations allow for a steady evolution, not only of existing species,
but for the development of new ones. Many people have the miscon-
ception that such evolution is always in a forward direction. This, of
course, is not true as, if the particular key characteristics cease to be
favourable, then that can be the end of the line for a particular variant,
or even a complete species. The archaeological records demonstrate
that a vast number of previously successful animals and plants have
died out. The reasons for such extinctions are very varied and include
climate changes in key aspects of temperature, rainfall, desertification
or sea level. Natural catastrophes ranging from volcanic eruptions to
meteor or asteroid strikes have also had a global impact. Changes in
the types of disease, predator and habitat similarly influence survival.

For the last few million years, most plants and animals steadily
evolved to accommodate slowly shifting climatic environments. Some
variations are quite spectacular, as evidenced by the variety of marsupi-
als in Australia after it ceased to be linked to the corner of the Asian
land mass. Madagascar similarly shows different evolutionary routes
once it separated from Africa. Survival has not been perfect as there
were around five major mass extinctions driven by natural events. The
most famous is the asteroid impact into the Mexican peninsula which

The Power of Imperfections. Peter Townsend, Oxford University Press.
© Peter Townsend (2022). DOI: 10.1093/oso/9780192857477.003.0016

destroyed not just the dinosaurs, but many thousands of other species. Survivors were few, but fortunately the major predators had died. Fossil records mean we are able to track the evidence on how these survivors adapted and branched into new species. Hominid, and our own ancestry, commenced much later down this evolutionary chain.

The sixth major extinction is in progress and is of a totally different character, and the cause is unequivocally humans. Not only did we destroy mammoths and dodos, and any other species we could eat, but we are continuing to do so, with ever more efficient hunting, loss of habitat and chemical treatments – as used in agriculture – to remove insects and native species that like the same crops as we do. In the process, we have consciously eradicated precisely the creatures which are needed for pollination, such as insects and bees. Looking at images of Asian farmers pollinating trees by hand, because of chemical destruction of natural pollinators, suggests we are heading into deeper problems. For example, the decline of the bee population can be cited quantitatively in the USA, where the Department of Agriculture numbers for bee hives fell from 5 million in 1998 to 2.8 million in 2018.

Global travel with ships introduced cats, rats and other destroyers of native creatures. Plus, we have imported pests, diseases, exotic animals and other life forms that have no native predators. There are of course natural losses of species, for example in islands which become inundated by changing sea levels, or weather pattern changes that created deserts. Current literature suggests in just the last 250 years, some 500 plant species and more than 1,000 varieties of animals vanished from the wild. However, when one looks at human interventions over a longer time scale, the extinction rates caused by humans are some 500 times greater than the 'natural' background, and for animal extinctions it is estimated to be at least 1,000 times. Maybe the planet would survive and recover with a new balance of plants and creatures if we all succumbed to a global virus. However, it needs to be very effective to eliminate eight billion.

Overall, there is considerable evidence within current lifetimes to realize species extinctions are taking place, but observing natural human evolutionary change is far more difficult to recognize. In part as we are self-centred and unwilling to consider that humans evolve, and quite critically we are unable to compare many generations simultaneously. Nevertheless, evolution of domesticated creatures, where we have introduced selective breeding, is apparent from farm animals to pets.

We have interfered with natural developments ever since we started hunting and, even more consciously, with the advent of agriculture and animal husbandry. Several thousand years ago it was recognized that selection of seeds, or young animals, could be controlled to produce subsequent generations which have characteristics that we found more favourable to us. The underlying science was not known, nor indeed was it needed, but the effects were obvious. Bigger horses and controllable farm animals, faster growing and more prolific crops, and animals which produce more milk or meat, are all simple examples of how we have capitalized on the natural variations in a particular species through selective breeding. In terms of the history of the planet, significant human intervention has existed over a relatively short period of time, of perhaps ten thousand years. In most farming examples one can recognize the changes, even if we have now bred out earlier characteristics. We can induce such developments (both positive and negative ones) for domestic animals, crops and pets as we are able to control their breeding, and start a new generation within a few years after birth. For plants, the selection may be as brief as on an annual basis. More obvious examples are seen in experimental biological programmes with trial species, such as mice or fruit flies, that have breeding life cycles measured in weeks. This allows a rapid view of evolutionary change over many generations.

Evolution of language

The nineteenth century recognition of evolution of species was culturally shattering, but it could easily have been proposed much sooner if people had recognized analogies between developments of plants and creatures with changes in language. Language evolution is obvious and speedy. Older generations are often effectively excluded from the current language of teenagers, and even when the same words are used their meaning and nuances can be totally different. Viewing old films or TV programmes immediately underlines how language, culture and social attitudes change within our lifetime. Extending this to, say, twenty generations (roughly 500 years) means it becomes incredibly difficult for the general public to read and understand the subtleties of an earlier text. Classical writers such as Shakespeare and Chaucer demonstrate that we may understand their broad plots, but totally miss the subtleties. Nuances, words and social attitudes have radically altered. Importing words from other sources, invasions and immigrations further generate evolutions of language but it is just more obvious than recognizing evolution of species. Both can evolve forward and be enriched by diversity, but equally may go into decay and become extinct. Classic languages such as Ancient Greek and Latin are partially retained in their modern progeny of Modern Greek and Italian, but many other languages have totally vanished. Aramaic was a major language of the Mediterranean Middle East 2,000 years ago, but this faded away, and just a few thousand people still speak the modern version of it. Aramaic script was widely used, and it became a forerunner of modern written Arabic.

Language unification can be a cause of dialect extinction. At the time of the French revolution (1792 onwards), barely 50% of the French population could understand French, and even for them it was not necessarily their first language. Political will forced acceptance and evolution of modern French, so that today probably fewer than 10% of the population understand any of their earlier regional languages. For a language to be still considered as live, the criterion is that a small group of people use it. Therefore, in the UK, Cornish is technically surviving, and there are efforts to increase the number of speakers. Worldwide, language loss is reported to be in the hundreds in the last century, or thousands if one includes dialect variants.

Changes between successive generations

I am a physicist, not a biologist, so I will use classic textbook examples which are well documented, and independently demonstrated, rather than those from recent literature. I prefer to do this as there is so much vehement religious (and secular) antipathy and fear of any discussion of evolution that using newer examples, especially if they are clear and spectacular and likely to arouse vigorous opposition, is best avoided. This is the typical pattern of anything controversial, or ideas that might upset the existing order or egos and status of existing 'experts'. The religious difficulty is that we have convinced ourselves that we are different, and superior, to any other creature, whereas evolution undermines this view. Biology clearly indicates our modest place among animals and astronomy has similarly changed our perspective of our importance within the cosmos. For example, only in the last twenty years have we been able to detect planets around many nearby stars, and it appears they are common features with some 2,000 already characterized.

Whether we discuss plants or animals (including humans), it is clear that a parent generation is never absolutely identical to the subsequent generations. 'Identical' twins and clones are not perfect and exact copies, but differ right from conception. Subsequent environment and nurture introduce further variations. A brief look at any plant species, or human population, instantly shows there will be an imperfect (i.e. non-identical) reproduction to the next generation. Effects that benefit from this include a range of sizes and ability to stand different weather conditions or new diseases. This adaptable diversity is essential for survival. The genetic blueprints for the next, and later, generations ideally should include a range of options which can be passed along to future generations and used as alternatives as required. Current agriculture and biology attempt to control genetic information have a severe weakness if future conditions for survival alter. Equally, the chemical changes driven by pesticides and herbicides invariably have follow-on effects which were not intended.

Familiar examples are natural plant seeds, which include the ability to survive under different conditions. If the same plants grow at very different altitudes, one single plant will produce seeds which are predominantly best suited to the altitude at which they are formed, but will include some variants which favour alternative conditions. If low altitude seeds are planted at a high altitude it means the bulk of the seeds will fail. There will be a minority which survive, and their progeny will be predominantly best suited for the new altitude. With slowly changing conditions this is an excellent survival strategy, but if new conditions prevail for a long time then alternatives will eventually fade away. Extinction can occur if there is then a reversal of conditions, as the species will have lost the necessary variants. When modifications progress and become permanent, it is deemed to be evolution of a new species.

I am stating the obvious fact that plants and animals do not breed identically from one generation to the next. For a market gardener this can be considered an annoying imperfection, but for long term survival it is progress. It is a simple fact, but the implications are immense. It means that genetic imperfections are the driving force that allows both survival under changing environments, and the ability to evolve into new species. The use of the term mutation effectively describes the same situation of changes between generations, but in emotive terms evolution is somehow considered a positive feature, whereas a mutation has the connotation of being a problem. Many mutants fail to survive, but others can open evolutionary pathways to new species. The medical profession is continuously battling such effects of mutations, which give drug resistance to bacteria and viruses as their mutation rates are major and rapid. When genetic information becomes corrupted it leads to new variations and, later on, I will indicate why this is such a common feature.

Modern genetic signatures offer many examples of developing diversity. For example, studies of different fish species in large isolated lakes show many hundreds of fish species clearly linked genetically to a common ancestor, (a parallel to breeds of dogs). The fish now differ in colour, shape, feeding and breeding habits. Time and environmental conditions allow such diversity. The rates of change are just slower for larger species.

Textbook examples of variations and evolution

The nineteenth century standard textbook presents examples demonstrating how members of a different species have evolved as a result of isolation of groups of animals and subsequent changes in habitat etc., in the Galapagos islands. Variations between finches and tortoises inspired Darwin to recognize they each had a common ancestor. Birds and tortoises on different islands have different feeding opportunities, and those best suited to such conditions bred most successfully to the exclusion of the others. Hence the balance of variants in their populations totally altered. The steady drifts away from the original common ancestors, in, say, shape of beaks of finches or shells of tortoises, are sufficiently large that new species emerged. This typically means they cannot interbreed, and/or their progeny are infertile.

Preferential selection can equally result from predators that eat one variant more readily than another. An oft cited example is the UK peppered moth. They live in woodland and are eaten by birds whilst they are settled on the surface of trees. The moths come in both light and dark varieties and the survivors are those which are best camouflaged on the trees. Prior to the industrial revolution in the UK, the clean, lightly coloured trees favoured survival of pale-coloured moths. Within the nineteenth century the pollution from the industrial revolution resulted in much darker tree bark. Hence the bird feeding habits preferentially reduced the obvious light members of the population, and the survivors were predominantly dark. Dark members tend to produce dark offspring, so the balance of the population changed. More recently, with cleaner atmospheres, more of the lighter moths are surviving and breeding. This is survival of the fittest from a population with alternatives.

Instead of predators one can pick examples of drugs and pesticides. Introductions of new drugs or crop sprays initially have a major impact on pests and diseases. Rarely do the treatments kill all the targets, as genetic variation means a small number are immune to the treatment. Survivors thrive particularly well as all the food etc. is available for them. An immune variant can then become dominant. Hence the genetic differences which result in variants of the species are effective in preserving it, at least in a modified form. The original version may never reappear, and after many generations the survivors redefine the species.

Many negative effects of herbicides, pesticides and antibiotics are well documented as rarely is their influence confined to the field, plants, fish or cattle that were initially treated. This was highlighted in a now famous book, *Silent Spring*, by Rachel Carson. She showed that side effects were numerous and, often, excess quantities were used that killed far more than was intended, and totally changed the natural fauna and flora. Further, in the case of humans eating the products and ingesting quantities, the marketing philosophy was that there was no danger as the levels were innocuous at concentrations of, say, parts per million. This is totally both false and misleading. Many of the chemicals become concentrated in different organs (e.g. liver) to parts to per thousand or more, and become as lethal to humans as to the natural insects they were intended for. The list of examples of such problems continues to grow with unexpected examples, such as recent examples of cow dung from antibiotic-fed cows destroying the soils where it is used as fertilizer. The drug content in farmed fish is similarly a reason why many drugs have now become ineffective, as the surviving variants of bugs are precisely those that do not respond to treatment. The others have, and so the minority of genetically 'super' bugs expand to fill the niche market that has been created for them.

Some genetic variations, which are not physically obvious, can confer considerable survival benefits (not just against pesticides). In the

Middle Ages, from about 1347 to 1350, roughly one third of the European population were killed by the Black Death. Statistics elsewhere are apparently less certain, but the plague effects may have been worse. The disease was initially thought to have been bubonic plague spread by rat or human fleas, but reassessments suggest personal contacts were more probable in spreading the plague than fleas. This model is consistent with a long incubation time. For the Black Death, the period over which the disease could be transferred (~37 days) was recognized, and Italian ports imposed a 40-day isolation period (hence our word quarantine). However, some sections of the exposed population did not succumb to the disease as they had a genetic immunity.

A classical example of this is known from descendants of a village called Eyam in Derbyshire (England), where during the London Plague of 1665 to 1666 the village isolated itself for fourteen months. A third of the population died, but some families were not affected. Genetic studies of their descendants revealed a gene called COR5 delta32, which not only resisted the plague, but is equally successful with immunity to HIV. This is not a universal gene of the human population and is identified in about 10% of white Europeans, 2% of Asians, and is absent in many other races across Africa, and also American Indians. Such diseases can resurface after long periods of time and several examples of bubonic plague were detected in mid-2020, and this was attributed to people eating uncooked animals that continue to carry it.

Other diseases such as sickle cell anaemia can offer benefits, in this case a tolerance to malaria. Yet again, one sees features where some imperfections can have a positive function.

Many viruses have been transferred by explorers and colonists, with consequent immense death tolls (e.g. in the European invasions into Central America). In return, other diseases were brought back to attack the unprotected Europeans. The only modern difference is that rapid international transport can carry new diseases around the world whilst they are still hidden in a symptom-free condition (apparent for the 2020 rapid global spread of the coronavirus). Of obvious concern is that if in the sparsely populated communities of the Middle Ages, 30% or more could be fatally attacked, then for the current ~8 billion who are packed tightly together, predominantly in large cities, anything similarly aggressive might eradicate a vastly higher percentage of these urban populations. A speculative perspective of future viruses is presented later in this chapter. They evolve remarkably rapidly, far faster than we can identify them and develop protective vaccines. Evolution is a reality for every living entity, not just for large animals. Viruses just mutate much faster.

The existential threats from disease and human-generated changes are almost certainly far more probable than asteroid impacts and similar large scale natural traumatic events. The only consolation for humans is that if the Mexican impact had not destroyed the dinosaurs, and other major large creatures, we would never have evolved from the small creatures that survived that mass extinction.

Human attempts at controlling evolution

Breeding of farm animals, or racehorses, is visually well documented via paintings but, for countries where 80% of the people live in cities, such imagery have little impact. Similarly, human-driven extinction of species may only be considered 'unfortunate' by those who have never seen such creatures and plants. For creatures that we never meet in our daily lives the majority of us may feel sad at their loss, but do not really understand the consequences of what is happening. By contrast, even city dwellers will be familiar with dogs and will follow fashions in what one will own, and look for in size or aggressiveness etc. Pet selection follows these changes as life expectancy of a pet, and its replacement, is perhaps around a decade. This evolution, driven by us, is on a recognizable time scale. Kennel Clubs effectively confirm this as they continue to acknowledge and characterize newly evolved breeds. More generally, breeds of dogs can be genetically decoded and the relationships between them tracked. The amazing fact is that they all appear to derive from an extremely limited number of female Asian wolves. Everything from Great Danes, to Chihuahuas and Asian wolves are fairly close cousins in genetic terms.

Clearly interbreeding is not always feasible, and many of the variations have significant genetic flaws which means that they could not survive without human care, feeding and frequent veterinary interventions. Some have breathing problems as their muzzles are too short; legs may be too short relative to body size; others are prone to specific diseases or types of cancer. Many genetic problems (caused by humans rather than natural selection) are now identified, such as a link between deafness in Dalmatian dogs and the size and number of their spots. This is an interesting genetic feature as white hair streaks in dark hair of humans can also be indicative of deafness. The same pattern underlines genetic characteristics that may well have originated in some common ancestral creature. Many canine redesigns may look attractive, but many are significantly inferior, and could not survive without us.

Our interference is equally obvious for plants and vegetables. Many roses and flowers are now grown to show large, long-lasting blooms with good symmetry. Often this has been achieved at the expense of losing the scent. Supermarket apples may look smooth and identical but can lack texture and flavour. The reason they do not have any deformities or attack by bugs and insects is not just the host of chemicals with which they were sprayed, but also that they are so tasteless that no insect would attack them. My garden apples are tasty, and highly desirable both to me and to bugs. Modern foods can be genetically modified to be resistant to certain diseases and, since genetic coding for one characteristic may also influence others, the improvements needed by the market may not be compatible with those of the consumer.

There are inherent conflicts between food production, marketing, quality, profits and benefits to farmers and society, as well as our future survival. A very obvious case can be described in terms of rice, a basic crop that feeds billions of people.

Natural and commercialized rice production

Rice has been a staple diet in Asia since Neolithic times and there were a vast number of varieties (termed landraces), almost certainly in excess of 100,000 variants, with genetic differences. Other claims of higher numbers often reflect that regional naming was different. These rice varieties adapted to their local environments such as soil, terrain, temperature, frequency of rapid changes in water levels etc. Clearly survival of the fittest justified this diversity, and it is an excellent example of successful natural selection. From the farmer's viewpoint they were locally ideal and not only provided a food crop directly from the local soil and weather conditions, but also gave the seeds for planting the subsequent crops with yields that might be harvested several times a year. Overall an ideal situation. The farmers were equally aware of the performance and differences in care needed for their local varieties. A further bonus was considerable understanding of their different medical benefits.

This was clearly too good to be true, and by the 1960s new strands of rice had been developed by food scientists which initially appeared

to offer a higher yield. These were effectively forced on farmers to displace their indigenous varieties. The commercial species, however, are
not perfect as they were not adapted to local conditions, they now need
fertilizers and pesticides, and in many cases are sterile. Consequently,
the farmers are forced to purchase both new seeds and fertilizers each
year. Quite typically, as reported for many crops that require pesticide
and fertilizer routes to give higher yields, the enhanced yields fall with
time as the soil becomes altered, and often within a decade the yields are
below the original pre-treatment values. From the viewpoint of farmers, they are in a far worse situation as they have lost their indigenous
varieties, have ongoing expenses to suppliers, and finish with reduced
overall yields. Equally, they have lost the skills needed to recognize how
their local varieties of rice are performing. From the opposite viewpoint
of the agrochemical industry, everything is perfect and highly profitable. This negative progress is encouraged by government legislation
aimed at achieving only a few varieties which are deemed important,
whilst the rest may not be used or allowed in mass marketing.

Once the 'imperfections' of having a vast number of different species
with adaptability and survival is destroyed, the extreme danger is not
just economic and loss of varieties with different properties, but exposure to a situation where a disease can wipe out all production,
rather than just one of the species. This worldwide move by commercial agricultural industries to monoculture uniformity applies to many
staple key crops. Monoculture proponents have failed to recognize the
example of the famine in Ireland in 1845–1849, where their reliance on
a single potato variety was destroyed by potato blight (*Phytophthora in-
festans*). It also destroyed the economic stability of Ireland. Within Asia
a repeat of the Irish monoculture farming problem is growing, as the
pesticides and chemicals kill the native insects which had previously
naturally removed pests such as the rice hispa and brown planthopper,
that are now depleting rice fields. The commercially engineered rice
may equally be unable to cope with changing climatic patterns of heat
and rainfall.

Mendel and peas

Attempts to understand the mechanisms and predict evolutionary
outcomes have been advancing for 150 years. The first classic, and
quantitative, example of studies of inherited properties was made by
a monk named Gregor Mendel in Brno using the reproducibility of

peas between different generations. Over the course of seven years he meticulously followed the success, or failure, of particular characteristics being transmitted to each new generation. He carefully controlled this pea breeding with nearly 30,000 plants. As sketched in Table 16.1, he recognized that some characteristics are dominant (i.e. are highly likely to appear), and some are recessive. Since there are two parent plants then this determines the relative probability that a particular trait will emerge in a future generation. The sketch indicates that, by the second progeny generation (i.e. grandchildren level), one in four showed dominant features, one in four recessive ones and the other two were hybrids.

Each successive generation combines one gene from each parent. Conventional labelling uses capitals for dominant genes and lower case for recessive ones. The table implies many characteristics can skip a generation (or more).

Scientific enthusiasm for his work was negligible during his lifetime as his publications were cited perhaps as few as three times before his death. His research stopped because he became the Abbot of the monastery. Nevertheless, his work was appreciated and recognized some sixty years later as a classic example of the pattern of genetic inheritance which applies to all species. The major difference is that we now understand more fully how it operates, and even some of the detailed molecular structure which causes particular characteristics to appear. The important underlying idea is that each property is encoded by adding features from each parent. We now know this to be the formation of chromosomes which are paired structures formed with one input from each parent.

The pattern works in both peas and humans. I had friends with four sons of very similar build and appearance. One parent was white and the

Table 16.1 Mendel's rules of inheritance.

Pure bred initial generation	Parent A Dominant		Parent B recessive
Gene options	D – D		r – r
First progeny options of pairs	Zero	4 of type D - r	zero
Second generation possibilities	25% of type D – D	50% of type D – r	25% of type r – r

other dark brown. The sons demonstrated Mendel's pattern perfectly as there was one white, one dark brown and two light brown.

Part II Encoding information with DNA

DNA is the shorthand for a highly complex structure called Deoxyribonucleic acid, which carries the information that determines the design, not only of us and all living organisms, but also of many viruses. In engineering terms it has brilliant simplicity. There are two backbones with a sugar-phosphate structure in-between to which are attached a sequence of four chemical units, which we label as C, G, A and T. One end of these is on one backbone and the other is covalently bonded to units on a different backbone. The possible pairs are not random as an A links to a T, and a C links to a G. Each backbone is on a helical structure and intertwined with the second helix running in the opposite direction. This makes replication very easy to understand as, effectively, the DNA can duplicate by unzipping the two helices and then each section can reconstruct by adding matching pairs backed by the corresponding other helix. A simplistic sketch of the structure is given in Figure 16.1. The caption offers the formal names for these nucleotides. Note also that most of the DNA structure is involved in the mechanical construction and support, not coding. The replication strand is termed RNA. More complex and detailed images are readily available on web sites and in articles. The truly amazing feature is that by merely defining the sequences with these four building blocks, we can encode all the characteristics and behaviour of life, both animal and plant. Not merely appearance, but behaviour, illness, intelligence and everything which makes us unique. Uniqueness is marvellous, as we differ from one another, solely because of imperfections in the copying processes at conception.

It is also worth remembering that life depends on near perfection in the copying processes, but this step can be modified accidentally by chemical reactions, viruses, or UV exposure which variously allow bond restructuring, or cell death. The UV example is obvious in the generation of skin cancer. Failure to make repairs implies ageing, decay and death. Most of us forget that cosmic rays from space result in ionization and cell damage. Approximately one cosmic ray per square centimetre passes right through us every second. Hence cell ionization, loss, distortion or mutation are inevitable features which we cannot avoid.

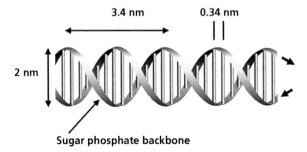

Figure 16.1 A sketch of part of a DNA chain. The links between the backbones are formed with A-T or C-G pairs. These units contain nitrogen nucleobases, Adenine-Thymine or Cytosine-Guanine, which are joined via double hydrogen bonds. The backbone chains unravel in opposite directions. Nanometres (nm) are one billionth (10^{-9}) metres.

Alternatively, they can generate new types of cells and DNA which may be relevant for evolution. Some estimates suggest that, each day. some 500 errors occur within cells just from normal background events, so it is amazing that such a complex mechanism can make repairs and survive at all. However, the repairs will sometimes result in a modified structure. Appreciation of the numbers of cells involved in every phase of reproduction, growth and developments merely underlines the fact that imperfections in molecular structures are inevitable. Whether we like it or not, this is evolution.

Changes between generations

If we progress to the present time then the underlying reason for inherited traits is much more obvious, but inevitably far more complex as we start to understand the details. The subject has developed its own language and terms which, to a non-biologist like myself, is very daunting. The decoding and understanding of the components are of such complexity that researchers will focus on specific features, and I strongly suspect that a total overview is beyond the majority of most people (even those with considerable expertise). For the non-biologists I have tried to consider, in Table 16.2, an analogy of the genetic terms with familiar features of writing. My list ranges from a library, down through books, to paragraphs, sentences, words and letters. The fact that these also differ in content, layout and even typeface, inks and paper is equally valid as it underlines the complexity of the task. Inclusion of the names and chemical structures of the molecules and the building blocks is totally intimidating (even for a biologist), so they are omitted. If this attempt is unhelpful, ignore it.

The modern level of understanding shows that Mendel had managed to extract the pattern of the influence of parent genes and he realized that each generation had a pair of determining factors, with one inherited from each parent. The details of chromosome pairs and the way genes are encoded along the spirals of DNA were experimentally separated and are becoming understood. To have any hope of recognizing how different genes confer particular properties, and how some genetic traits are linked, means it has been necessary to try to identify the pattern of the nucleotides A, T, G and C over long segments of the DNA molecule. This is a remarkable feat of chemistry, physics and biology, as almost the entire genome has been characterized for many cases. The scale is phenomenal as the entire book of instructions for a human genome is around 3 to 3.4 billion base pairs of DNA, partitioned within the 46 chromosomes (i.e. within 23 chromosome pairs). There are some 20,000 to 30,000 genes which have been identified. This may seem a relatively small number if it is the total genetic code for all the development and functioning of a human, but that would assume that each gene acts independently. If, as suspected, some features result from combinations of genes, then the human blueprint has much greater complexity than we currently envisage. It must also be noted that large blocks of the DNA may serve to function in ways we do not understand, contain obsolete information, encode latent information or be acting as safeguards for future needs. Quite surprisingly, whilst we vainly assume that we are unique and at the top of biological development, our diversity of genes is not exceptional and, to a large extent, is very similar to very many other creatures.

Replication requires that the double helix unzips and, since A must reform with a T and G with a C, the rebuild of the new helices should in principle exactly match the original. Different chromosomes are limited to different parts of a cell, so this minimizes mismatch and errors in the duplication stage. If all goes well the sequences of the ATGC coding should appear identically in the new DNA. If 'perfect' then each generation would be identical to the previous one. Fortunately, this is not the case or there would be no diversity, variations or evolution.

Table 16.2 A very simplistic view of some terms used for genetic coding.

Key words	Rough description	My analogy
Genome	All the genes which exist in a particular organism	A library
Chromosome	The packages which transport DNA into cells Content DNA and proteins use this to control its functions	Sections of book types
DNA	An enormously long molecule which is in the form of a double helix (like a long double staircase in the French Loire chateau of Chambord)	Sets of volumes in a section
Gene	A segment of the DNA which specifically encodes particular characteristics	One book
Nucleotides	Stairs which links the two parts of the helices There are only four types of these, labelled A, T, G and C They always pair in the same way with an A to T pair or a G to C pair (a fifth, uracil, replaces T in the RNA)	One chapter per nucleotide
Base pairs	Because nucleotides pair, biologists only count the stairs	Catalogue scheme
Allele	Variants of genes that give different outcomes, for example they define red, black or blonde hair	Similar books, different authors
Codon	A genetic instruction formed from a sequence of three nucleotides (e.g. TTT, GAC etc.)	A photocopier
Intergenetic regions	The material along a DNA chain between identified gene coding, purpose and value is mostly unknown	Books we do not understand
Mutation	A copy of a length of DNA that differs from the original Caused by numerous factors, these imperfections can open pathways to either improvement or extinction	Multi-lingual or inter-disciplinary books
Genetic engineering	Attempts by humans to control and modify a section of DNA	Invented scripts or languages

Even for identical twins, which can occur as the result of splitting of a single egg, the DNA pattern of the twins will differ, albeit they may appear very similar. Characterization of the DNA will certainly show family traits and enable identification of parentage in cases under dispute. For forensic science, the DNA equivalent of fingerprinting is extremely powerful and is now a routine technique to associate criminals to a particular crime scene, or indeed the opposite, of establishing innocence.

Perhaps the most surprising feature is that within half a century and with an immensely complex structure, we have progressed so far in our understanding of the functions of the genes and the coding system. Less unexpected is that human DNA has many detailed similarities with a vast number of mammals (certainly suggestive that we have evolved from common historical ancestors). Nevertheless, just a single gene difference between species can be remarkably obvious. My naive assumption was that we have great similarities with pigs. They are a similar size, definitely intelligent and, crucially, they can be used in medical transplants. However, the transplant success was originally blocked by a solitary gene termed gal-transferase. This causes rejection of pig transplants by the human body. An interesting example of just one in around 20,000 genes being critical. For medical transplants the solution was to modify pigs and remove their gal-transferase. In reality we have produced a new pig species for medical transplants.

We are able to create mutations of species, including humans, that have different responses to diseases and other key defining factors which might be used, not just for medical repair purposes, but to define future generations. A first response is that it has the potential to offer immense medical interventions that might avoid, or cure, a wide range of inherited or subsequent diseases. This is possible as, of 2020, genetic errors resulting in around 1,500 childhood diseases have already been identified. Equally, there have been successes in the chemistry of making repairs in the genome (e.g. by a technique termed CRISPR). Unfortunately, even where some experimental successes have been documented on rare inherited diseases, the pharmaceutical companies have ceased drug manufacture when there is insufficient profit – despite having conducted all the requisite research and development. Interest in more lucrative diseases is, of course, ongoing.

Why are there imperfections in duplication?

Appreciation of the numbers of cells involved in every phase of reproduction, growth and development merely underlines the fact that imperfections in molecular structures are inevitable. The very first step in human growth and development involves maybe 200 million sperm at one time. Only one of these can reach the objective of fertilizing an egg. Whilst the probability of a healthy unit succeeding may be greater than for damaged sperm, it nevertheless does not exclude such events. Nor does it imply that all 'healthy' spermatozoa are identical. Hence, even at this first stage of human conception there is an element of chance, imperfection and diversity in the outcome. The success rate for any one individual spermatozoa is no better than the odds on winning a National Lottery (i.e. very small, but finite), but with 200 million candidates, success is probable.

A very unfortunate consequence of the selection of winners in the sperm race (to fertilize the egg) is that they need not be the most perfect examples. This is seen in the statistics of major and minor birth defects. Really serious birth defects may result in the death of the progeny at an early age, even with the best of medical attention. Other defects will be visibly and physically obvious, but the children may reach maturity and still be visibly, or detectably, incapacitated in some ways. The estimates of such problems vary between different geographical regions, but it is cited with numbers as high as 5 to 8%. A comparable number of children will have genetic coding that makes them particularly susceptible to specific diseases (e.g. hereditary versions of breast cancer or glaucoma) or hereditary tendencies to mental or other health problems. Accurate estimates in such fields are difficult to make, but unfortunately it is all too apparent that many children commence life at a serious genetic disadvantage.

Medical genetic interventions and engineering

Well intentioned interventions or designer objectives carry risks, as we have minimal understanding of how co-operative gene responses occur, and what may seem to be a valuable advance in one aspect may have unpredictable and disastrous effects elsewhere, even if they are not immediately apparent. Examples of chemical contamination can lead to high incidence of cancer in grandchildren, but skip the first generation (e.g. as noted for those exposed to Agent Orange that was sprayed widely in Vietnam). Unfortunately, human behaviour and the financial market opportunities both guarantee that we will continue with such attempts at improvements and mutations, despite the fact that we only understand a fraction of the billions of lines of coding. These influence not just conception and initial appearances, but continue throughout the life of the species in future generations.

We have long experience with simplistic breeding programmes of domestic animals and agriculture, but previously we could not contemplate targeted genetic engineering. Within this century that limitation has changed. The initial medical motivation is well intentioned as it may offer the possibility of avoiding many genetic diseases. Nevertheless, the advances are so rapid that the engineering will be able to define many human traits. In principle, future parents may be able to select, not just the sex of a child, but also the colouring, size, shape, intelligence etc. This is clearly a disaster as, in many societies, the first consequence will be an excess of big, strong males, and too few mates. The 'idealized' body shapes will mean genetic links to the parents are severed and this, in turn, will mentally isolate the members of the families. Currently such changes are mostly in the realm of science fiction, but the reality is that this can be substantially realized and humans, as we now see ourselves, will be superseded in families rich enough to make diagnoses and engineer foetal changes. Family ambitions will inevitably wreck current social stability, and undermine the diversity which is essential for coping with threats to our survival. Many commentators have also emphasized that current knowledge is centred on data obtained for affluent whites, and there is far less understanding and research linked to other races, even though white people are a minority in the total world population.

Part III Viruses and pandemics

The 2019/20 appearance of the Covid-19 virus had the potential to generate a pandemic that would attack and destroy a vast number of humans. It is, therefore, useful to consider if this is an unusual virus, is it really as lethal as feared, and, in hindsight, have we learnt how to cope with future variants of it, or other viruses. The origin was not in humans but from animals, where it has existed for a long time. The route to us may have been direct or indirect. This is a typical example of zoonosis, where diseases sometimes move from their normal animal hosts to humans. Mechanisms of transfer between species are not always obvious, as they may include intermediaries between the normal host and humans. Once they reach us, we are the key factor in their dispersion around the world. This will be true for both viruses and bacteria, but here I will focus on those driven by viruses. 'Focus' is scarcely a good choice of word as viruses are incredibly small and can only be imaged in a frozen, dead form by electron microscopy, not by any optical techniques where (with bacteria) we can see live behaviour and how they evolve, divide and replicate. Viruses also differ from bacteria as they need to invade a cell nucleus in order to change its function to become a factory to make more viruses. Viruses have no way of reproducing themselves except by taking over target cells and using them as a mass production line for more copies. Part of the strategy is to make imperfect copies so that there are new variants. This is a key requirement as the version needs to be effective in a new species, but not to self-exterminate if they kill 100% of the new host.

Some cross-species routes are obvious, such as eating animals that live, or are kept, in very unhygienic conditions, died in the wild or were cross contaminated when in open markets. Widespread dispersal can frequently occur via animals such as bats, which can be carriers, even if they have immunity. They have large flight ranges, they also relocate, and their droppings are similarly widespread. Clearly, they are not the only creatures that can do this. To reach human targets many examples have been shown to have passed from, say bats, to intermediary animals (pigs, chicken, horses etc.) before reaching humans. Note the emphasis on bats is perhaps unjustly skewed by the fact that many studies have been made of them, and not necessarily equally of alternative species. Mosquitos and other insects can be extremely effective carriers and inject viruses into the bloodstream of people. In some cases, they are the major transmitter in diseases such as human malaria and myxomatosis in rabbits. Other creatures may be equally effective transmitters, and these include pets of cats, dogs and cage birds etc. Humans are, in fact, the optimal dispersal carriers as our travel is rapid and global, and there are eight billion of us. To initiate a virus pandemic the source number need not be large (e.g. one!) if there are opportunities for multiplication in many hosts.

Many viruses that have been long established in jungles have not been well studied because (a) we have not explored there and tried to find them; (b) some hit so effectively and rapidly that they totally destroy local human, gorilla or simian populations; and (c) collecting evidence and samples can be incredibly difficult and may need an extremely rapid response, which is not feasible for truly remote areas. Hence, the spread to humans may be unusual unless we inhabit the same area where the virus is prevalent. Transfer to humans can be initiated via contact with infected animals through cuts and grazes, killing them, or, quite commonly, by using them as a source of food. Several past examples were clearly linked to food from unhygienic markets where there has been cross-species contamination. Intermediaries may or may not show symptoms. An equally crucial factor which, in the past, has been overlooked is that we are rapidly destroying natural habitats, both locally and in primaeval jungles. This displaces both native animals and the viruses that they host. Alternatives, such as humans, can offer an excellent replacement host as viruses are not selective, but their mechanisms for reproduction rely entirely on a host and they continuously, and rapidly, adapt their genetic code to suit the host. Finding an ideal home of eight billion humans who can carry, develop, and quickly spread the disease, is an excellent survival opportunity. Therefore, as we deplete jungles and forests, we can definitely expect a steep rise in previously unknown viral types. Routes of eating 'bushmeat', 'game' and road-kill, or bites from animals, deer ticks or insects are all effective propagators, not only for viruses but many other diseases.

An extremely worrying factor as we encroach into new territories is that virologists believe there are literally millions of unfamiliar jungle creature viruses which have not yet crossed into humans. It is not that they do not have the potential to do so, but merely that they have not yet encountered the opportunity. Jungle animals, as do humans, have genetic traces of past viruses and appropriate antibodies that hit earlier generations. In extreme cases, just a few members of the population survived (as for the Eyam example).

We are already carriers of many viruses and, for them, we presumably have developed antibody responses, nonetheless we may transfer them to other creatures. In a new host with no prior antibodies, viruses hit extremely hard. They are endemic and always hungry for new hosts, hence finding a carrier and convertor that allows them to move between species is part of their survival strategy. In human examples the diseases taken from Europe to the American continent, by explorers such as Columbus, killed millions of the unprepared natives; in return, his ships brought back their gold and viruses that killed millions of Europeans (and the same diseases are still rampant, 500 years later). The immediate death rates were enormous. In the Caribbean, the short-term death rates were perhaps 80%. In Mexico, estimates of the population in ~1519 were 30 million people, there was a vast immediate decrease and, even by the 1570s, the numbers had only recovered to around 3 million. Note the original infected crew members on the ships may have been as few as 100 or so.

The replication unit of a virus is very small and it uses the target cells to produce and multiply. Often this is by attaching to the RNA units, which are both designed as templates for growth, and tend to encode errors in the replications. Hence, they are both a 'willing' copying device and tolerate, and perhaps encourage, mutations that are then better suited to a human or new animal host. Viruses which are likely to attach to the DNA chains are called retroviruses. They mutate more slowly. A classic example of this is HIV. It has been detected in stored human tissue from as early as 1908, and although it differs from the virus versions of a century later, it is not radically altered since, as expected, the DNA repetition errors will have been low. Intermediary carriers may show few symptoms but help in dispersion. An animal example was the UK Foot and Mouth virus, which obviously infected cattle, but was equally present in sheep, but with no symptoms. It could be transmitted by contact, or in windblown air, and had similarities with the poliovirus found in humans.

Morbidity rates

The scale of potential morbidity from viruses is quite variable and will generally differ between alternative generations of nominally the same virus. This is highlighted by the Australian use of myxoma to reduce the rabbit population. It had been detected in Europe in around 1890, and then deliberately used to control rabbit numbers in Australia in 1950, and in France in 1952. Rabbits were not native to Australia and had been introduced there, but without natural predators expanded exponentially. The virus was intended to bring numbers under control. As is typical, there was not just one variant, but at least five. In virus terms this was essential, as type I killed nearly 100% of the infected population (and so became extinct). Virus strategy of variants II to V allowed

an ongoing host to exist. The death rates were only (!) about 50% in
one case. Mosquitos acted as the carrier between rabbit colonies. So far,
this particular virus does not seem to have re-engineered to attack humans, but HIV certainly took a century to establish itself in humans and
cross-species adaptations and transfer can never be discounted. HIV has
currently killed towards 40 million people. Despite forty years of effort
to control or vaccinate, it is still not medically preventable.

The focus on new viruses is skewed; those which have clearly dreadful consequences, and/or a high morbidity rate are the hub of attention.
We may be overlooking many that are pernicious and either creating
long term illnesses, or still mutating to forms that are not yet obviously lethal. Names familiar to the general public may be limited to
those as seen on TV, and named for the region where they were first
noted. A short list of new or exotic viruses might therefore be Avian flu,
Ebola, Hendra, HIV, Lassa, SARS etc. More familiar, rapidly spreading,
examples include the annual variants of influenza, which are certainly
not minor in terms of numbers of people infected, but evolve and mutate into new versions every year. The list of well established, clinically
familiar viruses is of course enormous. To add some perspective, the
typical English and Welsh death rates from winter influenza are around
10,000 per year. Additionally, Scotland has a smaller population, but a
higher percentage of influenza deaths, which in some years has added
a further 5,000 to the mortality statistics. These numbers are extremely
modest compared with the influenza pandemic at the end of the First
World War which, from 1918 to 1919, infected around a third of the
world population, and killed between 20 and 50 million people. Adjusting the numbers to the current world population, one should multiply
them by about five. For just the USA that would be approximately 3.5
million. I am citing the USA as that variant originated there, probably
through contact with farm animals. Many soldiers coming to the war
died in the Atlantic transport ships, but this fact was suppressed for
political reasons. It was called Spanish influenza as the Spaniards, not
being belligerents in the war, could afford to acknowledge the extent
of the epidemic. The troops that contracted the disease then carried it
back to places as distant as India, New Zealand and Australia. It thus
became a global pandemic with many susceptible war-weary people.
The seriousness of influenza is also illustrated by the late nineteenth-
century examples of around a million deaths from Russian influenza
(1889), leading to the early 1890s epidemic which killed around 30,000 in

Britain and a million worldwide (relative to modern populations, these numbers would scale upwards by eight times).

Pandemics are not rare, with several in each century, but rapid international transport makes them global, rather than localized. In the twentieth century, 90% of plagues from viruses originated in a limited set of regions in Africa which lacked population flow to other regions. This was extremely fortunate as the variants of the dreadful corona-type Ebola virus have, locally, caused a morbidity between 50 and over 80%. It appears with symptoms very rapidly, which inhibits people travelling once infected. By contrast, influenza and Covid-19 have incubation times enabling carriers to travel across the world, and also mutate. Covid-19 appeared in a virulent pandemic form in China at the end of 2019. In the UK only a small percentage had precisely the Chinese version, and the majority of infections came from variants that developed in Europe. This was not surprising as that mirrors the pattern of people travelling to and from the UK, and new variants appear on the time scale of weeks. Subsequent analyses in both Spain and Italy indicate that there may have been a milder version in the spring of 2019, where it had probably been diagnosed as influenza. Mutations are not rare and, in 2020, Russian scientists believe they have identified around 100 different variants of it!

17

Hints for a successful scientific career

Part I Satisfaction from science

Having had several challenging chapters, the following will be very positive, and potentially useful advice on entering a scientific career. Personally, this has been, and still is, both enjoyable and productive, and encompassed commercial industrial activities plus scientific research in a variety of establishments and academia. In Part I the comments may be of general interest, and include some unexpected conclusions. Part II focusses more on gaining science funding, fame and collaborations once the career is established.

No career is perfect and there will always be some compromise in terms of responsibility, income, stress level and freedom to control what you do, plus job security etc. My viewpoint will, inevitably, seem prejudiced as I have always had a mixture of an academic environment with close involvement with real world problems. In many ways this probably offered the best of both worlds, and I realize I have been very fortunate.

For scientists there are several main options. These range from straight forward industrial work running from basic research, development and on to production. Academic institutions range from teaching in schools, to colleges and universities where there is usually a mixture of teaching and research. There are also national and international laboratories with a variety of objectives from standards to new energy sources, particle physics, astronomy etc. Military objectives can additionally exist as separate entities. The level of involvement of this mixture of education to original research is highly variable, and even in universities some people are almost totally at the extreme ends of these alternatives. Some academics are excellent at teaching, others research, some at both and some at neither. Within my university career, there has been a major shift from purely academic objectives to any activity which raises funds for the university administration. Indeed, the salary

The Power of Imperfections. Peter Townsend, Oxford University Press.
© Peter Townsend (2022). DOI: 10.1093/oso/9780192857477.003.0017

balance now favours administrative posts, which unfortunately attracts skilled researchers and teachers away from their specialities.

The global overview should be that science is an absolutely key part of modern society. Therefore, it is feasible to find a mixture of features that can allow individuality and/or teamwork over a wide range of ability and satisfaction.

Industrial collaborations are stimulating and, for academics, are undertaken by choice in areas where it is feasible to contribute, plus a key fact that they often provide funding which underpins more speculative 'academic' aspects of research. Money is the root of all research, from salaries to equipment. Competition for research council money, and other funding bodies, is partly a lottery and can be particularly hard for genuinely new ideas, or thoughts which challenge existing work. Slightly less obvious benefits of academia are that one has continuous contact with intelligent young undergraduate and research students, post-docs, plus multi-national collaborations. Many of these daily contacts are with people in their twenties. Permanent association with them is invigorating and I feel that this is still my own age group. The mirror has a different view, but mentally it is a superb environment to feel the enthusiasm and the confidence of youth, and be accepted by them.

A second unexpected benefit is that, whilst academic salaries may not match top industrial ones, the actual work is by one's own choice. Civil service statistics indicate that (as expected) higher salaries result in a close link to longer life expectancy. However, for equal pay, the life expectancy is noticeably better for those who were in charge of their destinies compared with those who were directed from above. I'm happy to note the life expectancy of professors of science is high.

I have carefully not mentioned colleagues in one's own university or research centre. Having worked in close collaborative research with more than a score of Universities, plus another dozen major research laboratories across the world, I have found an interesting pattern that I am always able to get total support, a willingness to share ideas, and have joint programmes and publications. Further, during such links there has often been successful participation between different competing groups in each site. I am a visitor and outsider, and therefore not a threat to promotion and other factors in the local politics. Since scientific publications are used as a monitor of success and international prestige, a bonus for both them and myself, is that we all increase our scientific output and advertise our skills. More interestingly is that, once I have left, the various groups in our shared experiments return to being separate and no longer collaborate. This is a familiar, and understandable, pattern that probably can be worth exploiting in all types of work.

There can also be benefits for students in such multi-national links. Some universities do not allow their students to graduate with a doctorate unless they have journal publications in which they are listed as first author. The implication being that they contributed most to the work. Reality is often different, and many publications have essential input from all those listed as authors, plus names of supervisors or group leaders whose input is sometimes quite modest. A caveat is that, if they provided the funding and facilities, then their role is still highly valuable. It is not uncommon that some supervisors will list themselves, or others, in the pole position, and by doing so consciously gain more years of help from very able students. With weak students an early 'first author' slot might be used to remove them from the laboratory. Fortunately, good students can often have visits to other labs and in several cases where I have seen this unfair practice, I have deliberately published items with such exploited students and listed them as 'first name'. This has then allowed them to submit their thesis and graduate. I know their supervisors have been unhappy, but none have ever told me so directly. For prospective postgraduate students, it is a point to note when choosing a supervisor.

My own policy with publications is to put the first name as the person with the most critical input (if obvious) as the first author, and the rest alphabetically. Since my name begins with T, I frequently am the last on the list. There is actually an unfair bias as the result of this type of listing, not just in publications, but also in job interviews etc., as it favours names that begin with initial letters of the alphabet. Some countries have journals that insist on alphabetical listings. A friend from such a country told me that she selected her first husband precisely because his surname was at the beginning of the alphabet (her maiden name was not). The marriage did not survive, but she continued publishing with the name.

The alphabetical name problem is not just in science. For classical music I am sure that most people could rapidly offer the names of ten composers with the names starting with the initial B, but would be more challenged with those later in the alphabet, of say N, R or V.

(Citing ten of the Bach family is probably cheating, unless you know their first names as well.)

Career advice for young scientists

Part of my aim in this chapter is not just to emphasize that science is a very rewarding career, but to offer some advice on how to succeed and exploit the imperfections in the way we behave. There are many social imperfections that create differences and define our existing peer group and established ideas. Once we understand the details of this, we may be able to capitalize on their weaknesses to improve our social and career status. In principle, a major step is to choose the right parents and start in the top 10% of society. This should guarantee good schools, the best universities, and enough contacts to benefit from nepotism in jobs, friends and general success in life. For the other 90% of us, we need to find a different path, or be accepted by the peer group we wish to emulate. The following plan is intended for an academic science career, but undoubtedly the essence of the strategy is the same elsewhere.

A potential scientist who wishes to move up the ladder of fame and fortune (maybe for fortune read career) will do well, as step one, to attend a top university and ideally gain a first-class degree. In reality, this is only proving that you are good at dealing with the exam system and understood the material. Excellent though this is, it is not necessarily a guide to future creative thinking. Indeed, I have worked with incredibly able colleagues with imagination and insight, who only focus on items that interest them and, in degree terms, were unimpressive.

For the next degree level with a research phase, and doctorate, be extremely careful in the choice of supervisor and topic. The three or more years for the PhD in the UK (much longer in some countries) are a little like a marriage (and lasts as long as many marriages), but similarly has output. In this case the 'children' are the publications and skills which are with you for the rest of your life. As mentioned earlier, the majority of academics may never significantly change course greatly from the work of the PhD. So, if you pick the wrong topic at this point it is a serious setback, or a dead-end. The aim is to find the most prestigious group and supervisor in a field that you like, who is doing work that is popular, accepted and well funded. However, beware of excessively famous people as they may spend their lives travelling, or be away on national and international committees etc. Equally you might not be properly credited for your own input. Once in the group, fit in totally, including dress style and apparent working habits (but actually work harder than you appear to do), so you are readily accepted (no sign of misfits or imperfections). This should ensure a rapid doctorate and good references.

At that point you will need to find income with a new post. Many will remain for a further few years in the same research group as their skills will be valuable there. Group leaders may not want to lose you, but in terms of an independent career this can be limiting. Once having found a job elsewhere, you need to establish your independence and do not forget to cultivate contacts and friends from the peer group of administrators and department heads (academic or industrial). In career terms this is faster than the science. For academics, join national and international committees, help to organize conferences and become well recognized. This similarly is still more important than the work, as it will help bring in contacts and research funds later on.

Whilst I did not have the benefit of this advice, I chose excellent intelligent parents, a very good PhD supervisor who allowed me to suggest new ideas, and I planned my work carefully, so that I submitted it whilst still 23. This early submission perhaps gave the impression I was incredibly bright (unfortunately false). However, I failed to appreciate the benefits of links to administrators and key name people. Equally valuable will be the links, and making the friends and contacts that are possible from attendance at conferences. A postgraduate student may at this stage have some results, or be a name on a paper or poster that is being presented. Exploit this advertising, as many people will offer useful comments and suggestions.

There is a hidden protocol in such meetings. For example, I realized that, at conferences, graduate students are at the back of the hall and the top professors and speakers sit near the front. I naively assumed this was deference to experience and wisdom. I am now one who sits near the front, and realize it is not because of status, but because advancing years ruin your hearing and eyesight. For most people, any student who moved into the front group would be assumed to be there from ability. Such a general perception is the first step to status. An alternative strategy is to attend a conference which is somewhat peripheral to your main research. I have frequently done this as I have a very diverse range of interests. Here again there is an interesting bonus. In meetings of the 'foreign' field my background may be weak, but relatively

I am an imported 'expert' from my own field. Hence there is instant approval, status and recognition of my assumed expertise. This feeds valuable comments back to colleagues of my own area, so it is a win-win situation.

Self-promotion, meetings and conferences

Back in Chapter 5 I mentioned that conferences can be an inspiration for new ideas and reveal the true leaders. I stated that most fields are driven by, say, 10%, but with valuable input from another 40%. The rest may be supportive, but some will not understand new aspects. This is life. Recognizing the key people is normally straight forward, but strong egos often exist, especially in weaker members of the group. I am not making a critical comment about scientific conferences, only underlining the fact that scientists are human and 'experts' have a wide spread of expertise. The same situation applies to all other aspects of life. The ability to track performance of, say, surgeons, shows an equally measurable spread in ability; financial skills of bankers and stockbrokers are highly variable and quantifiable. Beware of blatant advertising and self-publicity, not least as we remember our successes far more readily than the failures.

Within any type of meeting, delivery of material is important. It needs to be simple enough to be understood by the bulk of the audience, but with appeal to experts who can have subsequent discussion. Definitely do not try to shine by making the work seem complex. You will confuse 90% and not be impressive for the other 10%.

There are genuine, very significant and different problems associated with conference lectures, committees and presenting science to the general public. The obvious ones occur if the talks are ill considered in either the level of the content for the audience, or are too long for the allotted time. The temptation, with say PowerPoint presentations, is to include too many gimmicks, which detract from the content, or we fail to recognize that what looked impressive on the computer screen may not show up on the poor laser projector of the lecture room. Room acoustics, microphone problems and projection can have various imperfections but these are definitely not what I am trying to discuss. The more subtle imperfections are those which come from features which, in other circumstances, might seem like attractive differences. These include using normally familiar emotive factors of accent, language or dress, and use of colloquial phrases. In home territory they may be perfect, but to a multi-national audience at a conference they can convey totally different impressions. Churchill's comment, in Washington, of saying about the UK and USA *'Two nations divided by a common language'* is still valid, but it is applicable much more widely.

The more polished performers know that delivery and timing is important, but even they often miss the fact that men who speak with a deep voice are treated with more respect (and it has more sex appeal). Equally true for cattle as well as humans! A further oddity is that better dressed lecturers have students who get higher grades. So mundane statements presented with a deep voice whilst wearing a smart suit are applauded, but world-shattering comments in a squeaky voice, from a man in torn jeans are ignored. Recognizing this pattern early in life is a bonus which can have a major impact on not only one's career, but also social and love life.

A further problem is that scientific conference presentations tend to be like a detective story, with facts at the beginning and hope that there is time to draw the intended conclusion at the end. This is very different from the needs for the general public, industry or newspapers, where the practice is to hit the keynote headline at the beginning, add facts and reiterate the headline. Imperfect presentations thus include those prepared for the wrong audience.

Handling questions is essential, and some skill is needed to choose the best emotive words, and add the politician's spin to the science. For example, if I'm asked a numerical question about the accuracy of the data and I think the answer is around 10%, and just say so, the audience will realize that I guessed. With such questions from difficult people, I have normally made up a number, such as 8.5%, as this stops further questions. It is not a simple round number, so they assume that I had detailed evidence. Words are paramount. A politician would never say to the public that a new process, factory or material has 'toxic' properties. Instead the spin would emphasize that there is some 'powerful chemistry' involved. In the same way the semiconductor community replaced the term 'impurities' with more positively sounding 'dopants'.

The problem of discussing imperfections

As a physicist spanning Physics, Materials Science, Engineering (and more), I attend conferences concerned with defects and imperfections.

At such meetings no one disparages the topic and the choice of words. Imperfections are our life. Despite this, many conferences still have image problems along the lines of a conference poster or badge such as 'Defects and Imperfections in Materials' which is then abbreviated to DIM. Walking around with a badge claiming you are DIM is not good publicity, nor good for the ego. I tried to point this out in an overview presentation at one conference but said my review talk was to be about love, magic, war and marriage. The emotive panic response of the audience (especially the organizer) was very satisfying, until I admitted that these were acronyms. (LOVE was Lots Of Valuable Experiments; MAGIC was Manufacture And Growth of Insulating Crystals; WAR was Ways to Advance Research; and MARRIAGE came from Manufacturing, Advertising, Research Realities, Innovation, Applications, Growth and Exploitation.)

The word 'imperfections' is a problem (even to scientists, but hopefully not to any reader who has reached this point in the book) so I have often replaced it with euphemisms of 'advanced structural lattice engineering'. 'Empirical data' has emerged as results of 'optimization of a multi-variant structural manifold'. Such techniques have the benefits of helping to deflect negative comments and gain political spin and positive emotive feedback. The only danger with such examples is if one starts to believe them oneself.

The public attitude to words and phrases is extremely fickle, and both phrases and ideas which offer good marketing can change in fashion. For example, in the 1950s the energy associated with radioactive decay was seen as something very positive, and many mineral spas and drinking water were actively sold as having a high radioactive content. Similarly X-ray machines in shoe shops were used to check that shoes gave a good fit. The idea was fine and it helped to sell shoes, but the possibility that this could (and did) cause cancer in shoe machine operators was ignored. The opposite response occurred during the 1970s with the extremely powerful medical imaging technique that was, initially, called Nuclear Magnetic Resonance Imaging. 'Nuclear' had gone out of fashion and to the public 'Nuclear' meant weapons, atomic bombs and dangerous radiation. Patients rejected offers of such diagnosis. A rename to Magnetic Resonance Imaging (MRI) then brought it into fashion. It is expensive equipment, so having such a tool is also prestigious for the hospital. There are no obvious detected dangers to the patients. The only difficulty is that, in an EU directive, it was claimed patients should not be exposed to large magnetic fields, a ruling which, if enforced, would destroy this most valuable of medical analysis techniques.

By contrast, X-ray imaging is still promoted as a safe method for the many uses from dentistry, assessing broken bones or screening in mammography. In reality, better imaging needs higher radiation dosage and so more cells are ionized and destroyed (such events are highly undesirable). The cell damage is a proven potential precursor to cancers at a later stage. There have, of course, been serious efforts to improve the sensitivity of detectors and, for example, the X-ray doses used in mammography screening are now reduced to perhaps 10% of those in the early 1970s! In hindsight, many estimates suggest that the high doses used caused as many cancers as they detected. In terms of diagnosis benefits, relative to cell damage and potential cancer production, the X-ray images are considered simple, positive and beneficial, but definitely not perfect, and are still cited as the source of many subsequent cancers. The risks exist but need to be put in perspective of immediate diagnostic benefits, plus all the other radiation sources to which we are exposed, including those from rocks, pottery and aviation.

A scientific aside is that the radiation from a medical X-ray is often cited in terms of dose relative to that of annual background radiation. This varies from, say, 3 to 8 mSv (milliSieverts) per year, depending on location and natural background radiation from rocks such as granite, and flights at high altitude. Chest X-rays to CT scans range from 0.1 to 10 mSv. The soothing description is that these are trivial doses. Unfortunately, in defects studies with ionizing radiation the effects, damage, and long-term consequences depend critically on the ionization rate as well as the total dose. However, there are 31.5 million seconds per year and, if a 1 mSv X-ray takes under 1 second, the dose *rate* could be 31 million times greater than the total annual background. Ionization damage in insulating materials, and tissue, is strongly dependent on intensity. This biological sensitivity to intensity is very obvious as if we consider our sight, we know we can cope with sunlight throughout the year, but if we saw all those photons in a single flash it would be a million times stronger. The result would be instant, and permanent, blindness.

Tissue response and damage from X-rays is similar, and therefore it will be vastly different from the simple annual total dose from background sources. A fact never mentioned by the medical community to patients. Similarly, if X-ray mammography has not resulted in a cancer within five years, it is considered safe. However, many doctors claim the effects actually take longer to emerge, but the five year rule is again a comforting statistic, even if it is not universally accepted.

During most of the twentieth century, any small collection of metal particles was called a colloid. Such particles have a long scientific history and, for example, are the basis of the early Venetian ruby glass. The ancient glass contains small clusters of gold atoms. In more recent years the use of the word *colloids* to describe a small cluster of, say, metal atoms embedded in a solid, went out of fashion and the replacement was *nanoparticles*. In order to gain research funding, we must be constantly aware of in-phrases and currently favoured words.

Fashions change, and the skill is to sense when such words date the text. In proposals to the European Commission I know of successes where blocks of the proposal were lifted from the Commission's own statements of objectives. Most obviously it implies the proposal is in line with the priorities of the Commission and is very effective when the reviewers complete their EC check sheet. The time delay between writing the directives and assessing proposals is large enough that exact phrases will be forgotten, but sub-consciously the assessors will respond to the words they themselves may have written. The method works equally well in areas as diverse as student essays and town planning applications.

An alternative approach from a good friend gained him a high rating from UK research councils. His strategy was to seek money for unpublished work he had mostly completed. He not only outlined the aims he had in mind, but also made quite detailed predictions of the likely data that would be recorded. Inevitably, not only did the funded work succeed, but his predictions implied he really was very astute (he was). This, in turn, led to more funding.

Part II Scientific publications and grant proposals

Any research programme needs money for people and equipment. This is as true in industry as for academic research, and so a key skill for a successful scientist (and administrator) is the ability to raise money. For some this is an alien challenge, and so for such people it is essential to be in a team with good fundraisers and to give them full credit. For academics there is a further need to succeed by publication in respectable journals, as this maintains one's credit rating. 'Respectable' in this case requires some critical level of peer review. A strong publication track record helps grant applications, so is a key factor. Unfortunately, having papers published is often easier for work which is 'more of the same', rather than something original. Referee's critical comments seem far more prevalent if the work is novel, or with particularly impressive improvements on earlier publications. Some research scientists find this very difficult. My own response has been fairly pragmatic. I take the critical comments of the reviewers of the journals as evidence that perhaps I was not as clear as I intended. Then I make adjustments and try again. If that fails, the less stressful route is to consider other journals and submit my scientific offering to them. In part this is because reviewers can become entrenched in their views. They are anonymous, so are less tolerant and rational than they would be in a face to face discussion.

Viewed from the opposite side of the process, one can of course find authors who totally reject the wise suggestions we make when we are the reviewers. In one example, I saw the same paper four times on behalf of different journals. It was totally unchanged although I had spotted, and pointed out, a fundamental and significant error in the article. The authors made no attempt to correct it but just passed it to different journals. The process is imperfect as I eventually saw the paper published in a fifth journal, still in its pristine erroneous version. To some extent publication can be a matter of persistence.

This is not the case with research funding as there are fewer opportunities for success. Those of us who have written many research grant proposals (a sort of up-market begging letter) are, of course, always critical of referees who fail to recognize the brilliance of our research. There are normally several assessors and often the different referee reports will appear to relate to totally different quality work, quite literally with summaries ranging from 'outstanding' to 'rubbish' for the same proposal. Nevertheless, the chance of gaining funds is finite, historically with perhaps a one in ten chance being typical in, say, materials science, but probably now worse. Skill in proposal writing is essential as the system is less forgiving than publishing articles as there are rarely alternative funding bodies, and sometimes the first attempt must succeed. Unfortunately, it is essential to simultaneously convince all referees why your proposal should be accepted, rather than be in the 80 to 90% who are unfunded. Grant writing is a tricky business and plagued with irrationality. During four attempts to gain funding for a novel project of optical cancer detection (mentioned in Chapter 7) I had several very favourable referees on each occasion, but always a fourth or fifth who was negative. In grant terms this meant no money. On one attempt, one referee wrote about my suggested improvements for the photo-multipliers '*This is a forty-year-old problem. If there were a solution it would have been found – Reject*'. The next resubmission also failed because a referee said '*This will obviously work, so it should be funded by industry*'. Fortunately, a proposal to the EEC succeeded, as did the project, and the improved performance was 50 years ahead on the preceding improvement trend line from international companies.

Attempts at assessment

The subsequent section will deal with the errors and perils associated with trying to grade people, whether in exams or grant proposals. This is an example of imperfections with few redeeming features and frequent injustice.

Publication of exciting results is clearly something that one wishes to do, unless it is an idea that might have good commercial potential. In which case, patents might be the first objective. Many people enter this new industrial pathway and then launch start-up companies and discover that the commercial world is totally different from their academic experience. Some are successful, but many start-up companies struggle to survive, whereas a few winners have spectacular success. For the majority who remain in the safety of academia, publications are used as a monitor of performance. Not just by the number of publications, but by the prestige rating of the journals and the number of times their work is cited in other journals. At first sight this seems a reasonable approach, but it is deeply flawed for a host of reasons.

The academic communities, administrators and the world in general, are now obsessed with attempts to quantify the quality of scientific work and the status which reflects back on their institutions. To some extent it is clear which journals attract the best quality work and, consequently, articles in mainstream topics in these journals attract many citations. This is a self-perpetuating system, as the status of the journal is raised, and so more people want to put their best work in such journals with a high 'impact factor'. For advancement at the start of a career, citations and publications in high impact factor journals are desirable. However, less obvious is that any publication that includes a

Nobel laureate co-author, will gain citations disproportionate to the effort involved. I saw this happen very clearly when a friend of one of my students used our equipment for a modest experiment. We included the name of his supervisor in the list of authors (both polite, plus he was my friend, and he agreed with the content). He happened to have a Nobel Prize, and suggested a far more prestigious journal than I had considered. The paper was *immediately* accepted and published. It then had an immense number of citations, not at all commensurate with the importance of the article. My friend said that prior to the prize his previous publications drew few citations, but soared afterwards.

Merely counting papers and citations is moderately sensible if one is working in a very popular field with lots of other groups around the world, and the results are routine and not challenging existing theories. During submission it also means the references in the paper are likely to include potential reviewers, who in turn will be happy about this and, perhaps subconsciously, more likely to accept the article and subsequently cite it. Publication difficulties escalate if the ideas and data are challenging existing theories and equipment performance. Firstly, human nature instinctively rejects new ideas, and especially if they are critical of earlier ones (no matter how flawed they were). Secondly, data acquired with equipment that is far superior means no one else can duplicate it and, by default, there will be no citations, even if it is accepted.

The playing field is far from level between topics. Medical journals gain more citations, as do fields where there are very few journals. Solid state physics and materials science have a diversity of journals, and this spread of publications can reduce their impact factor. Similarly, journals which used to include conference publications were undermined, as conference proceedings are cited far less than regular articles. Many journals now have a separate section for conferences to try to overcome this.

Prejudice is equally detectable in reviewing as, in experiments conducted by leading journals, articles have been sent to reviewers with alternative source addresses, such as Cambridge University or a fictitious Backwater Community College. Guess which versions were accepted and which rejected. Ditto with names attributed to the papers.

The greatest imperfection in gaining recognition and meeting the citation criteria comes for those who are imaginative and branch into new territory. If the work is well done, others find it hard to compete, so the work is not cited. Leading a field of one, or being too far ahead in a topic, is bad career strategy. As an experimentalist I have seen this happen with people (including myself) who have built equipment which far exceeds the performance of the rest of those in the field. The articles are fine, but unchallenged and unquoted. A possible option is to find a friend who is willing to use the same world-leading equipment and publish independently. You can then both work on something where you can argue in public, and so both gain papers and citations. I have not tried this, but I'm sure it would succeed. A further route to gain citations is to write long review articles. Really long articles are probably not totally read, but always in a publication it is wise to cite background references. If, when writing a new article, you recall some immensely long review then a citation to it will probably cover all the papers that you have not actually read yourself. The review writer gains citations and is then happy. As a writer of numerous reviews, I know they rarely contain original work, but they have been excellent for focussing my own thoughts on a subject (and gave many hundreds of citations!).

The preceding thoughts on developing an academic scientific career strategy should rapidly result in a Professorship and good financial support. At that point start to do original work, as reputation and contacts will help to offset the prejudice of referees about new ideas.

Can one quantify performance?

In all aspects of education, and life, there are attempts to make assessments of performance and ability. Whilst the sentiment is reasonable, the realities of how such judgements are made are highly variable, often deeply flawed, and almost always very subjective. We do this continuously when watching TV, discussing politicians, listening to music or talking with other people. In these examples there is no doubt that the judgements are personal and subjective, and we recognize it. However, throughout education (from school to academic activities), interviews for jobs, and even sporting activities, we assume that these can be quantified and trusted. The consequences are profound as they define how we advance through life, the types of job that we have, the leaders and governments we elect, and even whether or not we engage in wars. In every one of these examples there are intense weaknesses inherent in the assessments. In this chapter I am concentrating on education and science and will rapidly indicate some of the variations that we accept, or perhaps have never considered.

Starting from the most basic school education example of examinations and grading, there is an immediate distortion factor from the quality of handwriting. Good, clear, neat writing means that we read it easily, feel friendly towards it, and mark it generously. This leaves a deep psychological impression on the students as to how they relate to the others in terms of grading. Their expectations and ranking are immediately skewed by this very simple but inherent difference. Having taught, graded, and been involved with university examination boards, it is equally apparent that legibility of written work is a significant factor in the assessment of all subjects. In experimental trials where the same text has deliberately been written by different students, there is a correlation between grade and legibility. At school level, a teacher who has to mark thirty papers will find it difficult to be equally fair in terms of concentration, and a legible tidy script automatically is favoured. Similarly, if they are marked in student name sequence for each question, the concentration level by the last ones will be less generous than for the initial ones. Personally, I always have marked scripts in a different sequence for each question in an attempt to reduce this bias.

From my university experience, it is not extreme to suggest that the legibility differences from hand-written examinations are worth at least one degree class. The response to writing quality is also highly variable between people. My own handwriting has never been good, and has deteriorated, so I am deeply grateful to be able to use word processing. It primarily gives clarity so even I can read it, as well as the bonus of easily making alterations. My first book was hand written, and before being typed had deletions, changes and many odd scraps of paper attached to the original pages. The option of many drafts was very limited. The reliability of degree assessment can further be perturbed when there is a choice of options that are marked by different teachers. Once again, individuality can mean that what one person considered an excellent level of understanding, may be assessed as mediocre by another. Examination board discussion can struggle with such issues.

Additionally, some universities have exchange programmes where, for one semester, students may take courses in a totally different university, or industry, often to benefit from experience with a foreign language and/or culture. An excellent concept, but if grades are to be transferred between them it is essential that all the organizers understand that different cultures use different marking schemes. For example, I discovered that my university had not appreciated the scale of such alternative gradings. On one occasion we had given a top-grade mark of, say, 75%, but a foreign student then came and asked why he had done so badly as, in his home university, 75% was failure (!) and a top grade was over 95%. So as an exchange student between different sites, check that the administrations understand this. Internally, multi-discipline courses can display precisely the same variations.

A related student problem emerged when we accepted foreign post-doctoral students without realizing differences in degree nomenclature. We had foreign applicants with undergraduate grade degrees who wished to become postdoctoral students. Their grades ranged from one to three, so of course we accepted those at grade one (a UK first is our highest level). The students were very weak, and I then discovered that the country had level one as the lowest (not the top). Subsequent grade three students were absolutely fine. The administration did not care as their primary aim was the fees.

Having been amazed that no one had spotted this use of alternative schemes I was further surprised when I was on two different research council committees. One went with a grading of one down to five, and the other from five down to one. Apparently, committees randomly selected which scheme to use and assessors were not told which scheme was in operation, so they may have used the opposite gradings! There was some embarrassment when I pointed this out, and a unified scheme was introduced.

Assessments and comparisons continue throughout life as they define promotions, job opportunities, status and friendships. For my focus on science careers, administrators and funding bodies would obviously like simple quantitative indicators to help with promotions and grants. Tangible inputs may be linked to performance items such as (i) how much funding has the person raised, (ii) has their work been published and discussed by the wider world community, and (iii) has the person been invited to speak at major conferences or at other establishments, or generated publicity through the media. This may be clutching at straws as it is an incredibly difficult challenge because of the diversity of topics, patterns of work and their inherent problems. Established teams with commercial experimental equipment may generate many items per year, but only add more examples to an understood topic (e.g. the same experiment with a range of different materials). By contrast, a theorist is likely to work alone, have a very low publication rate, and in many cases have a tiny world audience able to appreciate the new theories. Hence, there are major variations in simplistic monitors of performance by administrations and funding bodies which are far from perfect, and basically unquantifiable.

Counting publications is easy but, in my view, of very limited value as it is too variable between topics. Note also that many journals have charging schemes for publication and this will penalize authors who lack finances. Further, as mentioned in Chapter 5, there are around 2.5 million science and medical publications per year, which dilutes the readership and reduces citations to work that may be of excellent quality.

Citations of journal publications in other journals is a further possibility, which is favoured by some administrators and funding bodies as it offers a number. One scheme is termed the h index (h, as it was suggested by J.E. Hirsch). I will outline how the index is derived, but since I have little confidence in it as a general tool, I will emphasize potential weaknesses where it fails to deliver a fair assessment. I admit the aim is sensible and have no idea how to offer a better alternative. Consider a graph, as in Figure 17.1, which displays the number of citations per article against the decreasingly cited number of published papers. The point where the publication number drops below the number of citations is the h index. In the figure h = 5, as by paper number 6 and beyond, there are fewer citations.

Note that people with established careers, and a steady publication rate, may have a higher index rating, say for h = 25. This fails to indicate that their total list of publications could extend to several hundred, with potentially all of them with 25 each! Therefore, such authors are seriously penalized. An alternative suggestion is to sum the total

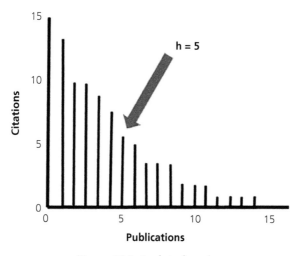

Figure 17.1 An h index plot

number of citations (N) and define an h $\approx 0.54\sqrt{N}$. This is obviously a total guess as I do not see how one can justify a precise 0.54 when making a simplistic approximation, except to make it appear as though there is some scientific basis. ($\sqrt{N}$ is the square root of the number N, e.g. $\sqrt{49} = 7$). For my unfortunate example with an h $= 25$ plus many more also at 25 citations, the N variant could easily deliver an h value approaching 50.

The h-index method is partially successful in the USA where, for mainstream physics, an h > 12 might be the case for an associate professor, > 20 for a full professor and > 45 for a member of the US National Academy of Sciences. Hirsch suggested that h > 60 would be truly exceptional. To support my personal view that it is a very poor monitor, because the problem is so complex, I need only quote that Albert Einstein reached an h-index of just 44, and Richard Feynman a mere 37!

With a computation in terms of N they would fare better, except that they gained fame over prolonged careers and Feynman is also famous for his highly stimulating books of undergraduate lectures, compiled with Leighton and Sands. In some applications of the h-index, only journal articles are counted (not books) and/or neither are citations more than ten years old! The logic of these restrictions escapes me as many books are highly cited and new ideas can take years to germinate and take root. So, neither Einstein nor Feynman would have achieved even their ridiculously low index ratings.

I have read that, for physicists, perhaps as few as 1% have a total of 3,000 citations (using the N value this gives and index of $\sim$29). Comparing physicists with those in medical research or other topics is totally unreliable as citation patterns and number of published works vary considerably.

I am clearly not in the Einstein league, but some of my books and publications have around 1,000 journal citations each (others are in hundreds). Equally, there was nearly a ten-year delay between our publications on ion beam implantation for optoelectronics and it becoming internationally fashionable and a major applied technology. I therefore view h-index ranking with considerable suspicion. The h $\approx 0.54\sqrt{N}$ estimate is perhaps better as it allows time for ideas to percolate into the literature. To put the spread of these numbers in perspective, one of my former graduate students is both excellent and highly organized. She has a large research group and an N value over 20,000 (i.e. an index in the mid- 70s!).

Further assessment problems are that there is no feasible mechanism where one could weight the input of different authors on a multi-name article. Many group leaders insist on being on every publication. Some major astronomy and particle physics labs have articles with several hundred authors listed who contributed to the project in some way. If they all individually cite that paper in just one future publication, they instantly have an extremely distorted N value. Further, brief letters count equally with long overview articles.

Other factors to note are that some names are quite common and search engines that construct these index data can easily be confused. (I have experienced this as there is another P.D. Townsend and we work in related fields. I am quite happy as he is a highly respected physicist.) Mistaken identities are especially true for Chinese authors, since there are only around 200 monosyllabic surnames. Finally, we should not ignore the possibility that citations may not be praising the work, but criticizing it. The classic example here must be for Pons and Fleischmann, who were two respected electrochemists who believed they had evidence for a process of cold fusion. If correct, it had immense potential across the world for low-cost energy production. Hundreds of people dropped their current projects and tried to duplicate the experiment. There were around 200 citations within a year and maybe 6,000 in total. None were able to duplicate the experiment, but from this one article alone they gained an N index of ~48 (i.e. above Einstein!).

Difficulties with acceptance of new ideas

An unfortunate topic to mention is that, whilst one can have brilliant ideas, they may be totally rejected, or destroy a career if they are too far ahead of their time. In passing I have indicated several examples. My earliest involved Galileo, where he was promoting evidence that the Earth (and therefore humans) are not at the centre of the universe. Locally we are just a planet circling the sun. At that stage there was no understanding of the scale of billions of both stars and galaxies. Unfortunately, conflicts between science and religions have frequently been a barrier to promoting new ideas.

For example, in 1805 Philipp Bozzini, in Vienna, made a precursor of an endoscope which initially was approved by the local medical committee. However, government and church intervention banned it as the church view was that he was being 'too curious' about the workings of a living body. He was then forced to move to Frankfurt, but had difficulty both obtaining a licence to practice and obtain citizenship.

Slightly later, as the result of experimental work in electricity, Georg Ohm published a book in 1827 which included the now familiar equation linking voltage, current and resistance ($V = IR$). However, German scientists felt that experimental work demeaned science, and thus the German minister of Education declared that 'such heresies meant he was unworthy to teach science'. His experiments were in conflict with the mathematicians Fourier and Navier. So, he lost his post. Fortunately, by the 1840s, his views were accepted.

By the twentieth century one might assume attitudes had changed. In 1911, Alfred Wegener recognized, as most of us will have done as children, that on a map the outlines of Africa and South America looked as though they might fit together. He improved the fit by looking at the continental patterns, rather than just the coastline, and found very good evidence of matching geology and fossils that confirmed an earlier link of the two continents. Whilst he could not explain why, he proposed the model of continental drift. This was ridiculed and he lost his job. Confirmation and models did not emerge until the 1950s.

In Chapter 8 I mentioned the idea of Hedy Lamarr, in the 1940s, of frequency hopping to offer secure communication to wartime submarines. This was totally rejected by the US military, but when they eventually used it, they robbed her of the patent as she was not a US citizen! The technique is still absolutely crucial to modern communication technologies.

I cited a personal example of rejection of grant proposals in around 2000, because it was a forty year old problem, and 'therefore insoluble'. It suggests our mentality has not greatly altered. Equivalent stories from many other areas of science and technology are too numerous to list, but could readily justify an entire book.

Part III Patterns in scientific progress

My examples of valuable imperfections have spanned many technologies, from flint knapping to optical biopsy but, in all cases, it is clear that not only are the materials and technologies controlled by imperfections, but the way that science has advanced is frequently haphazard, even when there has been a steady drift forward. Progress is slow as scientists are as fallible as all other humans; they automatically reject novel ideas and have a negative sensitivity to ideas which challenge their status. Often progress was from experimentation and empiricism, and this was followed by detailed theory. Novelty was actively impeded in some cases (e.g. optical fibres) as it displaced those with outdated expertise. Even with semiconductor production there was a reluctance to move from adding dopants by thermal diffusion (i.e. cooking) to the now universal route of ion beam implantation. None of this is surprising as the changes in techniques were accompanied by changes in expertise, investments and personnel.

An associated difficulty is that, for the general public, there is no excitement to hear of minor advances, although they expect products to be steadily improving. Consequently, both the press, the media, and scientists prefer newsworthy items where a flash of genius has caused some great leap forward. Mostly these 'leaps' are the result of long and hard contemplation and experimentation, but for publicity a description as a flash of inspiration is preferable. Some such inspirational events are probably fictitious but, for the non-scientific journalist, it is easier to write in terms of some accidental event which inspired the big step forward. The Eureka moment of Archimedes in the bath considering density and displacement of water had the correct press value. Newton explaining gravity by describing a falling apple has a similarly nice common touch for the public, but it was probably pure advertising fiction. There is also a tendency to denigrate many advances that apparently included some fortuitous event, and say they were just examples of serendipity.

Separating science and serendipity

To a non-scientist, a first impression of serendipity is that there was a random event which just happened, and this gave the lucky scientist fame and fortune. Certainly, to the outsider, or competitor, it may be considered unfair good luck. In reality, this is rarely the case and progress through serendipity is the demonstration of skill, planning, insight and a willingness to interpret an unexpected event. A quick web search of examples of major scientific advances which are attributed to serendipity produces an immense list over a vast range of discoveries, as well as numerous Nobel prizes. Perhaps the more unexpected feature of such searches is to count just how many examples and famous people are being quoted. Many Nobel Prize winners have hit such fame and fortune as a result of accidental experiments and observations. Indeed, the wealth of Nobel himself was based on his semi-accidental discovery of how to safely handle nitro-glycerine (i.e. dynamite). His brother was less lucky and was killed in an explosion.

Serendipity may also just mean recognizing that a product intended for one application is valuable in a totally different role. Very early studies of chemicals used in photography also produced celluloid. In a lecture I would say this success was loudly trumpeted, as the first beneficiaries of this were elephants. The new material was used in the manufacture of billiard balls and, since it was much cheaper than ivory, it resulted in the shooting of fewer elephants. Serendipity linked to medical advances from side effects are legion, and of popular interest are contraceptive pills and Viagra type products, which started out with quite different types of medical role. Some scientists will even claim that all life forms and evolution have developed from a sequence of fortuitous accidents.

I think the test of a good scientist is not that some accidental new important result happens, but that the event is recognized and then repeated and developed. I am grateful that most of us have had such leaps forward in our research. In less positive moments I also wonder how many major possible advances are missed because we are preoccupied, or not in the laboratory when the oddity appeared. This type of oversight is extremely easy if tired, or inexperienced, when faced with an unexpected result that might have been an equipment error. Experience is essential. By good fortune I once saw a graph in our lab waste-paper basket which one of my team had thrown out, because there was a major anomaly on the set of data which he had never seen before in other experiments. Fortunately, I recognized that it implied we had heated through a phase transition of trapped impurities. This accidental event opened up an exciting new research topic that has resulted in many insights, publications and invited lectures. The oddity was actually a brilliant gem of information. That first example was very satisfying as it was of a new type of optical fibre fabricated by my daughter in her laboratory in another university, and she was wondering why it had some unusual properties. Such events really add to the pleasure of hands-on laboratory research. Examples we missed I will, of course, blame on time spent teaching, or administration and writing grant proposals, rather than my own blinkered view of new data.

Recognizing serendipity

Serendipity can be exploited at many levels and the most useful feature is to recognize that, throughout our lives, we are continuously exposed to information in forms which are not explicit. In school, the teaching was offering us new information that had been considered and digested. We knew this was intended to be informative, but because it was presented to us (whether or not we were interested) we probably failed to fully consider it or see more widespread patterns of how to analyse new information. Serendipity was therefore just an extension of advances made by famous people. What we really need to remind ourselves is that we are continually hearing, reading, or seeing facts that we can put together and recognize patterns of behaviour. We need to recognize we are intelligent enough to process such inputs to make our own conclusions. Perhaps we should not call it serendipity but think in terms of intelligent observation. The results may not be fame or wealth, but they can be a happier life and a better choice of friends or partner.

Success with partial understanding?

The answer is obviously going to be yes as, if we had total understanding, then there would be no fun or motivation to pursue further knowledge and research. I am a fairly typical research scientist in that I have a reasonable, but limited, expertise in areas where I work, plus a slightly larger general knowledge. As a graduate student I had assumed I did not fully understand my subject because I was inexperienced, but that given a few years, I would have deep scientific insights. A lot of time has passed, and as a rather more mature professor I still do not understand in depth, but realize this is the norm and in career terms it does not matter. Indeed, this is a positive imperfection as it is a stimulus, and the impetus, for further effort.

Equally, most scientists would like some degree of fame and international accolades. Fame will mean different things in different careers. In scientific terms, film star type recognition by the general public and frequent TV appearances is very rare. It also requires different skills from those of pioneering science. Within the research community, some view aspects of fame as being invited to conferences as an 'eminent' speaker, or to write review articles in prestigious journals. The reality of such invitations is that they favour good speakers and clear writing. These skills sometimes do not appear with the most knowledgeable scientists, who can too easily ignore the fact that most of their audience will be less well informed (or lack the necessary background or intelligence). However, being realistic, most reviewers, key-note speakers, or referees of articles and grant proposals are likely to be less ignorant than many others in their chosen field. Partial understanding of a developing subject is the norm, and the more critical experts recognize weaknesses even in well-established theories and models of how scientific processes occur. Imperfect analyses and partial understanding are acceptable as a pragmatic way of discussing a subject and attempting to make some progress. Limitations of understanding are more likely to be admitted by the very best scientists, as they feel secure in their work and themselves. This is encouraging both for younger scientists as well as for any non-scientist reader. The key conclusions are that we can survive and advance with an imperfect understanding of detail; even the greatest scientists are not perfect and we should not be inhibited by praising advances made by serendipitous observations.

We may assume Einstein was a great mathematician, but in his violin playing he was harshly criticized with comments such as '. . . count Albert, 1, 2, 3'. I know the problem, sometimes playing the violin notes and counting are conflicting.

Small science versus big science

Many students are seduced into thinking that the only important science involves vast expensive projects in major laboratories, such as CERN, astronomical projects or NASA etc., plus highly complex mathematics and models. Indeed, they can be interesting and they attract extremely able people and consume vast quantities of money. However, from the world view we have lost very able, intelligent people who might have contributed significantly to progress for mankind, rather than some esoteric science with extremely expensive toys. Governments support such projects since few politicians will understand enough of the science to be critical, and so will not wish to appear ignorant. For politicians there is a bonus that the immensely expensive equipment offers good photo and TV opportunities, plus improved image status by associating with Nobel laureates and famous scientists. Indeed, the same may be true for many of the scientists involved, who will only have a partial understanding of the highly imaginative mathematical assumptions and theories. One drawback of these facilities is that one has to work in a specific site and so one's life is tailored by the work, with immediate contacts with people of a limited outlook and focus. This is a loss for family members.

My experience with multi-participant projects is that, for lower echelons, they often offer little personal satisfaction and recognition. There can be success in their specific tasks, and some personal satisfaction, but no real glory for a minor cog in a major machine. The big-name leader may be equally unhappy as the top job will involve so much administration and financial worries that the pleasure of conducting the research is lost. For myself, personal satisfaction from success on a project with five people feels ten times better than with a group of fifty. Maybe there are hidden laws that personal satisfaction times the number of participants is a constant, and an inverse rule that smaller groups collaborate more closely and have more impact.

Computer graphics and journal publications

Computer graphics with the use of colour are now routine in science journals, but I often disagree with younger colleagues as the graphs may look more sophisticated, but frequently they hide essential data. There are several inherent problems. The first is that a computer screen display will have a good colour sensitivity, whereas in a journal the colour printing can be markedly inferior. The second, and less obvious weakness is that some 10% of Caucasian males are colour blind to some degree. Note 'colour blind' is a seriously discouraging misnomer as it may only mean their colour response is not typical of the general population. Hence anyone with a wider spectral sensitivity will fail the standard colour blindness tests. Only a very small percentage of people have no colour sensors, and usually they are compensated by having more of the simple rod-shaped detectors, which are some 100 times more sensitive than the cone-shaped colour sensors. In low light levels their light sensitivity is far greater than usual.

Science journals vary greatly with graphical items and some include incredibly complex graphs constructed to show how different types of data are interlinked. Such figures may be detailed, but for me, unless they are on a topic of specific interest, I invariably just ignore them, as the effort involved to understand them can be considerable. Such a response undermines the numbers of readers who might have been interested in a simpler, and clearer, presentation.

Other journals recognize the conflict between fashionable coloured graphs and clarity and so present both colour graphics in their articles, and have 'on-line' access which includes additional views of the data in 'old-fashioned' black and white alternatives. To highlight how much these views can differ I will use Figures 17.2 and 17.3 from a recent publication made with three friends. They are young and prefer the colour images, I am old and prefer the black and white. Figure 17.2 directly compares the presentation of a modern colour-coded intensity plot with an older style contour map. The data represent optically excited luminescence where the intensity changes as a function of temperature. For me, the number of contour lines offers a simpler indication of the relative intensities.

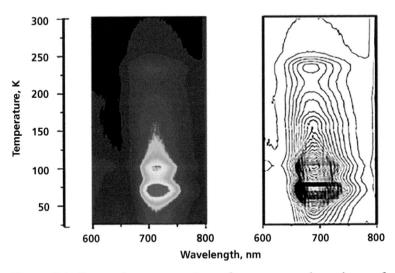

Figure 17.2 Contrasting presentations of temperature dependence of photoluminescence of pink sodalite. The colour intensity scale is indicated on Figure 17.3.

(a)

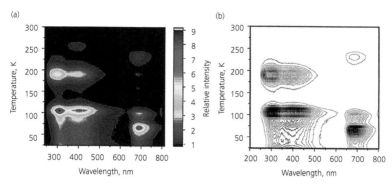

(b)

Figure 17.3a and b (a) is a colour-coded intensity plot, and (b) is a contour version of the same thermoluminescence data. Figure 17.3c (below) is a transparent isometric plot of the three variables which reveals far more detail.

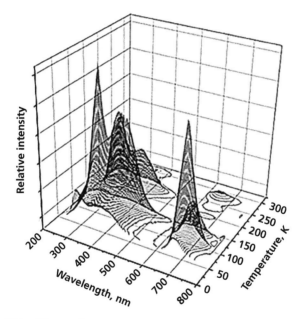

Figure 17.3c Thermoluminescence data from a mineral named pink sodalite.

Figures 17.3a and b present similar plots for thermoluminescence, where both the intensity and spectra change with temperature. The colour-coded map locates the peak positions, but disguises their relative intensities and any fine detail. The contour version is slightly better, but with such a very wide dynamic range there could be an improvement by using a logarithmic intensity scale, so as to include both weak and very strong signals. However, the final figure (17.3c) is far more valuable as it is the isometric plot of the three variables (intensity, temperature and wavelength) and it is very easy to visualize and separate the component features and gain an insight which (for me) is totally lost in the colour-coded intensity view. The transparency of the grid even allows one to see through strong signals to weaker ones in the background. My view is based on the fact that, as a young graduate student (pre-computers), graphs were plotted point by point by hand, and one was immediately aware of subtle details. Within the next decade one may be able to display them as three-dimensional holograms, which would be truly valuable.

Science and the media

I have not discussed the public aspects of promoting science for a mass audience via TV. For the natural world, programmes one can immediately cite superb presenters who deliver excellent visual footage and an empathy for both the subject and the audience. These are a real pleasure to watch. Astronomy is perhaps the second most successful, although it is far more challenging for the presenters. Perhaps 80% of TV audiences live in cities and cannot see the stars in the night sky. They therefore lack any real connection with the topic. Medical sciences present highly imaginative colourful graphics. For example, the images, colours and animations for the coronavirus are totally graphic art (not observations and measurement). Eye catching certainly but, in reality, they are not imparting facts and intuition into the subject for a typical TV audience. Engineering and other types of practical science may fare better as the audiences they attract are self-selected, because they are interested. The same is true for those reading science magazines or books (such as this one).

By contrast with detailed journal publications, the mass media items invariably involve sensational events, meteorites blazing across the sky or earthquakes and tsunamis, and they are all highly visual, and accompanied by a tiny piece of science. Extra interest is gained by adding scenic locations. Dramatic claims by the medical profession are marketable. By contrast small advances, or facts which discredit the former spectacular interpretations, are never headline news, even if mentioned at all.

The net effect is a public image of science which can be quite negative, or encourage an assumption that science must be difficult. Additionally, many scientists do not recognize the immense gap between them and a mass audience. They fail to fully remove jargon and formal scientific terms that interfere with communication for informative and popular entertainment. Even the image of scientists can fail to show them as perfectly normal people. Not least as in TV interviews in laboratories, one can be asked to put on a white lab coat to show we are scientists. (I have refused to do so.) It does not help that some scientists are remembered for abysmal dress sense, a very poor barber (Einstein) or film villains (Dr Strangelove).

I believe a wider acceptance of scientific methods would be beneficial as I am still idealistic enough to think that this could benefit humanity, rather than just making better bombs, computer games, or the continuous electronic communications so that we no longer appreciate genuine direct human contact. I understand the problems for marketing of TV and newspapers who want attention-grabbing novelty. Murder and catastrophe are excellent in this respect, but a carefully considered advance in science which has developed over several years is rarely considered newsworthy. Even less newsworthy in media terms is evidence that proves some great exciting claim (e.g. of a super drug) was actually less successful than originally claimed, or even that it is a failure with bad side effects. There are also cases where TV programmes offer extremist views (which may be in major conflict with the majority of documented scientific research), as this attracts attention and helps the TV channel ratings. Claims that they are being impartial are clearly rubbish. This is equally true of mass electronic media where facts are not deemed important, electronic images are doctored, and so-called fake news is rampant, as is the intense enthusiasm for conspiracy theories.

Anyone who can counteract such negative factors will have a real future in the media. They must create excitement in science, and clearly delineate what is fact and what is opinion. To do this, and be recognized and appreciated by the mass market, not just scientists, they will need to be exceptional, and they are desperately needed.

18

Science in the realm of opinion

Thoughts on viruses, evolution and astronomy

I am moving away from hard and substantiated scientific facts into areas where, despite data, interpretations and acceptance are strongly controlled by emotive, historic or political opinion. Their main difference from the preceding science is that the perception of the ideas does not progress, even when information is available. The topics, nevertheless, are firmly linked to imperfections in our lives and continue to have impact in areas from disease and pandemics, to religious beliefs. All have been the cause of conflicts, wars, human atrocities and immense numbers of deaths. Understanding their background, and recognizing our past mistakes, could therefore be beneficial. If we fail to do so then our future is not merely compromised, but could lead to our extinction.

Part I Responses to viruses and pandemics

Control and/or inhibition of virus epidemics is ongoing, and for well-established viruses we tend to develop antibodies, have genetic protection or immunity (i.e. useful imperfections). Medication and vaccinations can help but, for a randomly evolving virus, these are moving targets with long development and production times. Vaccination is not the panacea presented by politicians as their effects may only be temporary, or counterproductive. Equally, the body defences may focus on the vaccination, and in so doing be less effective at coping with new variants. This means we exacerbate the problem. Attempts to use anti-influenza vaccinations have not eliminated the disease. They are effective for some people, whereas the same strain has induced illness in others. This is not a criticism of vaccination, only the realism of dealing with a highly complex and changing problem.

The Power of Imperfections. Peter Townsend, Oxford University Press.
© Peter Townsend (2022). DOI: 10.1093/oso/9780192857477.003.0018

Anti-vaccination activists point out that to develop a vaccine is incredibly complex and it may involve more virulent versions and concentrations. This is certainly recognized and extreme measures are taken to cope with this. Pharmaceutical accidents can happen, and long-term storage of samples always carries a non-zero risk of accidental release (or, indeed, an intentional malicious or terrorist act). There are no such things as perfect developments and containment. This is life, but without the courage to take risks we would never have travelled, explored or developed any type of technology. Examples of miners, fishermen, police, medical workers, and all the military actions throughout history, emphasize risk is part of life. A population that has a lack of courage is doomed.

Many vaccination programmes can be highly effective, but need to be correctly administered. They have suffered from bad publicity when there were mistakes in the production or distribution (e.g. underfunded programmes, where needles were used many times). Needle re-use by drug addicts highlights this type of spread of infections. Viruses are a moving target, and a vaccination developed for one strain may be ineffective a year later (influenza is a classic case). They can work well, e.g. in eliminating smallpox (both versions). Nevertheless, simultaneous (multiple) vaccinations, designed for the benefit of drug manufacturers and distribution, have the potential to overload small children, and intuitively it would seem preferable to have separate vaccinations as, delivered separately, they can be excellent.

Controlling infection rates is a key factor in the ability to limit the spread of diseases. Fairly obviously they link to social conditions of the density of population, general hygiene and close contacts. This is highlighted by the emphasis on the reproduction factor, the so-called R zero (R_0). In reality this is very difficult factor to estimate. In principle, the sentiment is that if each infected person can infect several others ($R>1$) then there will be a rapid spread and an epidemic. However, if they only transmit to less than one other ($R<1$) the disease will fade away. High population densities and many contacts favour $R>1$. Hence the strategy of using isolation wherever possible is temporarily helpful. If there is no restriction and contact with many other people (e.g. as in modern cities and public transport) then the initial spread is exponential in terms of numbers. For example, if $R = 2$ the pattern goes 1, 2, 4, 8, 16, etc. ... Hence, in major cities with high density tower blocks and transport, R values rise. Any strategy to reduce contacts can temporarily reduce the R value, but it will not eliminate the disease. This fact seems not to have been understood by many nations coping with Covid-19. The 'lock-down' approach is effectively a delaying tactic, by reducing contacts in public transport and social gatherings. The economic downsides of closure and collapse of businesses, vast rises in unemployment etc., may actually cause greater long-term damage to society than the disease. This is certainly a relevant concern with Covid-19, as an extremely high percentage of people tested for the disease appear positive, but are asymptomatic.

The political view is that there is a bonus in reducing hospitalization of severe cases. However, better planning could address this and, at least in the UK during 2020, there have been extreme failures in this respect to protect medical staff and anyone else unfortunate to have been admitted to hospital for other reasons. Indeed, some 10% of deaths were linked to patients in hospital after accidents etc. Similarly, some 30% of care home deaths were linked to Covid-19 active patients being sent out of hospitals into care homes that lacked isolation facilities. Deaths and disease of patients who could not receive any hospital treatments have been seriously increased by the closure of hospital departments with their narrow focus on Covid-19.

Hidden in any estimate of communication of any disease are questions of (a) how long can one be infected before there are any symptoms; (b) when does one become contagious; (c) having had the disease is a person still able to transmit it; and (d) can there be re-infection of the original strain, or of a modified one. The pattern will differ with every variant of every virus.

Ebola hits extremely quickly, so far mostly in small village communities, it is fatal to between 50% and 80%, and, whilst locally dreadful, for geographic reasons it has stayed limited in the original community. The few exceptions have been medical staff who went to help. However, surrounding populations of gorillas and other simians have been totally annihilated over large areas, as they continuously travel.

By contrast, influenza does not have immediate symptoms and one can spread the virus whilst still feeling relatively normal. Recovery may seem fine, but one is still susceptible to re-infection, or new strains. HIV, and many other virus types, allow the carrier to survive and continue to infect others before and after symptoms are obvious. Note that the HIV retrovirus does not yet have a guaranteed vaccination, even after forty years of study (and around forty million deaths).

Both government and medical attempts to deal with a new virus pandemic are, therefore, a mixture of guesswork and panic as invariably there are inadequate data available for the uncertainties (a) to (d) listed above. The politicians' claim that they are using 'science' is a fiction, misleading, and undermines our views of genuine science.

Self limitations will occur if, once infected this leads to immunity, or death. Many diseases that are widespread in children effectively limit the number of targets, since most adults will already be immune. Vaccinations can equally reduce the R value. For a new virus there is no such factor. The number estimates are harder to assess if people can be re-infected, or continue to be contagious. Further, for some diseases our natural immune responses may cope with a small infective exposure, but totally fail with frequent exposures. Clearly a disastrous problem for inadequately protected medical and care workers.

There are genetic differences in response, as mentioned for the Eyam example in the case of the London Plague. Equally, some sections of the population are very vulnerable, with a disproportionate number of deaths. Elderly men were high on the initial Covid-19 list, people with diabetes accounted for ~30%; obesity was another factor, as were other pre-existing health issues. Statistics are further complicated as there appeared to be an excess of deaths in the so-called BAME group (Black, Asian and Minority Ethnic), particularly among health workers who may have been poorly protected. They thus include many carers, often with close extended family contacts, and in many cases living in more densely populated areas, and/or low income and jobs with high numbers of contacts. Genetics is a factor in a pandemic, and may be positive or negative.

The isolation route has severe consequences for everyone and the economic life of the country, and there are predictable further factors, both positive and negative. In the UK in 2020, air quality was vastly better, there were fewer road accidents and a reduction in cases of influenza. In parts of India, the reduction in road deaths was greater than those dying from the virus. Winter 2020 influenza in the Southern hemisphere dropped dramatically during their 'lock-down' against Covid-19.

By contrast, there are many unfortunate consequences of the isolation approach, including very major increases in domestic violence, divorce rates, mental stress, financial collapse of businesses, potential long term destruction of the music industry, an immense national debt and major economic impact on countries worldwide – with an increase of three million UK people becoming unemployed. Balancing these factors is a political nightmare and there is no solution that will please everyone. If we learn, we may do better with the next pandemic. Although it is not possible to stipulate which virus will cause the next pandemic, it is certain that there will be one, and following historical patterns, it will occur within a decade.

Part II Accepting the reality of human evolution

Observing and controlling evolutionary changes for humans involves two problems. The first is purely psychological, or religious, in that we may believe we are a perfect species, or the most advanced, and therefore do not wish to realize we are impermanent and changing. The second is a purely practical one that hominid evolution is very slow, and hence visually impossible to note for an average human, as we never sense more than parts of a few different generations. Even though we may know parents and grandparents, we cannot directly compare them, age-to-age, with ourselves or our children and under the same socio-economic conditions. A third factor is that there is no selective control of particular characteristics; mating and reproduction are mostly by personal choice, and strongly biased by individual attractiveness between males and females. It is precisely such variations between people that drive this crucial stage in selecting a mate. The children will not be identical with either parent, and overall new generations typically only appear after, say, twenty years. Hence, in a human lifetime we have a minimal glimpse of visible changes in the appearance and behaviour of perhaps three or four generations of humans from the same family. In the past, average life expectancy was much less, so we would have noted even fewer changes, and this would have reinforced our perception that we did not alter.

Evolutionary trends will also be significantly disguised by external rapid variations in fashions of clothing, hair styles, beards etc. Statues, paintings and photographs clearly show our idealized images of beauty and appearance can be very transient, and come in and out of fashion. Extreme examples of body building and cosmetic surgery further underline that we have varied tastes in what we see as desirable or revolting. There are further additional cultural differences in ideals. The obvious, rapid, and well-documented details in the evolution of languages, morality and variations of religious beliefs, are just as extreme as those of physical appearance. These aspects of human evolution are irrefutable.

Archaeological data are extensive and extend back to prehistoric ancestors (such as the Cro-Magnon period) who were intelligent enough to survive and had artistic human characteristics of making cave paintings. Bones offer a fragmented, but fairly continuous, record of alterations in the human skeleton over some 35,000 years. Viewing this much wider time span, there are very obvious differences. These are not just in terms of height, but also skull features such as the size of teeth, the angles of inclination of facial bones etc., which all reveal steady alterations with time. If confronted by a Cro-Magnon or Neanderthal we might assume we are of different species (but related). This is a spectacular result since we are still only considering perhaps a few thousand generations of development under conditions which did not actively encourage selective breeding. (Remember that dog breeders achieve visible differences within a decade.) Skeletal data may broadly indicate changes in brain size, but not more subtle properties, such as intelligence, which are not immediately linked to brain size. A search for evolution of intelligence is pointless, as it is what is needed to survive in the climate and culture of each lifetime. These conditions change.

The example cited by Darwin that finches etc. adapted rapidly to local environments on different islands, merely underlines the fact that environmentally driven evolution is a reality for many plants and creatures. This is in addition to those from climate variations, or from the more dramatic changes of meteor/asteroid impacts. Alfred Wallace and Charles Darwin only had visual external clues that various creatures had been closely related. Whereas the new possibilities to analyse genomes has added a totally new dimension to such studies and revealed shared, and often unexpected, links between many nominally separate species in terms of information coded within their DNA. Humans and mice share around 97.5% of their working DNA (as noted by analysis of chromosomes), and a slightly higher similarity (~98.5%) with chimpanzees. Differences between different races is far smaller at around just 0.01%. This is less than the diversity within an ethnic group. Clearly there is no scientific justification for racism. The DNA differences between people emphasize both that we are individually unique, but there are patterns within families and regions. The variations are sufficiently large that there is no way we can claim we are an ultimate, perfectly designed creature. Instead we can quantify that, genetically, our DNA variations are no different from any other creature.

Our major claim to being a slightly superior species is that we have evolved not just language, but tools and writing that pass on our ideas from one generation to the next. As a scientist I see this as incredibly valuable, as we can bypass the need to try and relearn everything within our own limited lifetime. The written records enable improvements in the accuracy of science, but as with all history, this is never perfect and is skewed by the fame and opinions of past events and writers. Nevertheless, with tangible records there is steady progress as ideas and information evolve.

The truly negative feature of historical writing is that it continues to perpetuate thoughts and attitudes so that, even when they are deeply flawed, they become entrenched in the way we behave. The egocentric belief that we are unique is just one example. Failure to recognize that we need to care for others, rather than exploit them for their land and resources, is another. Warfare and greed are traits in a wide variety of animals. We are the same, which merely emphasizes we are just another primitive animal, not something superior. The very positive aspect of humanity is that we have a moderate level of intelligence, and if we wished, we could use it to produce improved attitudes to one another. That would be a spectacular evolutionary step, which would justifiably raise us into a superior classification.

The net conclusion is that our understanding of evolution may be imperfect, but it is real, measurable, and is as valid for humans as any other creature or plant.

Religion and acceptance of human evolution

To those who totally believe in religious writings I would ask, what do we truly know about the authors (often unnamed), and what can we say about the generation and cultural climate at their time of writing? How much have texts, words, meanings and nuances varied in successive translations? Would we show the same total confidence in believing the texts of any modern politician? If we look at modern views on history over the last two millennia, we see immense numbers of disagreements and totally reverse interpretations of how the winners and those in power portrayed their actions and wars. This is a direct parallel with writings on religion, whether about sun worship, or the mythical gods that have now faded into folklore. As a scientist I have often looked at earlier science writing and textbooks of my own generation. I am frequently surprised how writers who were extremely prestigious in their day made statements that we now know to be totally wrong. If science were a religion, such errors would be perpetuated and unchallengeable.

With animals (except humans) and plants there is rarely any problem when discussing evolutionary patterns. Despite the archaeological evidence for human evolution, there are still many people who feel they cannot accept the concept of evolution for religious reasons. This is despite the evidence, and the very clear current situation that the world is populated with nearly eight billion people who are all different, who vary in size, shape, colouring of eyes, hair and skin. This is little different from the variations in breeds of dogs and cattle. Some features are merely related to nutrition and social background, and here rapid changes can take place. In the British army in the First World War, the officer classes were typically five or six inches taller than the lower ranks, merely from their differences in background and nutrition. This height difference reduced with the availability of more food and different social behaviour. There is now evidence in some current populations of an absence of wisdom teeth. This is a measurable genetic change.

Our key difficulty is that we are unwilling to accept that we are just another minor species that is steadily evolving, hopefully in a positive direction. Throughout written history we have assumed that we are not just different from all other creatures, but vastly superior and the epitome of species. This egocentric view has equally made us assume we are at the centre of the universe. In remote history we realized that we, and other creatures, were not in control of our destiny, and assumed that perhaps some greater power shaped our lives. In some cultures, this power was linked to animals, in others it was the sun (clearly our source of heat, light and seasonal variations). However, by imagining there were superhuman entities (gods) we both explained why events occurred, and simultaneously made ourselves special by assuming them to have similar appearance and qualities. We invented these gods in our own image and imparted human-like behaviour to the myths we imagined. The gods of Egyptians, Greeks, Romans, Norsemen etc., are basically no different from all the modern TV soap operas, or science fiction films. Any additional powers and abilities are just part of our imaginative licence, which totally ignore all the laws of demonstrable science.

Religions brought power and wealth to local individuals, who asserted themselves as being the representatives of these gods. It still does. These are powerful key factors in driving both religions and empires. The desire for power, wealth and control of people are entrenched in human societies, and religion provides all this for the leaders and upper hierarchy. The pattern is no different for non-religious doctrines or non-sectarian states. There are many obvious current examples. The pattern is repeated in historic and current regimes of royalty, emperors and dictators. The pyramids of power only offer a few top positions, and consequently religious, and political, schisms occur, with creation of new sects, cults or political philosophies. All bring power, wealth and influence for the new upper layers. (The difference between a religion, sect and cult is basically the numbers involved). The top leaders have always dictated what we should think and how we act, motivate wars and crusades, colonialism, persecution of alternative groups, and even try to define how we live our private lives.

In the case of droughts or floods, earlier leaders claimed the gods demanded human sacrifices, but they (who purported to have this inside knowledge) were never the sacrifices. Instead they gained more power by deciding who would be killed, without personal risks to themselves. Religious and ethnic wars are typically driven by inputs that are in total conflict with their teaching of tolerance and behaviour. For me, the difference between excellent humane concepts and realities of religious actions are intolerable and totally unacceptable.

Dominant and powerful people therefore entrench religious views, and market them by claiming we are a perfect, ultimate, unchanging, superior species. From this selfish viewpoint I understand the instinctive rejection of the concept of evolution, and why the antipathy can be so intense.

Humans are a feeble species so we have needed to develop and act together in groups. This is not unusual as there are many pack animals and other social groups, with examples from fish to ants, bees, monkeys, wolves and lions etc. Religions thus evolved as part of early tribal groups. Even variants of animal and sun worship have not vanished, as discovered by twentieth-century explorers. Overall, there are current estimates that perhaps 100 religions are being pursued with reasonably large numbers of adherents. Additionally, there are immense numbers of sects and schisms within the major religions (i.e. offering new top posts and power bases for leaders). The pattern is to persecute any variants of the same set of basic beliefs. This appears to have happened with virtually every brand of religion.

With a childhood Christian background, it is easy for me to cite a multitude of examples of Christian persecutions and massacres, by Christians. It seems unbelievable and is in total contradiction with the teaching of Jesus Christ, which is unequivocally saying we should care and love other people, no matter what their background. Two historic examples include the killing of 20,000 Cathar Christians, and the mass pyre to burn some 200 or more priests which was ordered by a Pope between 1209 and 1229. Mass murder on an equivalent scale took place in Paris on St Bartholomew's day in 1572. The Spanish Inquisition offers a further example. There are not just hundreds of historic examples, but one can cite modern ones, such as participation in the 1994 Rwanda genocide (an historical act that was mentioned with great regret by the Pope in 2020). There has been continuous ongoing Catholic and Protestant mutual torture and murder (even within parts of the UK). If the priests on each side actually had any real interest and belief in Christian teaching, they could end such a destruction of society, at a stroke. For me, their failure is unforgivable, as they are to blame.

At every level of society, from school to international politics, minority groups are always portrayed as villains and the source of economic failure, and therefore can be persecuted. These are demonstrations of the most primitive tribalism from which people have not yet evolved. The underlying logic of such attitudes is to acquire or maintain existing power structures, without regard for humanity. At the lower levels of society, there is fear and resentment from changes and threats that might worsen the current and familiar way of life. Locally this means a strong reaction against immigrants whose languages, religions and culture are different. These all bring changes in society and people will automatically feel fear and resentment, and focus on negative aspects rather than any benefits. Precisely the same resentment is felt by people who are invaded and subject to colonialization and/or exploitation. Attempts to proselytize new religions are often part of such invasions. In hindsight, the patterns have sometimes carried unintended benefits of new skills, music, or even an imposed administrative language which allowed a large diverse region to become a single nation.

Undoubtedly, we want the equivalent of tribal companionship, and this may be satisfied by association into religious communities of church, mosque or synagogue etc. Many people may equally find tribal solace via football crowds, bars or music. Even scientists like to congregate with others working in related topic areas. All such alternatives are familiar, and may even be genetically driven. Accepting the evidence of evolution does not undermine this social need in any way. However, it is a real problem for any religious leaders who believe we are unique, unchanging and with a special relationship to a Creator that only they, the ecclesiastical leaders, can interpret, and hence control the lives of the masses. Accepting our true, and very minor, place in the Universe is therefore difficult. Such a change in perspective does not remove the option of a Creator, but such an entity has to be for the entire Universe. It is definitely unimaginable, as the entity would be far greater than any version that would need, or demand, worship etc., from such an insignificant species as humans based on Earth. Crucially it is well beyond the understanding of our limited human intelligence.

Part III   Astronomy and our place in the universe

Our egocentric view as being the most important species has been seriously undermined both by archaeology, biology and astronomy. Astronomy has moved our view of being central to the universe to being just a planet around our sun, to a further downgrade of being in the outer fringes of both our galaxy, and the observable cosmos. Initially our only sources of light were the sun, moon and stars. At night we once could see a magnificent plethora of stars. Now, with polluted skies, and for city dwellers, such spectacles may never be appreciated by large fractions of populations. In the UK, the last time the stars and Milky Way were fully visible was during the black-out of the Second World War. Even then the skies were degraded from towns with chimney smoke and other atmospheric pollutants. The patterns of night light sources included a monthly moon cycle, a few bright items (planets) which moved in a predictable pattern, and an annual rotation of all the other 'fixed' light sources. Oddities included the occasional eclipse of moon or sun, meteor showers and rarities (from comets and meteors). Therefore, not unexpectedly, our egocentric nature naturally put us as the centre of all these light sources and assumed the stars were permanently fixed. Since the patterns matched the seasons it was a small step to assume these might be linked to the fixed star pattern.

Ancient visual astronomy was remarkably accurate and it was possible to predict many items, including eclipses. However, these were seen as bad omens that presaged deaths or catastrophes. In some societies they resulted in sacrificial offerings to appease the gods who had made the eclipses. An extreme example was where early Persian kings abdicated, were replaced by a stooge, who was executed after the eclipse, so the king could safely return. From our perspective this was farcical, as their astronomical predictions were so good that it obviously was not a random sign from the 'gods' but an accurately predictable phenomenon.

Astronomy, by the time of Galileo and Tycho Brahe, included telescopes, and so the paths of the planets (by now firmly named in honour of the gods) could be tracked. Relative to the Earth, they were moving in complex epicyclic paths. However, all these epicyclic patterns simplified into slightly elliptical orbits once we chose the sun as a focal point (not the Earth). Poor Galileo was criticized by the church as the associated simpler mathematics also defined Earth as being just another planet circulating the sun. Galileo caused religious unrest because, if the sun was the centre of the local region of space, then we were no longer the key planet. Humans were less important, and by definition it destroyed our religious centric view as us being the most important species at the centre of the universe.

Telescopes steadily added further downgrades to our key role as not only were there many more stars than could be seen by eye, but they appeared in clusters. Later on, distance estimates showed their scale was vast and stars and galaxies are also in motion. Within this telescopic range we, and our solar system, are just one speck in an outer arm of one of the spiral arms that rotate around our galactic centre. We are, at least, on the central plane of our disc-like galaxy. Our galaxy is far from unique, as other galaxies appear in whichever direction we look. Within this century astronomers have already detected several thousand nearby stars which have planets around them. The clear implication is that there are likely to be billions of planets, as there is no reason the rest of our galaxy should differ from the ones that we can see.

Even at the level of local stellar astronomy, recent Australian, international, and new sophisticated equipment have provided observations with higher precision. From these we have recently identified over 130 further minor planetary bodies in our solar system. Spectral analyses of the light in our galaxy have revealed a wide range of organic molecules which are assumed to be the building blocks of life and evolution. All such evidence undermines the justifications of our human assumptions regarding our importance and uniqueness. We now need a major change of perspective to recognize what we can achieve by our own efforts, not what was somehow ordained for us by a supernatural power.

Astronomy is totally fascinating and the immense progress over the last century has offered ideas for models on how our universe has evolved. The current fashionable model has it appearing from zero in a 'Big Bang'. The model is not perfect so there are imperfections and adjustments that include new ideas of dark energy and dark matter. These are conjured up to try to fix the differences between models and observations. Nevertheless, we are human and still looking for uniqueness, if not for ourselves, at least for our universe. If everything exploded from a single point, we are once again at the centre of this creation event (try to ignore that this logic applies to any other location). Our enthusiasm and limited intelligence trap us in the medieval human-centric view, even if we recognize that our observable universe contains billions of stars, we subconsciously want a model with a single creation event (just for us). Even astronomers will rarely mention that if space is truly infinite, and has always been so, then the Big Bang model will have occurred, and will continue to occur, over and over again and will still be happening somewhere in the infinity of space. Our detectable universe is just a speck in the unimaginable multitudes of universes that did, do, and will, exist. Rather than being the centre of everything, we are a totally insignificant, tiny, transient item.

A local world-centric view by astronomers can appear quite unintentionally. In the Big Bang model, a key feature includes measurements of a microwave background which is seen in all directions and is interpreted as residual evidence for the event. Effectively it corresponds to a background temperature of around minus 270 °C (~3 K). The intensities vary with direction, but classical textbooks and articles often cite the pattern seen looking out from the world by Robert Wilson and Arno Penzias as presented in 1964. Subsequent graphs with more data often preserve this display, which is effectively an oval like a Mercator graph of the globe. Wilson and Penzias were in New Jersey, so their local central axis (north/south) was in the middle, whereas the original Mercator world map was through Europe. In reality, such plots introduce huge distortions in size and perspective. (For example, a Mercator projection makes Greenland eight times too large.) The only satisfactory display is via a three-dimensional globe. Spherical displays are instantly more illuminating and, in the case of the microwave background, can suddenly reveal distinct patterns and anomalies that were previously hidden, since they do not align with the familiar New Jersey north/south axis. Various examples can be found on modern websites' images, where the centre is rotated on different locations, and enhanced, to indicate clear axial features, and/or intense minima. Interpretation is of course highly speculative.

Unintentional self-importance can never be totally suppressed in our imagination. Our own system is made out of particles that include electrons (negative) and protons (positive), which we term normal matter. These are the building blocks of atoms and everything else. In a Big Bang event (an ultimate imperfection) it is assumed there would have been an equal quantity of antimatter (i.e. where the charges are reversed) but, for some reason, the 'normal' matter mostly survived. With difficulty we can experimentally generate anti-matter items, such as molecules of anti-hydrogen, so they certainly can exist. For me this raises the obvious question that if equal quantities appeared in a Big Bang, and our galaxy is made of 'normal' matter, then there must somewhere be an identical mass of 'anti-matter'. Is this randomly dispersed? Are some galaxies entirely composed of anti-matter, or did the initial event eject matter in one hemisphere and antimatter into the opposite direction? The even 'unthinkable' suggestion is that anti-matter is more prevalent than the material that we experience.

Therefore, one can speculate that (a) there may be entire universes made from anti-matter that may permeate into our region of space, or (b) that in our current explosion event there was clear separation into regions of matter and anti-matter. Separation is not impossible as normal and anti-matter hydrogen molecules are paramagnetic and should differ in their response to magnetic fields. Conceivably, some of the galaxies we optically detect may be entirely anti-matter. To a non-astronomer it seems quite feasible that such concepts could potentially resolve the problems of so-called dark matter and dark energy that are currently discussed as being a major percentage of our current 'normal' universe. Mixtures of zones within our observable range may include both matter and anti-matter zones, with consequent energy generation where they interact. Existence of more distant universes may equally exist, which attract and cause ever faster expansion of our own universe. Highly imaginative ideas are often discarded as scientific ideas follow fashions, so making changes is a slow process, especially if new models differ from the current majority view. It would also need courage to further downgrade our importance in the total overview of matter and life by admitting that our entire universe is a mere blip in the pattern of universes over an infinite time scale.

For writers of science fiction anti-matter planets and galaxies offer many plot opportunities, as the arrival of a space ship or probe with an anti-matter origin would generate an annihilation explosion that would totally dwarf our Mexican asteroid event. Fiction writers have already exploited chemical chirality (i.e. whether the symmetry of a molecule is left or right handed) which means that whereas we can process the amino acids and proteins normally found in our food, we are unable to do so if they have reversed symmetry. These mirror effects were often used (such as Alice in Wonderland) long before their validity in chemistry was discovered. Changing the plot to photons from normal and antiparticle universes is an equivalent thought and maybe we will discover . . . ?

Inevitably, each universe will collapse and vanish into nominally nothingness; perhaps the conditions needed for further Big Bangs. For those seeking solace in religion, this ongoing increase in knowledge and perspective requires a total reassessment of the scale of a Creator, to a level which is way beyond our imagination. What we, as humans, must struggle to do is to accept just how insignificant humans are in this total celestial pattern. Once we accept this, we will realize that our future is determined by us and we cannot attribute disasters or successes on mythical gods.

A philosophical ant may think its nest is the centre of the world, with their life designed to worship their queen. In many ways we behave in a comparable pattern. Not least, because we seem, and indeed are, inherently totally unable to grasp the concepts of infinity in terms of space or time, and are further confused by the science that says space and time are one and the same. Even Einstein found it very difficult.

19

Improving our future lives

Part I Weaknesses in human behaviour

This chapter will differ greatly from the earlier ones as I want to focus on imperfections in life and our prospects of survival. It inevitably involves human behaviour and a vast range of views and opinions. However, general patterns can emerge, even though the same 'facts' will be totally differently interpreted. No doubt my views will be total anathema to many. I make no apology for this and am happy if my comments are contentious and stimulate genuine thought, but not if they are merely rejected without consideration. 'Truth' in human responses can be multi-faceted and will depend on our history, social context and personal background. My aim is to indicate where there are problems with our society, and encourage ideas on positive changes to reduce them. Equally as valuable as merely focussing on humanity, is how we are changing and destroying the entire planet, with the clear possibility that it will become uninhabitable for all species. This is not some far future problem, rather one that confronts us now and which could be irreversible within the lifetime of those alive today.

My criticisms of the current world will, of course, reflect my age group and the experience of very different past conditions. Rather than dismiss them as merely ramblings of a geriatric, please read beyond that and recognize that there are genuine issues, many of which only become apparent with experience. In this chapter I am being quite uninhibited in expressing critical views of social behaviour, but do so as I am well aware that there are many who hold similar, or even stronger sentiments. I also believe that it is feasible to alter life for the better. I fully admit that earlier generations were racked with unfairness, bias and mistakes. My hope is that the current world might be willing to make more effort.

The Power of Imperfections. Peter Townsend, Oxford University Press.
© Peter Townsend (2022). DOI: 10.1093/oso/9780192857477.003.0019

The challenge to positively exploit the defects of humanity

Imperfections are inevitable, inherent and present in every aspect of both science and life. I started with technological examples that were exploitable. In life we disproportionately respond to flaws and minor defects, from beauty to speech and appearance. These can be subconscious responses and we may not be aware why we have unease, no matter whether we are listening to music, looking at art, watching top sportsmen, reading literature or even watching TV programmes. In every aspect of life, we respond, and often focus on minor flaws from musical sounds that could have been better, to different artistic brush strokes or paint colour, athletes who are having bad days, story lines, plots and acting that have the occasional weakness. TV and film examples are, of course, the most evident as many plots have totally unbelievable events and scripts, include intentionally pure fantasy and/or just rely on computer graphics.

The truly positive bonus offered by starting this book with discussions of technology is that we have a perspective where, *despite* partial understanding and flaws in the materials and manufacture, we can often produce new products which are superior to earlier examples. Progress followed from understanding the difficulties. With inanimate materials experiments can be destructive, and the only emotive consequences are for people who had theories which were not correct. Exceptions exist with technologies related to weapons and industrial developments that displaced existing jobs or skills.

Whilst faults and weaknesses in humans and society can be described in similar terms, of having imperfections, it becomes far harder to accept or offer criticism if new ideas are alien to our past thinking. There can be vehement antipathy if they challenge current ideas and lifestyle. We need to cross this mental barrier to advance. In science a working hypothesis can be flawed, but we can discard it and try again. Human behaviour has entrenched views, so change may only be accepted by a later generation. Nevertheless, we make progress, so perhaps there is a glimmer of hope.

I ask your tolerance, and where our views disagree, do not take personal offence but at least try to consider why I am offering particular types of criticism. We now live in a culture where there is a new attitude and insistence on blocking controversial ideas under the guise of something called 'political correctness'. I am totally opposed to this. In reality it is censorship by a minority. I agree that ideas should be presented with courtesy and not as personal attacks, and that in very many cases people differ greatly in their ideals and prejudices. Rather than calling this a fault, I see it more positively as a factor that makes us interesting.

New ideas, and reassessments, are more likely to be considered when they are presented by eminent or qualified people. TV advertising of chemical or pharmaceutical products succeed when promoted by people wearing spectacles and medical attire. We are gullible. I have mentioned that for a TV interview about breast cancer detection, the camera crew were very unhappy that I did not wear a white lab coat. They said I was not giving the right image for their stereotype scientist. Such prejudice in reporting is not confined to science. In a sporting example, during TV filming of some advanced-level fencing, the interviewer wanted the people to make the sword play actions like those one sees in films, as actual fencing fights were too fast, and unfamiliar.

When experimental evidence and quantitative testing are lacking, predictions are very model sensitive. This is a major problem for computer simulations, which change with both model and input data and acceptance reverts to ingrained prejudice. A classic example is ongoing with the discussions of global warming and climate change. In many parts of the world, for example the USA, belief in the reality of climate change and the rejection of it happening seems to be split on lines of political ideology, rather than evidence. Underlying reasons (conscious or not) involve the fact that changes in behaviour will probably reduce commercial profits and force a need for more economical lifestyles, plus reducing the number of fuel-hungry luxuries of cars, planes, military 'toys' and an excess of food. Hard evidence in recorded temperature patterns, changes in rainfall, tornadoes, and the visible loss of many glaciers (including in Alaska) and images of the ever-increasing areas of the Arctic without summer ice cover exist as indisputable evidence. However, they do not impact sufficiently on daily life, and so are dismissed as mere scientific data. In parallel with this rejection there is a strong detectable antipathy to all science, despite that the wealth of the US (and most developed countries) is totally based on engineering and science.

Climate *modelling* of change is incredibly challenging and variable, and the details can be blurred by egos and prejudices of the people involved. Our human imperfection is therefore to take no action and dismiss the evidence. Unfortunately, ignoring the current window of opportunity to make change is likely to have irreversible and disastrous consequences in the coming decades. A decade is a timescale that neither politicians, nor the general public, seem to consider, and priorities are more likely to be measured in days. Maybe senior politicians are too old to worry about a future after their elected term, and only a minority of them understand any type of science, and possibly not even the prompt sheets of their assistants.

A worse imperfection exists with religion and political dogma, as there is never definitive quantitative and testable evidence available, and poor ideas can survive far beyond their initial proposal. Historically the results have been wars, social disorders and millions of deaths. The saddest aspect of history is that we do not seem to learn from our past mistakes.

Although religion is the one aspect of life where people claim there is perfection, both the stories and the consequences of belief in the supernatural are littered with obvious imperfections. Historic deities were given story lines no different from modern dysfunctional TV soap operas. This is a global pattern from early examples from sun and animal worship to the present. Despite the obvious hindsight that the Greek, Roman or Norse gods were totally fictitious, people were extremely devout, believed and prayed to them, worshiped and offerings to support their religious hierarchies, and willingly gave them control over their lives.

After several millennia we can view such historic events fairly dispassionately and easily recognize the underlying reasons why humans look for something beyond our experience. Factors include: (i) we are a very feeble and insecure species and seek an explanation of why events occur, (ii) we are articulate, contemplative, and assume that there must be some guiding force that has made natural events happen, (iii) we wish to be remembered, (iv) we are basically tribal which, effectively, means we want to be part of a community and have direction from people we think are our leaders, and (iv) a small percentage always exploit this pattern, want power, wealth and control of others. The techniques that support these social structures (and religion) range from intimidation, to selling ideas of an afterlife which will punish those who do not fit the pattern, or reward those who do.

A similar 'work or be punished' approach is sometimes used by teachers in schools. Politics (and religions) offer considerable control, power and wealth, which we unquestioningly accept, variously from monarchs, aristocracy, power hungry politicians, aggressive military or demagogues. A modern addition to the list has been internationally based commerce that uses products to exploit our lives and money.

We may be less aware that we mostly buy food that is driven by economics rather than free choice and quality, or upgraded electronics and science that is unnecessary but marketed as marvellous as it is commercially profitable. Other major items are developed as they are deemed to be key items of military importance. The fact that some major piece of equipment (e.g. an aircraft carrier) can take ten years, and billions to construct, seems hard to justify. Quite often they are outdated or obsolete when actually delivered, or could be easily sunk by a new type of low-cost missile against which they have no defence.

Overall, this sounds like a sad indictment of human history. Indeed, it is, and is precisely the set of imperfections that we need to consider if we are to improve our existence so that future generations will survive. The hard part is to admit we are responsible for making these changes, and we must not take the lazy view that it is pre-ordained by a government or mythical supernatural power.

Rather than be disparaging we need to recognize that, throughout history, there is an underlying sentiment that we have never been content with the world as we know it, and instead give ourselves comfort by assuming that somewhere there is the possibility of powers that we lack, and a future life where everything is better. Our objective is this nebulous idea of perfection. One cannot criticize hope and idealism. They are essential, as the world is still ravaged by poverty, illness, war and intolerance.

A realistic view of life is that there are many events that are totally beyond our control, and/or understanding. Weather, crop failure, harvests, health and plagues are obvious examples. Battles and wars may be marketed as essential and/or of religious necessity by the leaders who cause them, but for the participants they are just another example of the disasters. Tribal warfare is not limited to humans and clearly exists with many other animals, from ants to chimpanzees. The net view is that we cannot avoid imperfections in the way we have to live. This is precisely the same trend as is evident in developing all our different inanimate technologies. Materials and ourselves are imperfect. Even if we feel we are making progress, our understanding and theories may be flawed, and their implementation is imprecise as well.

Part II Generation-dependent attitudes

Each generation evolves from earlier ones and so do their objectives, mores and attitudes. History may highlight some of the more obvious step changes, but in general these do not strongly reflect day to day living for the majority of the population as, often, behaviour and new laws were (and are) generated by a limited ruling class. For example, the Magna Carta of 1215 is usually cited as being the foundation in Britain of power moving from a monarch, and influencing the legal system. It is described in emotive terms of current parlance of democracy, freedom and human rights. In essence this may be true, but at the time it was merely a power takeover by twenty-five Barons and the Archbishop of Canterbury to protect themselves from King John. The original document was rescinded more than once, and reappeared in modified guises.

A parallel oft-quoted document, the US Declaration of Independence of 1776, severed the US colonies from Britain. There are fine words about human rights etc., and innumerable interpretations that change between generations. From twenty-first century perspectives, one of the key issues was that there was an implied intent to abolish slavery. Certainly, many of the signatories were opposed to slavery, but in the culture of the time two thirds of the signatories were actually slave owners (some of the non-owners were too poor). Despite the hype about the document it is frustrating to realize that, after 250 years, racial equality is far from a reality across the US population.

These examples indicate just how difficult it is to judge and appreciate the attitudes in different centuries and realize that, for the masses, life was hard and without any major input to improve their conditions. Generations of the grandparent vintage survived, or grew up at, the time of the Second World War. Modern youth will find it difficult to appreciate and recognize the consequences of the extreme shortages of everything. There was rationing of food and materials for clothing etc. which worsened *after* the war and continued until around 1954. Therefore, there was no throw-away attitude to everyday items, instead there were political slogans such as 'Make-do and mend'.

Food was in short supply with very few items from abroad or other cultures (e.g. dishes from curry to spaghetti (in tins) were not widely available, or unfamiliar, and pizza was only served as a small quadrant of a tiny disc). The focus on survival had numerous benefits. There was a considerable desire to help others, and obesity from over-eating was very rare for the majority of the population. More surprisingly, the same types of British grown food were something like 40% more nutritious than the present-day mass-produced equivalents. Other contrasts with current times are that now even DIY shops are going out of business, as modern youth will throw away items and buy new, or spend large sums of money on a new smartphone because it is fashionable, even though their earlier model was working perfectly well. They never consider that a phone costs more than the *annual* income of half the people in the world.

In the 1950s, people were concerned about their appearance, but there were stringent restrictions with clothing coupons, a shortage of materials (and no charity shops with unused items). Therefore, imaginative dressmaking evolved to recycle older items. The contrast with the current world, where roughly one third of UK clothing purchases are never even worn, is extreme. Survival was a priority, and they could not indulge in the luxury of excessive concern about their personal image, so there was less criticism, self-harm was extremely rare, and far fewer people had mental health issues concerned with their looks.

This contrast between times within living memory is considerable and they reveal some serious self-generated defects in modern attitudes that stem from excess affluence, to technologies that bring electronic bullying and criticism, rather than any face-to-face comments that would require far more courage to make. Whilst the 1950s BBC was often elitist with clipped accents that now seem quite alien, we were not bombarded with TV soaps of totally dysfunctional families who spend most of their time in high outrage confrontations and obscenities. Inevitably, watching such items influences the way we behave in our own lives, and they are undoubtedly a factor which undermines personal relationships and modern marriages. Similarly, people had time to think and relax alone when working or travelling. The modern contrast of incessant communication (probably a misnomer) via phone and computer can easily undermine one's life and self-esteem.

The net effect for the 'grandparent' generation was less constraint on how they acted, or viewed their friends and neighbours, and allowed them to have aims that were achievable without incessant distractions and criticism. They were more independent and self-reliant. They had to be. In terms of friends, these were people they knew, not some nebulous list of electronic people who said they liked (or hated) them on contacts via social media. In reality, modern claims of having 200 'likes' is meaningless since mostly we will never have met the people, and certainly would not even remember their names. As a scientist, I and my group have published many hundreds of papers, created several new areas of study and applied technology, and have over 30,000 citations to our work. It is satisfying, but definitely not a measure of friendship. That category is limited to perhaps a dozen people. The rest are acquaintances, work colleagues, or people we meet during sport or musical events.

Part III   Imperfections in our behaviour

If we are trying to track behavioural patterns that evolve between generations, then we need to separate local conditions (as for the lives dominated by a war) from inherent attitudes because we are human. As a starting point we can try to identify the major flaws in our behaviour. There are vast numbers of scholars who have attempted this, but here a convenient example might be the Ten Commandments. They appear in two books of the Bible and reflect both Jewish and other Middle Eastern teaching. From the versions we read in English translations and/or re-interpretations, they offer some sensible instructions such as 'Thou shalt not . . .' commit murder, adultery, steal, lie, covet etc. However, from a teaching perspective these are poorly written, as saying 'Do not . . .' is far less effective than positive instructions on how to act. By contrast, the 'Sermon on the Mount' by Jesus is far better presented as it emphasizes the rewards for one's positive actions. For example, interpretations such as 'Blessed are the peacemakers . . .' or 'Do unto others as you would have them do unto you' are very clear and positive instructions to guide us on how one should conduct life and consider others. One does not need to be religious to recognize this and use it as a basis of an unselfish life. I believe it is essential to have a strong moral code on how we conduct our lives, but if people lack their own motivation then, for some, a background religious motivation may be useful.

Since I am trying to show that imperfections can have positive aspects, one may also cite religious sources in terms of the Seven Deadly Sins, as they include the familiar problems of gluttony, pride, greed, lust, envy, wrath and sloth. The negative aspects of these are evident in every aspect of our lives, from personal and commercial, to political and military activities. Nevertheless, without these inherent failings we would not have had the motivation and drive to expand our territories, develop new technologies, invent writing and mathematics, or move on to technological developments. The real problem is not the latent selfishness, but failure to limit it to activities that improve *all* our lives without destroying those of others, or the resources and other creatures on the planet. In these religious examples there is total agreement with my message from modern technology that defects have stimulated advances, but failure to keep them under control is disastrous.

We are highly selective, blinkered, and dishonest in our interpretations as to what is acceptable. An extreme example is murder. When we see reports of such events, we are variously appalled, horrified and outraged. Nevertheless, when we go to war and kill millions, or practice 'honour' killings, 'ethnic cleansing', or crusades and religious wars, the people involved seem to have totally lost all contact with the positive side of humanity, and rarely recognize these atrocities as mass murder. Merely by changing the word for destroying a life of another person from murder to euphemisms of warfare is completely false. We are equally hypocritical in that people who commit murder in Western society do not lose their own life, but are merely incarcerated for a few years (at a high financial cost). Whilst there may well be different grades of murder, I see no justification for allowing such people to survive if guilty of terrorist atrocities (i.e. effectively warfare), or first-degree mass murder.

We are equally blinkered in handling weaknesses such as gluttony and wealth. Over the last fifty years, developed countries have vastly increased their food intake. The ill health consequences of excess weight, and minimal physical exercise, are rife. They include medical conditions of diabetes, cancers and heart problems and many associated illnesses. Obesity is driven mostly by over indulgence and poor diet (note however, for some there are unavoidable medical reasons). When first drafting this section, I recognized that obesity is strongly linked to excessive eating, but only on careful reflection did I realize just how serious are the problems of self-image, and commercial pressures, which are driving such a high percentage of the population into this very negative situation and ill health. Obesity is not merely an unfortunate problem for a few per cent of the population, but, in some sections of the social mix, it is well over half the people, and extremely obvious in their children. For them, they may never enjoy a healthy life. The response of undereating is equally damaging.

The causes for obesity are quite complex and certainly major villains that contribute to it are excessive advertising by the food industry, oversized portions in items we buy, poor nutritional value of many products, excess of sugar and synthetic additives (that supposedly add flavour but which, in reality, deliberately interfere with our stomach to brain communications and convince us that we are still hungry). Finally, many people are unwilling to cook, and they merely buy pre-packaged foods. None of these factors existed for the war-time generation, instead these chemicals and drugs are 'benefits' of food research.

Commercial inputs are deeply involved with these eating disorders. Marketing works with skilled colourful packaging, intensive TV advertising and low-price promotions. Food quality is reduced relative to that of good cooking, by swelling the volume with bulk and flavour additives. The hundreds of books written about diet, or slimming, encourage fads of eating and nutritional claims that oscillate every decade between totally different extremes of what is good and bad. For example, fats were villains and carbohydrates were good, and then the reverse. Mono substance diets were proposed that initially caused weight loss, but our responses typically rebel and the apparent weight losses are reversed. There is, equally, peer pressure that people do not recognize obesity if their friends are a similar size. Advertising encourages unobtainable views of body image. An unhealthy lifestyle of watching computer or TV screens, instead of taking part in sports and exercise, results in poor muscle tone. It takes dedication, careful shopping and food preparation, to counter these highly negative market forces.

Obesity or diabetes occur for many reasons, but the modern increase in numbers is firmly linked to diets and factors encouraging gluttony. Consequences are not just poor mobility, breathing difficulties and heart strain, but obesity is often a root cause of most diabetes. The net effect is not just an illness, but a shortened life span. The 2020 UK Covid-19 data indicated that one third of fatalities were of people with diabetes. There are similar obesity links to the rise in dementia.

All such ill health generates immense medical costs, and just one side effect directly linked to diabetes is that there are some 50,000 amputations per year in the UK. If one assumes a subsequent life span of just ten years, then this implies the country has some half a million amputees as a consequence of a condition that, for many, were self-inflicted. Additionally, it is difficult to quantify, and put a price on, the selfishness of causing immense strain on friends, family and the medical profession. A significant error of the medical services is that their focus on treatment never significantly highlights the dangers and costs. Nor is there any great effort to confront people to induce them to make a personal effort and drive a real incentive to retain, or regain, fitness and a healthy body shape. This would also require aggressive action against the food industry and few politicians have the determination to even mention the need.

As part of this particular failing of excess enthusiasm for food, we only use about a third (or less!) of the crops that are grown. The fault lies with many people. Supermarkets always want 'perfect' crops and so a third of the output from farms is rejected, although it is perfectly edible. A minor caveat is that some supermarkets now include a small number of 'wonky' misshapen vegetables. The beautifully shaped foods in such shops do not necessarily have nutrition and good flavour. I have just mentioned that they can be only 40% as nutritious as the same crops eaten during the Second World War. Similarly, supermarket apples etc. may look perfect, but this is rarely matched by flavour.

We are induced to buy in large prepacked quantities and many (even most) households consequently do not use a third of the food that they purchase. This waste is reprehensible as we should not ignore the reality that more than half of the world population is starving, undernourished, or lacks clean water and sanitation. Here there is a defect situation which needs immediate response. Not least, as the total world population is increasing rapidly, primarily within the less developed countries.

I focussed on gluttony but within this same category of eating habits are many other variants, such as bulimia and anorexia. In each case people, both male and female, have intense mental problems with their own image and undermine their body in attempts to change it. This ability to be both self-centred and self-destructive in order to distort natural body weight (plus or minus) is a consequence of an affluent society where we can waste time and money on trying to conform to an ideal shape that has been marketed by clothing, and other industries, for their own profit. I unashamedly repeat, half of the world population are undernourished or starving – their ambition is survival, not some image driven distortion of the human form. Changing this global problem should be our focus. Obesity would then reduce to just those who have inherent medical or genetic problems, not self-generated ones.

Pride and appearance

All animals, including us, take care of appearance in order to attract mates. Methods range from fantastic plumage and dance routines for birds, to even more dramatic changes for humans. Simple methods to improve our 'perfection' and attractiveness range through hair styles, clothing, cosmetics, body building and gym fitness courses to more traumatic surgical attempts to change shape. This is a universal behaviour throughout history from most societies across the world. The only differences are that the 'improvements' fashionable in some groups are seen as disfigurements by others. The Schlager duelling scars from a sabre are a classic example, as they were seen not as a failure in defending one's face, but as a sign of the courage to continue in a fight. They were a clear indicator, in the nineteenth and twentieth centuries, that these scarred people were of the officer class who had attended important universities in Austria or Germany.

Hair styles, beards, moustaches, skin conditioners, creams, sun tanning etc., similarly all help to define the social group to which we belong, or would like to belong to. The implication across this entire spectrum is that we are constantly pressured to feel that our natural appearance is inadequate in some sense or other, and we need to override these imperfections to make us more attractive, or to define our social group.

Taking a completely unorthodox viewpoint I will say that all these imposed surface changes are actually imperfections added to our true appearance. We like them, add them, and view them as improvements. A perhaps cynical, but most certainly true, comment is that we are encouraged to do this as the cosmetic, hairstyling and clothing industries are immense multi-billion-pound industries that give employment to millions. We are, therefore, made to feel discontent with our natural appearance in order to support these industries. This unfortunate message is then amplified from our peer groups. We have taken many thousands of years to evolve to our present appearance and, in reality, this is likely to be a sign that we have reached a good approximation of the human perfection that is required to survive in our current changing world environment. It is not a unique model as we differ even regionally in appearance in terms of height, build and colouring. The changes can vary noticeably over small distances within a small country such as the UK. They equally differ between social classes. Modern city societies, with a multi-ethnic and regional mix of people, will define future evolution into different examples of natural 'perfection'.

Selective breeding of cattle and other animals manipulated their characteristics to meet our ideals of appearance, or usefulness. They are certainly no longer creatures that would survive outside of our domesticated environment. We may not admit that the same has happened to humans but, in reality, only a small percentage of current populations could survive the conditions that existed before we had help from technology.

Rather than focus on evolutionary changes, which were outside of our control, we have worked on superficial aspects of appearance, dress and surface decorations. Advances in medicine and biology have altered these possibilities and, in recent years, our approach to engineer changes to our appearance has included being willing to undergo surgery to alter our structure. The nominally permanent changes in appearance of adding tattoos and cosmetic surgery are now examples of expensive modern industries which are based on imperfections that we are adding. Both these methods define our peer groups, and are major sources of employment for various groups of people who are exploiting our insecurities of not being able to accept the way we appear naturally.

Purely cosmetic surgery may employ the same sophisticated skills needed for medical surgery, but it relies on exploiting and encouraging us to feel inadequate. It is a highly profitable industry. In the case of initial breast implants, the scale of the number of implants per year (in 2019) was around 30,000 in the UK and 300,000 in the USA. I have carefully inserted 'initial' as in a business model one can expect repeat business to replace, or remove, the implants (albeit at a higher cost than the original work). Initial breast implant costs ran up to £8,000 on average (2019 data), and somewhat less in the USA. If all goes well the silicone ones will last from ten to twenty years. This is far less than the life expectancy of the women who have been implanted, who are often as young as twenty. With a UK life expectancy of some seventy-five years, removals, replacements and/or repairs are a certainty for most women, at costs which will match or exceed the original numbers. The majority may be successful, but scarring, pain, swelling and implant rupture, plus chronic side effects, are well documented and not that unusual.

There are several types of implant and structured examples which are less likely to rupture, but they variously have been linked to a wide range of major problems and are now described as BII (breast implant illness). Loss of sensation is also common. The restructured breasts feel different in terms of contact with other people (i.e. they feel solid and unreal), and the extra weight often results in sagging with time. Not surprisingly there are many media and web sites from women saying they were a mistake.

Parallel comments apply to cosmetic changes to other sections of the body, and in facial work there can be loss in expressiveness. This is very unsettling to observe and may significantly reduce the number of friends.

Trying to emulate some idealized shape set by film, TV or other fashionable people cannot be successful as there are too many variables involved. We are unique, precisely because of our natural imperfections, and aiming to be a clone of some currently fashionable star cannot achieve 'perfection'. Surgery is only superficial, it will not alter our innermost thinking and will only change the views of our most superficial friends (not necessarily positively).

The very unsubtle message I am offering is that if one looks at art in the form of statues and paintings over the last two millennia, then the 'ideal' female, and male, forms have varied greatly with time, and even in a single generation may differ considerably. Glamour photographs from both Victorian times and the twentieth century relied on actual body shapes, and include a diverse range of desirable figures (precisely because there are highly personal views on what is attractive). Modern electronic images are far less honest as they will have been adjusted with software to remove imperfections or reshape various parts of the images. If your friends do not like the way you look, then change your friends. Real friends will accept you as you are. Surgery may alienate the ones that you have. If they only care for your appearance, they are not true friends.

Tattoos

Bodily decoration comes in many forms where some are permanent, as for tribal scarring, some are totally removable as for lipsticks, hair colouring etc., and a third group includes tattoos. This is an ancient form of skin decoration which has variously come in and out of fashion. It has often been associated with particular groups such as sailors, or in some Far Eastern countries it is symbol of particular groups of gangsters. Therefore, wearing tattoos can have very different meanings and responses if one is travelling across the world. Previously, tattoos were indelible and a lifetime fixture, but there is now some limited success in removing them with extremely short bursts of high intensity laser light. Basically, the colour of the laser light is chosen to be absorbed in the dye, and the power is such that a dye spot explodes from the surface. In multi-coloured tattoos the different colours require different pulsed laser wavelengths. The removal is slow and expensive, so before having a tattoo it is wise to be sure that you will be happy with it as a permanent feature. Not all colours can be perfectly removed. A final caveat is that in some countries the coloured dyes used are potentially carcinogenic.

Points to consider are that earlier style tattoos such as 'I love Mary' may be regretted if you no longer love her. 'Mary' at least is a fairly common name so there is hope of finding a replacement. More exotic names are problematic. Young people should also consider that a tattoo on a firm young piece of flesh may look good, but as one grows older, we lose muscle tone, and skins become wrinkled and flabby. The tattoo of youth will be both dated and ugly. Because it is a permanent change of body image it is variously driven by a need to conform to one's peer group, an example of self-harm, or a result of a very poor self-image and a need to have some decoration to improve it. The concept of needing tattoos is definitely in the category of imperfections for many people. Conformity is undoubtedly a driving force as in the UK it has moved from being quite rare, to being the norm for around about half of the millennial age group. The role models are a vast number of popstars and footballers. Nevertheless, the tattoo removal industry has boomed in the last five years. The costs of such processing indicate there is clearly spare wealth for many people.

For my purposes, tattoos fit nicely as an example of an imperfection, which if selected carefully, and for well-considered reasons, can be a decorative accessory that appeals to a peer group. Nevertheless, it may be an alienating factor if one then works with a different set of people. Equally, poor designs, bad workmanship, or changing friends and social groups can make them undesirable.

Part IV Good and bad aspects of extreme wealth

Whilst considering why we can justify some of the Seven Deadly Sins (in moderation), there is the perception that greed for money and excessive wealth are always negative. Poverty is definitely a disaster if one does not have enough income to live and survive. Definitions of wealth are variable as they will depend on the type of work, lifestyle, friends, acquaintances and expectations. Wealth and financial hardship therefore mean different things. Even using data such as 'median household disposable income' is only a simplistic guide, since the value is shifted upwards by all the highly paid. Back in Chapter 5, I explained the difficulty of citing some type of average number where there are a wide range of incomes. For 2019 numbers, median or average London incomes were from, say, £25,000 to £30,000, whereas the high-income tail of the distribution had many UK company executives and directors above £90,000 per year, with a top end above £4 million per year. This totally takes the focus away from the millions who are struggling to survive and buy even the most basic necessities.

Diversity of wealth may be greater than we imagine. In 2019 there were some 150 people in Britain in the billionaire category. Many others had moved in order to save a few billion in tax. Globally, high wealth examples exist not only from industry, but also from criminal and political activities. With such extreme numbers the temptation is to say this is a highly imperfect society in terms of the distribution of wealth. Absolutely true, but in many cases, the wealth supports large numbers of employees, medical foundations, health, education and welfare programmes. We can also look at historical examples of very successful businesses that have led to philanthropic support for culture, museums, infrastructure and charities. Life is not simplistically in black and white, with extremes of good and bad.

With enormous earning potential in industry and commerce, one may wonder if we attract the very best people in to running the country. Financially this is unlikely, as in 2019 the basic salary for a UK member of parliament was around £80,000 and an additional sum lifts the salary of the Prime Minister towards £150,000. By contrast, in academia a university Vice Chancellor can receive £500,000; some BBC presenters are paid far more (e.g. £1 million). If we are looking for defects in the system then it is highly probable that we are not necessarily attracting the most able people into politics. The political monetary issues are actually more difficult to assess as Prime Ministers have a substantial life pension, and some past ones have become multi-millionaires as a result of influence, consultancy and lecture fees.

Financial excess is not just a problem in the UK, but it is a world-wide disaster undermining many nations. Greed is widespread, and real wealth has always been concentrated in the grasp of a small fraction of the population. Unfortunately, the divergence between the rich and poor is steadily widening, i.e. completely the opposite of what should happen in an intelligent society. I am amazed by current estimates that fewer than 100 super-rich people now have greater wealth than the sum of the poorest half of the eight billion on our planet. To emphasize this injustice, I have a set of scales in Figure 19.1.

One might assume that this does not apply to the richest nations, but in the USA the 2020 numbers show the wealth of the poorest *half* of 330 million people is matched by just three super-rich individuals. This nominally rich nation has a steadily increasing gap between rich and poor. Further, the country had a 15% unemployment rate, despite having the world's highest GDP (Gross Domestic Product) of $20 trillion. Many economists state that the GDP does not measure the wealth and economic health of all of a nation, but instead it is merely monitoring a very skewed top end in the distribution of wealth, such as for banks, stock exchange and financial services.

Such imbalances have happened throughout history and, as usual, the overview is that there are far too many poor; slavery continues (both in name and effect), and many jobs are rewarded with unreasonable sums, but whilst our system of recompense for work is highly

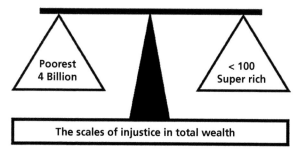

Figure 19.1 An indication of the spread of wealth between the poorest half of the world population and fewer than 100 super-rich individuals.

flawed, there are a few philanthropic exceptions. Nevertheless, fewer philanthropists and significantly better conditions for the lower paid would be preferable.

Difficulties of communication

In Chapter 5 I offered a warning about our difficulty in interpreting data, graphs and tables which results, not from intelligence, but a lack of education with such information. Therefore, even many highly intelligent people who do not use such presentations in their daily lives may either not fully appreciate what is being presented, or indeed just skip over them. The group certainly includes administrators or politicians, who have only been tutored in non-mathematical topics. Scientists, such as myself, will of course feel we are superior in this respect but, in reality, it is all too easy to see a pattern in a set of data and believe it supports some preconceived view. For example, I saw a graph when I was searching for any correlation between excess weight and a reduction in life expectancy. In one example, data were offered from many countries across the world and the trend line showed that countries with a higher percentage of obese people had a *greater* life expectancy! Numbers ranged from around fifty years in a few countries, up towards around seventy six for the USA. The flaw in the listing and interpretation was that it is not possible to make such a plot when there are many variables. It was actually saying that people in very poor countries will not be obese, and most will die young.

Taking a single country (USA), and plotting data by state showed a totally different result. Life expectancy fell from around eighty two down to seventy from Hawaii to Mississippi, with fewer obese people in Hawaii, and towards 40% in Mississippi. The temptation is thus to say this is proof of obesity reducing life expectancy. However, the same correlation emerged in a plot of age versus *median* income with Hawaii having the greater life expectancy and being richer. In this situation it is not feasible to unequivocally link life expectancy and obesity without specifying equivalent income levels (and not *median* values as the data sets have both a different spread in different states, and differ with income). So, continue to beware, and separate where the data are consciously skewed in their presentation, or where the statisticians failed to recognize it. Something obvious to them may not be understood by others!

Failings and thermodynamics

Our human failings may have driven progress, but clearly change is essential, every generation says this. As a physicist I am familiar with the laws of thermodynamics and there is a well-documented example that says in all reactions/activities the quantity called entropy can only increase. Entropy is a measure of disorder, which if phrased in more emotive terms is telling us that, no matter what we do, there is an increase in background chaos. Or more bluntly, things can only get worse. This is not our intuitive response as we normally are delighted that we can make superior metals, or grow large gemstones that seem to have very few imperfections compared with natural stones. Whilst true, as soon as we take into account the energy costs of the total processes involved, we see that 'perfection' in one area is offset by far more disordered processing during the manufacture. We are extremely prone to focus on one aspect of a topic (for both scientific and social problems) and we totally fail to view the entire picture. This is not just an act of individuals, but one which permeates society from politics, to the media, and unfortunately very often from well-intentioned campaigners for change.

In a sensible desire to reduce pollution from petrol and diesel we claim that electric cars will be our salvation. Discussions rarely mention that energy conversion is variable and often no better than around 33%. Rather than directly burning the fossil fuels (with limited supplies and lots of pollution) we wish to replace motor engines with battery power. Citing efficiency is hard. The literature is full of conflicting values from optimal, realistic, or downgraded from followers of alternatives! To generate a charged battery for the car or bus there is an initial step of charging that did not exist for petrol and diesel. This source electricity requires energy. None of the steps are perfect. Idealized systems cite charging examples at maybe 66% efficient (or more), whereas other situations of a rapid charge can be as inefficient as around 33% (or less) and has dramatic effects on the working life of the battery. The net result of generating the electricity, a fast transfer to the electric battery, and then driving, is that the *overall* efficiency may have dropped down towards a mere 10%! We have used more *total* energy and, in some cases, involving two or three times as much. Even this is a short-sighted view, as additionally we are solving one problem but causing many more and, instead of depleting oil resources, we are instead creating extensive mining requirements for materials such as lithium or cobalt for batteries. Both exist in limited accessible quantities. Oh, we forgot that the mining uses vast quantities of diesel and destruction of land areas where the minerals such as cobalt exist. Perhaps, even worse, is that there are plans to mine cobalt and manganese etc. from nodules on the seabed, and all the consequent contamination and chemicals will be left to spread within the oceans. Indeed, we may reduce pollution from the city streets, but the cost globally may well be far more energy usage and a great deal more total pollution. As I mentioned, the disorder of entropy only increases, chaos rules.

Part V   Future human defects and how to anticipate them

The unfortunate situation we are in, where we have advanced because of our latent mental and social imperfections, is incredibly difficult to change. Virtually any example of entrenched behaviour that I discuss will not be believed, or met with strong resistance to any criticism or suggested changes. I will take a simpler option and discuss major flaws in society which are still at the development stage. It will be clear that the potential future problems are extremely serious, but the imperfections are not yet so embedded that we might not be able to make improvements if we admit they exist.

Past generations had major faults, but they are history and we cannot change them. Focussing on them merely wastes current energy and effort that could be directed towards current problems. Whereas modern generations, such as the millennials (born in ~1980 to 1995) and those born after 1995 show discreet differences. Both groups will define the future, so indicating patterns of failings, and previously non-existent dangers, may help to avoid them in their children.

The millennials grew up at a time of relative wealth, where parents would remember the shortages and hardships of their own parents and grandparents during the Second World War. This will undoubtedly have influenced attitudes to offer protection and generosity to their children. Excess protection is not ideal as exposure to dirt, playing in the open-air etc., which were the norm of both earlier country and urban life, means the protected children do not develop immunity to many allergies, such as hay fever or, oddly, things such as peanuts. In adult life these over-protected children suffer more medical problems. 'Immunization' applies equally to independence of thinking and survival. I have read that some US states have passed legislation that parents may be prosecuted if they allow children to play in a park without adult supervision, or travel alone on public transport! They have clearly lost the balance between sensible protection, and stifling in cotton wool and freedom to interact with other young people without adult interference. Young people need freedom to establish how to cope with life. This is seriously undermined if all their activities are always monitored by adults.

A consequence of over protection is that the young often lack self-confidence and find difficulty with challenging situations. Often they are referred as the 'snowflake' generation, to imply that they are transient, without real substance and can easily collapse and melt away. This may well make them pleasant people, but employers say they want continuous praise, and if the job is difficult or tedious, they are very likely to quit and move to look for another one. A remarkable newspaper comment from a UK police recruitment officer said new millennial applicants were not willing to work in evenings and weekends, and were highly opposed to anything confrontational. An equally perturbing comment from a London hospital was that the new potential doctors have extremely poor mechanical skills as they have not played much with toys, or built anything (but spent time on computers and phones). The result is that when they commence, they are not able to perform delicate surgery! The hospital is running special training courses to try to offset this. Perhaps with a special emphasis on thumb surgery as, since the advent of smartphones, there has been a leap in the number of people with thumb problems resulting from extended phone usage.

There is a trend to easily take offence so that in, say, university lectures or debate, students complain if anything is presented that might upset them, or a view that they find contentious. This means they ban discussion on non-favoured topics and controversial comments, rather than have the courage to hear them and then engage in proper debate. Because of their attitudes, numerous university administrations have taken an easy, spineless route and demanded that no lectures or discussions will contain such material.

In precisely the same way I have read that, in 2019, two notable UK universities have banned meat from their cafeterias because of student pressure. Students have total freedom not to eat meat, but definitely not the right to limit the menu for those who to choose to eat it. The further pressure to stop meat production is bizarre as this would remove all other products such as butter, cheese, ice cream, leather

etc. and lead to synthetic substitutes which often have poor flavour and, many consider, are actually unhealthy, and their chemical production is very high in energy consumption and generating greenhouse gases.

This is an excess of political correctness. Personally, I find it totally unacceptable and a ban on freedom of speech and action, which is precisely the opposite of what stimulating discussions and ideas are supposed to generate. There is a saying that 'Freedom of speech includes the freedom to offend'. This is a healthy viewpoint and allows strong opinions to influence our views. The military equivalent for success is 'Know your enemy'. For this book you will guess my version is 'Know your defects'.

As quoted in writing by millennial people, their background may indeed make them content and friendly, but they often seek an anodyne environment that is totally misguided as it will just undermine discussion, logical thinking and innovative ideas. I see this excess of political correctness to be no different in principle from suppression of views which, historically, were disastrous. They are the key techniques for dictatorship, military or religious rule. We may no longer burn witches, but the spirit of intolerance is precisely the same.

Listening to, and considering, critical comment is an absolute necessity if we are to advance in knowledge and personality. As one who is fairly successful in scientific research, I know that part of the reason our group is effective and has the ability to have innovative ideas is that I am happy to have students and colleagues come to me and say 'you are wrong', and tell me why. On one occasion I remember being very moved by a young foreign student who pointed out an error in a paper I had published (he was right) and said that, in his home country, he would have been expelled from the university for criticizing a superior. The failure to listen to criticism will destroy both contentious ideas, and progress in separating the good from the bad. The current excess of political correctness is a major and pernicious defect.

Post 1995

For the younger group (post 1995) there are some major flaws from the ways their parents behave, plus weaknesses of development in the youngster's social interactions, that arise from advances in technology. The results are clear and, once recognized, could be avoided. My aim here is to offer hints on how we should modify current and future parenting. This is a perfect social example of identifying imperfections and then using them to our advantage.

Step one in raising well balanced sensible, caring, intelligent future adults is in the choice of parents. If all is well, they will be fit, healthy, mentally well balanced, and able to provide for their progeny. Avoidance of drugs, alcohol and other items that will influence the infant are fairly obvious. Slightly less recognized is that, even in the womb, the baby hears the parent's voices and these voices imprint as important. Sounds, such as music, define the type of music the child is likely to enjoy. Noises of conflict, aggression and arguments (even from the television) produce negative reactions on the development (both emotional and chemical). Many soap operas and films contain intense and continuous conflict which influences baby responses (as well as the adult viewers).

If possible, one should try to avoid passing on serious genetic imperfections. This is a delicate decision if one is aware of them before conception. Many defects emerge unexpectedly that are a challenge to all members of a family. Nevertheless, there are many examples (both past and present) of people with severe physical problems who have very successful lives, and contribute greatly to society in fields as diverse as art, science, music and technology. The example of the Paralympics very clearly shows that physical disabilities can often be offset by dedication and training, to produce even sporting abilities far beyond the performance of many able-bodied people.

A baby needs as much stimulation, care and affection as possible, particularly in the first two years. Failure at this stage can lead to permanent poor development in all aspects of life. More surprisingly it cannot be generated by greater care when the child is older. Here a modern imperfection emerges that did not exist for earlier generations, which is ignoring the child because of other simple technological distractions.Older prams had a baby facing the mother (or other adult) and recognition, eye contact, speech etc. were continuous. Modern seating, certainly by pushchair age, has the baby facing away from the mother and, since the turn of the century one sees that, rather than talking and stimulating the baby, many mothers will spend much time on the smartphone chatting to all their contacts, and ignore the child. By about the age of two, many babies have now been given a tablet with images and pictures. They will play incessantly without looking at their surroundings, and totally miss all the education and life that they should have been experiencing. Even at meal times I have seen families in cafés where children are playing electronically and missing out on any social adult interaction with their family. This is a key step to being trapped into reliance on electronics and a major inhibitor of development of originality.

The urgency is that not only are the babies and young children often ignored by parents using their phones, but there is a steadily increasing trend to give children a tablet on which to play games, and later on, a phone. The scale of these electronic devices varies with age, but Figure 19.2 is sketched based on data prior to 2018. The fine details are unimportant and the pattern will vary with region, country and social class, wealth and religion, but the trend is universal in developed countries. It is probable that the numbers for the very young children are more significant than shown here. By their teens, phone access is certainly over 90%.

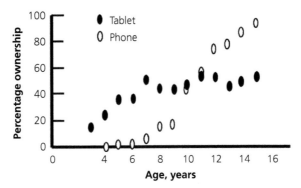

Figure 19.2 A sketch of studies on the ownership of tablets and phones by children. Some have devices as early as two years old and, by early teens, more than 90% will have a phone.

The fantastically useful advances in communications, from mobile phones to computer-based systems, does indeed offer lots of speedy access to information (both true and fake) and to a large extent results in a new philosophy that, rather than learn and remember, many IT (Information Technology) experts say we can forget learning and rely on web sites. I agree that web sites are useful, and I certainly exploit them but, without storage in our own memory, it is a big hindrance to having original ideas if the facts are not in one's personal little brain. A further weakness of this reliance on electronics is that it needs a relatively minor natural, accidental, or terrorist action to disrupt electronic communications. Examples of power failures and government action to block access to some web sites are well documented (even in nominally liberal countries). So far in this computer age we have not had a major hit from a solar emission, but as assessed from earlier examples, it could wipe out satellite and most other types of electronic communication, for large sections of the developed world. Power loss has often occurred as a result of minor solar emissions (as well as from terrorist attacks), but a major event would be catastrophic and result in many millions of deaths in advanced societies that rely on electronic communications and electrical power.

Yet another imperfection is that the temptation to play computer games is very strong (most of us do it to some extent). Computer games do not help development for real life, and many are just warfare and destruction games which subconsciously undermine a sense of morality and social values, and definitely mean we do not recognize the pain and long-term damage from violence and warfare.

Children who commence learning in this electronic age lack the experience of older generations and the recognition that friendships and opinions of others are shaped by face-to-face discussion, whereas online comments can be made without sensitivity, be totally false or just plain offensive. It is much harder to be rude to someone in person than via an unsigned electronic post. One cannot be critical enough of cowardly cyber trolls. This is something that must be communicated to children at a very early stage, even before they go to school.

The consequences of electronic communication via social media, as cited from medical evidence, is quite alarming. Examples now include teenagers with a higher incidence of loss of self-esteem, anxiety, sleeplessness, poor concentration on schoolwork, an 80% rise in mental health claims (between 2014 and 2018), a significant rise in child suicides and self-harm, a lack of self-confidence, greater aggression in young girls and 'psychological fragility'. In part, such problems will be directly associated to living in the virtual reality of an electronic world, where communication is totally different from direct personal contact. One cannot underplay the fact that real communications involve not just words, but all the visual clues that body and intonation offer along with the words. Even for the words, the frequency response and quality of sound from a smartphone lacks both the original frequency response and dynamic range of live conversation. It is fine for trivia and unimportant chat, but it is inadequate where one is making a carefully nuanced choice of phrase, and equally it will fail to convey our subconscious modulations of tone and timing of words that we use, depending on our sentiments and emotions. Not only do we miss this key information, but those who spend hours of their life glued to the phone may no longer recognize such information in face-to-face encounters. They are isolated from the pleasure and information in the subtlety of language.

Visual content is highly critical for emphasis and cultural communication. For example, in Italy people will still use hand gestures whilst on the phone, as if the other person is present. UK gestures may be more restrained but are equally important and many patterns have been documented and discussed. The examples used by Desmond Morris in his original 1967 book *The Naked Ape*, together with his city version *The Human Zoo* (1969) may well need new insights to include electronic communication as used by later generations, but the wealth of clues still exist for those who want to find them. For example, some people tend to look up to one side (or turn their head away) when they are not being honest, others will over compensate by looking too directly. Whilst many people learn how to control their facial expressions, girls in sandals should know that this does not extend to their toes, and these will often be curled when being less than truthful. Men have similar foibles, and when two men meet and shake hands the dominant one tends to look away. However, beware if both look away as this may mean future conflict. Similarly, in some jobs people are trained to override their natural actions (e.g. salesmen who convince you they are being subservient). Other familiar examples are TV presenters who use an excess of hand movements because they find talking to a bleak camera lacks any real-time response, as they would expect from an audience. We need, and expect, some feedback to what we are saying or doing, not just in conversation but in acting etc. Having given many lectures and talks I know it is a direct and valuable monitor on the success, or not, of the delivery.

Less obvious, but real, is that we use speech and sight in communication but, even for city dwellers with a reduced sense of smell, we are very sensitive to pheromones and other chemical clues about reactions of people to one another. Often this is subconscious but, certainly in more intimate relationships, we are strongly aware of the signals we are receiving. In the countryside, male moths can detect female moth pheromones at a kilometre away, dogs are several hundred times more sensitive to smell than us, and so are used in police work for tobacco and drug detection as well as searching for lost people in mountain rescues, and, more recently, in very successful detection of cancer. We cannot match such sensitivity, but many of us recognize changes in attitude via these non-verbal signals as well as sexual clues.

Non-communication can be just as apparent in a group as sitting in front of a computer or phone screen, if the people are not actually looking at one another. I have very often seen groups of teenagers in a café, sitting at the same table, and all speaking or texting on their phones, but ignoring their 'friends' at the table.

Isolating young children from direct contacts means they will not easily communicate with grandparents and older generations, not least as the older generations may be less skilled with electronics or prefer not to use such contacts. Addiction is unfortunately an appropriate word to use. Children who spend a long time in front of screens tend to continue to do so in later life. Many teenagers now watch TV or computer screens for at least six or seven hours a day. There is a strong link between such usage and lower school grades. In my list of imperfections that need addressing this is high on my priorities, as these very same children may be equally vulnerable to cyber bullying, insecurity and making inappropriate social media links. Further, they are the ones who will shape the environment and lives of future generations. They need help now, whilst it may be effective.

Final comment

In this chapter I have attempted to apply the techniques of materials technology to identify faults and failings of life, rather than materials. That was easy. The extremely hard part is to recognize and admit the reality of these imperfections, and then have the determination to modify our future behaviour and so reduce their negative impact. For materials, if we fail it does not matter as we can always start again with new samples. In life this is totally different. Children are unique and irreplaceable, and there is only a single chance to guide and direct them, or they will continue with the errors which have entered their education.

My idealistic views on how we might achieve a better world, which is equitable for all people, is in conflict with 10,000 years of history saying it has not happened. This is a glaring major failure. It is feasible to make change, but it needs thought, great care in the detail, and an enormous amount of effort and charisma to motivate people to aim for a more acceptable world. All we currently have is positive sentiment, but seem not to recognize that we are, and have always been, exploited at every level from politics to commerce.

20

The greatest challenges that we face

Historical weaknesses

In principle, I would like in this final chapter to look at the imperfections in the way we humans have interacted with one another, and with the planet, over the last ten thousand years. My aim is to see what we should do differently in the future. There are no shortages of examples of really serious mistakes either by negligence or intent and, as expected, they clearly demonstrate all the underlying human faults of greed, selfishness and desire for wealth and power. On the positive side these fuelled technological and scientific advances, but often with a total disregard for both others and the planet. We focus on short-term gain with limited foresight regarding long term consequences and planning. In the past, when the world population was smaller, our impact on destroying animal species would not have seemed apparent; mammoth meat might have been tasty and would feed a tribe for a long time. The fact that we were exterminating an ancient species might not have been obvious when each generation survived for perhaps only thirty years, and the total world population would have been counted in tens of thousands. However, when such creatures were limited in extent and matured slowly, they could not survive human predation. With a modern population approaching 8 billion, our food consumption has soared far over a million fold, so we will require ever more food and other resources.

We cannot reverse our past mistakes, but we might learn from them. With our current behaviour of environmental destruction and increasing numbers, we are probably going to add humanity to the list of extinct species. The time scale is likely to include our children and grandchildren. To change and switch from destroying the planet to genuinely caring about it, so as to preserve it for both us and all the other creatures and plant life, is an enormous uphill challenge. We may

The Power of Imperfections. Peter Townsend, Oxford University Press.
© Peter Townsend (2022). DOI: 10.1093/oso/9780192857477.003.0020

already be too late. In terms of useful land surface, we currently occupy around one quarter with farming, and one quarter by buildings. The pressure is such that even deserts are being watered from deep artesian wells to make arrays of vast circular fields. This can only be a short-term option as, not just for the deserts, we are drawing on subterranean water that has been there for many thousands of years. In the UK, it is a consequence of the last Ice Age. It is not being replenished, and the quantity is measurably decreasing.

Primaeval forests are being destroyed for cash crops, animal feed or cattle. We should remember that a minor climatic shift in North Africa stopped the annual monsoons and rapidly converted fertile savannah into the Sahara Desert within a few decades. We are destroying Amazonian and Pacific forests and this might have the same consequence. It is a dangerous, potentially disastrous and irreversible route, not least as such forests are the great generators of oxygen. Changes in attitude and solutions are required immediately.

A new pattern of longevity

A contributing factor to population growth over the last two centuries is that life expectancy has suddenly risen after, 10,000 years of remaining below forty years. I was surprised just how recently this began to soar as the result of fewer child deaths, better living conditions and medical treatments. Figure 20.1 sketches the pattern for past UK data and later predictions. The First World War projected a dip in the early 1920s, but typically after such events of war or plague there is a baby boom. By 2020 the projected number had risen above eighty. Women have a slightly greater life expectancy than men. For a country such as the UK this increasing survival rate adds visibly to the total population. It also raises the percentage of people who are not necessarily working and contributing to the economy, but instead, require money and resources for health care. This is a broad pattern and there are, of course, very considerable differences linked to relative wealth, type of job and social class. At present there is no obvious plateau. Numbers of those surviving to greater than 100 have risen over the last seventy years from around 100 to some 14,000 people. This is a tiny percentage of the population, but a noticeable shift. The current trend for patterns of death rate is for a steep exponential decline in the nineties, where only half those aged ninety make it to ninety one, half of those to ninety two, etc. This sets an upper limit to the predictions of longevity.

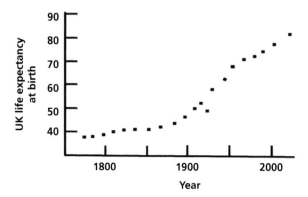

Figure 20.1 UK life expectancy at birth. Note, a dip was predicted after the First World War. Numbers for men, women, and background wealth, job etc., are all subsumed in this overall estimate. It is merely a guide reflecting past data plus tentative future extrapolation and, realistically, it probably should be a plateau for the more recent dates.

Figure 20.1.should be viewed as defining an upper limit for developed nations. It is very dependent on country and the social mix of rich and poor. The USA has a shorter life expectancy than the UK, despite spending far more on health care. However, this spending is only of benefit to a limited section of the population. The disparity is strongly linked to wealth and ethnicity. Of global concern is that the vast population in Africa has by the far the lowest life expectancy, with current numbers in the low fifties in many countries. Their total surviving population will rise steeply from better health care and include more elderly people. This requires infrastructure to cope with increasing geriatric numbers.

Figures such as Figure 20.1 need extremely careful interpretation. As with the figures showing how electronics and communication rates were initially doubling at a steady rate (e.g. Figures 11.3 to 11.5), such plots were only relevant when they were not limited by other factors. The 'transistors on a chip' hit an upper limit once the dimensions of interconnecting sections could not support the power being dissipated in them. The solution was then to move to several processing chips. Whilst obvious, and predictable by electronics engineers, an inaccurate marketing hype was used to impress the general public for as long as possible. Figure 20.1 has similar flaws. It will be used by the medical profession and governments to claim that they are achieving an extension of life expectancy. In reality I suspect it will turn out to be a significant over-estimate, not least as it cannot include consequences of future wars, pandemics, or other world events such as economic collapse in different nations.

During the 2020 Covid-19 event, the UK claimed that there must be extreme attempts to 'save' the grandparent generation, who were hit most critically among the general population. The implication and reality for the rest of the population is that there were immense costs incurred, and an explosion of unemployment from a collapse of a vast number of businesses. This is clearly unrealistic. We all have a finite life expectancy. If it happens to increase whilst fully preserving the quality of life then it is acceptable but, unless this is the case, then a shorter life is preferable. As one who is past retirement age, with many friends and acquaintances of similar, or older vintage, there is a considerable disagreement with this particular political claim that we must be preserved at all costs. We have all seen relations who have considerable frailty or illness as they age, and far too many suffer from forms of dementia, plus physical problems of mobility and/or incontinence etc. These are extremely common and demoralizing. None of this is the image that we wish to leave to our children, grandchildren and friends. For many, death, whilst still independent and fit, is far preferable. If it happens to arise from a corona or other virus that is absolutely fine (that is life!). It also means we may leave some residual wealth, rather than spending it on care homes and nursing, or being a major liability and worry for our families who may be looking after us, rather than enjoying their younger years. The overall state of the country, and the world, is more important.

Our mature view is completely contrary to the medical and political pronouncements on why we should be forced to exist. An unfortunate comment that one might add is that many of the medical profession appear to be only interested in extending life, regardless of the quality. Care homes vary greatly and from ones I have visited, some are excellent, whilst others focus more on the numbers of residents in order to have a profitable business.

A minor point is that apparently the elderly cope better than the current millennial generation when under stress. Indeed, older people will remember wartime and times of shortages, these strengthened their will to be independent and survive. In reality they are equally stressed, but are more resilient. A second aside is that the 2020 Covid-19 death rate was particularly apparent in men over seventy, and overall ~80% of deaths were of people with pre-existing medical conditions. Figure 20.1 indicates that, relative to the predictions at their time of birth (~1950) the septuagenarians had already exceeded their projected life expectancy. One should not apply current views retrospectively. Similarly, figures such as 20.1 are based on historical input and cannot account for totally different lifestyles, such as the more recent escalation of obesity and all the consequent health issues.

Reducing overpopulation

Our major challenge is how to reduce the world population. We need to plan this extremely difficult task internationally with very clear steps. The targets are not just to reduce the rate of growth, but actually to go steeply into reverse and drop the total population by many billion! Preferably not as the result of a world war or a significant pandemic. The crunch problem is that reduction is completely contrary to all the international industrial and business strategies, which are entirely driven by the need for ever expanding markets. Not just with more people, but by convincing us that we need new products, even if they are no better than former ones. Repairs and long-term usage are frequently discouraged. Waste and destruction of limited resources are seen as desirable. I scarcely need note that the lack of political actions to control use of resources is so entrenched in all major cultures that it will be incredibly difficult to motivate a reappraisal of government and industry, even though millions of individuals can recognize the need. Idealism of conservation, recycling etc., are valuable in terms of emphasizing what must be attacked, but the resultant scale and impact of such activities is still minimal. and non-existent in some nominally advanced countries.

In terms of waste, one can take a non-contentious example of plastics, which emerged in the 1950s. These are used universally, even to supermarkets as packaging for vegetables that are naturally designed with protective outer layers. Some UK cities do attempt recycling but many millions of tons of plastic contaminate the streets, countryside and oceans (~20 million tons each year from the UK). Photos of beaches after hot weekends underline the selfish disregard by those who used them. The plastics, string and other rubbish kill millions of birds, sea mammals and fish. The rubbish accumulates in ocean doldrum zones with up to a million fragments per square mile. In total there are many millions of tons of such material in both the Atlantic and the Pacific oceans. Interestingly, recent analyses (2020) on a finer particle size scale show that the fragmented particulates are pervasive within many of the bodies of fish that we eat. The total ocean plastic content is, in fact, at least twice the original estimates. It is having a significant impact in reducing the quantity of fish and crustacea that we can harvest. Not only do ocean fish have plastic particles, plus particulates of glass for plastic stiffening, but they appear in root vegetables, apples and fruit.

Response to death of animals is mostly just sympathy and it motivates minimal political action, whereas threats to humans may actually result in a response. I had wondered if more refined studies on humans would reveal precisely the same contamination. Unfortunately, since drafting this chapter I have now found 2020 data saying not just that we permanently contaminate many human organs with plastic and other residues, but they can exist in fragments far smaller than the width of a human hair. Particle sizes range down from millimetres to a mere 10 nanometres (10^{-9}) which can easily cross the blood/brain barriers of any animal. The inference is that it is a factor in human illnesses such as brain degeneration and dementia.

On the scale of imperfections our excessive, and rising, world population, plus the destruction of other species and the planet, are by far the greatest that we can consider. The hard questions to face are 'what can we do?' and 'can we implement adequate change?' Altering growth rates and total population will be strongly resisted. Commercial opposition will be intense and, in many cultures, there will be great opposition to limiting the birth rate. The major difficulty is that there is no obvious way of doing this within, say, fifty years. Failure to achieve population reduction peacefully, and under control, will mean it will occur naturally during this century with far greater dramatic consequences of global wars and mass starvation.

I see a two-step agenda. The first is to find a way to feed the current population, who will in many cases exist for another eighty years. Step two is to find moderately acceptable ways of inhibiting and reducing further expansion. The first stage is possible if we can reduce the immense excess of crops that we grow and, instead of throwing them away, actually use them. Non-use is variously cited between 50 and 75%! UK household wastage is > 30%. A serious attempt to reduce this excess

Table 20.1 Targets for change and survival.

Government intervention to reduce food wastage

Improve efficiency of distributing excess food to those who are starving or homeless

Cut back on global de-forestation for crops and cease growing non-essential crops

Reassess the dubious value of monoculture farming

Improve equality of conditions for all people, independent of sex, race and ethnicity

Education, for both men and women. This is a priority (it typically lowers the birth rate)

Make intense efforts to induce people to care for their own health and fitness

Improve the reality of democratic government, not just the current token examples

Ensure these improvements are driven globally, not merely in rich nations

Change human nature (!!) by reducing greed for land and power, warfare and dictatorship

---

production and purchasing could support some of an expanded future population. However, it is just a holding stage whilst we co-ordinate long term measures to actually reduce growth rate and total numbers. The first step is essentially in the hands of current adults. The second phase is even more important for their descendants.

Suggestions on how to make the requisite actions will be variable and/or contentious and, as a prompt for this major activity, I have added Table 20.1 which includes both targets and some ideas. Hopefully, better and fully considered methods will be discussed and implemented. The first step on the list must be actions which mean current and young populations do not starve. We need to forcefully minimize our waste production in every aspect of life, and cease destroying the natural resources. These are not idealistic aims but essentials for survival.

The UK has hundreds of charities which use unused supermarket food to feed local poor and homeless people. Truly excellent and valuable, but this opportunity should be exploited by government along all parts of the food chain to help local farmers, modify the wastages from imported food, add more realistic sell by dates on supermarket foods, and raise the awareness of the general public how much they waste (or merely goes to waist). The scale of rationing in the Second World War may be unbelievable to millennials, but people survived, were not overweight, and seemed very healthy. Those numbers were extreme, but an intermediate scale would have immense benefits to people, health and reduced wastage, land usage and demands on the health services.

Concentrating on aspects of food and land usage, one can highlight total destruction of forests to add crops such as palm oil and cattle feed. Not forgetting that such forests generate the oxygen in the atmosphere. Palm oil may be excellent but is added almost automatically to products which definitely do not require it. Beef is fine in moderation but it, and other meats, are eaten to excess. In US restaurants I have been offered the choice of 400, 800 or 1200 gm steaks (~1, 2 or 3 lbs). This is unnecessary and incredibly bad for one's health, as well as land usage. (In the Second World War, the UK total *weekly* meat ration was less than 400 gms.)

I have already mentioned that the commercial practice of monoculture crops, from wheat to rice, over vast expanses with pesticides and fertilizers is, invariably, only a short-term gain for farmers. Farmers are trapped into dependence on agri-business chemicals, lower yields, a reduced income, and total destruction of native animals and insects, etc. which fertilize and maintain the soil. Most millennials will not know of the US dustbowls this caused, or be aware that such problems were being highlighted even in the 1960s (e.g. Rachel Carson), nor that that the extinction of bees means many entire orchards are now pollinated manually.

I am convinced that if we had the political will, not only could we feed more people as the numbers expand, but do so without increasing land usage, with the very positive bonus that we might finish with a fitter, healthier and slimmer population in advanced countries, and reduce starvation and undernourishment globally. This would buy us time to decide how to minimize both the rate of population growth and address the harder task of reducing it.

The other obvious, well-documented factor is that where there is an increase in education across a community, particularly where this is for both men and women, there is normally a decline in the birth rate. Contraception is also effective. This applies both to underdeveloped and industrial nations. Japan is possibly an extreme example where women are educated and have become firmly entrenched in business activities, and there are far fewer children. In industrial terms the country is highly successful and also tops the league of life expectancy.

There has, indeed, been a drop in global fertility from around 4.5 live births per woman in 1960, to around 2.5 in 2020. Extrapolating this change into the future is extremely difficult and opinions vary, both on the rates and the desirability. For example, I saw one official presentation where the 2100 projection was as low as ~1.5 and it was portrayed as a disaster. To emphasize that particular prejudice, the graph for the future was distorted and foreshortened with a change in date spacing. This made it appear to have a very rapid decline. Incredibly, the authors suggested making efforts to increase the birth rate and/or encourage immigration in order to maintain commercial markets. The same source also suggested that the population of sub-Saharan Africa would treble to around three billion.

It is easy to make a wish list of changes that are required globally. Implementing them and the detail of how they would operate is quite a different matter. Nevertheless, we are at a point where we must move from paper idealism to hard implementation of a change in the way humans have functioned up until now. My list will prioritize economies in food and consumable usage so at least we can feed a fast-growing population. Stimulating people to care for health by their own actions would greatly reduce the overload on national health and medical services. Steps must include reduction in self-abuse of drugs, overeating and lack of exercise. Education is paramount in all this and it must address every level of society, with total equality for men, women and every individual, irrespective of class or ethnicity.

The really challenging steps will be to drive industries, of all types, to be in accord with efforts to reduce wastage, and governments to focus less on weapons, war and expansionism, and use their energies to resolve domestic and world issues. Table 20.1 lists a few of these idealistic thoughts. We must feed an even greater population, and simultaneously find a route (preferably peaceful) to slow and reduce the total. I view many of the thoughts not as idealism, but as absolutely essential necessities if humans are to remain on the planet.

My aims are clear as, if we fail, there will be total destruction of the natural world that existed prior to humans, a collapse of society through lack of food and water resources, and almost certainly this will trigger revolutions, political chaos and, potentially, a major worldwide war. Although the threat of a nuclear war is perhaps less than a few years ago (e.g. as considered in the book by Martin Rees), rational thinking has never been part of humanity. On the contrary, there has been persistent greed for land and resources via wars and exploitations.

One could believe I am deeply pessimistic, but in reality, I am being both idealistic and hopeful that if we could drive political will, and reduce commercial pressures, we might be able to achieve the contractions in population and use of resources that would enable us to have a sustainable planet.

Natural threats to human survival

Natural events have always played a role in locally reducing the human population. The normal destructive actions of earthquakes, volcanoes, tsunamis and major storms are of course very serious and, for each of them, there are typically several hundred events each per year. However, their global impact is normally very limited. These 'routine' events may indeed destroy a city or island, but in terms of eight billion people (and rising steeply) the reductions are insignificant. Nevertheless, some volcanic eruptions have contaminated the atmosphere, caused crop failures, starvation or revolutions, and their effects lasted several years.

Variants on these natural phenomena with more dramatic impact have been featured in many TV programmes, both factual and serious, or merely melodramatic fiction. They are often described as mega-scale natural events. In reality, although they can exist, they have happened so rarely that they will never motivate us to make any preparations. One slightly probable spectacular event is the production of a mega-tsunami, created by a landslip and collapse of one side of a former volcanic mountain in the Cape Verde islands. A section that slipped some 73,000 years ago set up a wave estimated to have been some 240 metres high. A repeat performance is feasible as the unusual geology of the volcanic island will cause it to recur. Minor splitting has recently commenced, and the major separation of one side dropping into the Atlantic will eventually follow. The subsequent tsunami will destroy much of the East Coast of the USA and parts of Europe. Bad news locally, but *globally* it will be minor in terms of population, although there would be widespread economic effects. Similarly, a second big natural event could be eruption of the Yellowstone volcano, this has happened around every 700,000 years; incredibly dramatic as it would wipe out the USA, but currently viewed as highly unlikely in the next few thousand years.

Asteroid impacts have had immense publicity as the one that formed the Chicxulub crater, in the Yucatan peninsula in Mexico, wiped out both the dinosaurs, and virtually all life on the surface of the Earth. Again, such a rare and unavoidable event that we can ignore it, not least as there is nothing we can do. Technology allows us to monitor many incoming asteroids and, perhaps, may offer some ability to intercept and try to divert their trajectory so they miss the Earth. Not quite science fiction but far from a feasible reality in terms of protection.

Immediate dangers and contingency planning

Moving to events that could happen immediately, or statistically will certainly occur in current lifetimes, world governments need to recognize their importance and attempt to have advance contingency preparation. One such would be one of the very many, ongoing, major Coronal Mass Ejections (CME) from the sun. As I mentioned in detail in Chapter 11, they are continuous and obvious as they create the spectacular and beautiful events in polar regions. The northern one is the Aurora Borealis. It is worth repeating here as we have moved into a totally sophisticated age of electronics, which would not survive a major electrical disturbance on the scale caused by a direct hit from a large Mass Ejection. A minor event might merely destroy satellite communication. Even this would have an immense impact and Government strategies to try to cope with such an event must, therefore, be well thought out in advance. By contrast, a major CME that hits the Earth would cause not just loss of satellites, ground communications and optical fibre systems, but also the electrical grids. Their destruction would totally destroy our current ability to survive over large swathes of the Earth. Based on previous major CME events (as in 1859) a modest estimate is to assume that, in the northern hemisphere, it would only (!) extend down to, say, the 40th parallel. Globally this roughly would define a zone including Washington, Beijing, all of Europe, and everywhere north of them. The impact would be extreme within affected cities since all communications and services would cease. Some estimates suggest it could potentially kill over 80% of city dwellers, and a lower fraction of rural communities. World trade and economics would collapse. A prepared strategy of how to cope is incredibly difficult but it is needed as not all events will be so extreme.

Rather than this collapse and high death rate being a solution to the excesses of over-population and land destruction, I suspect the reality would be economic collapse followed by wars, and a rapid invasion of the vacant northern territories, with a typical post-war boom in birth rate. Within this century we missed one such CME by nine days. We may be a small target, but the sun is continuously active in generating the flares with the radiation and energy that drive the ever-present Auroras. Their intensity merely varies, and an occasional direct hit from a large event is a certainty. They have happened frequently in the past, but in a pre-electronic era there were negligible power and communication implications. The difference for us is our new, and total, dependence on electronic technology. This is our Achilles heel.

At the other extreme end of the size scale, sources of catastrophic events are pandemics induced by tiny viruses. The Covid-19 variants have demonstrated just how unbelievably ill-prepared we were for such a global attack, and many government responses were amateurish. Pandemics are not rare, they just vary in the scale of their consequences and, with rapid world travel, they speedily encompass the globe. In terms of the potential lethal possibilities of a virus, the first variant was relatively trivial, with global death rates typically below 0.1% of the total population. Even these unfortunate people often had prior conditions that made them particularly vulnerable. This is not an insensitive comment, and I fully recognize that for those families that were affected, the results were very sad and tragic. However, within a wider view of virus pandemics, we were fortunate, as many have variously killed from a third to well over 90% of the populations they attack. The obvious problems are that modern global travel allows rapid dispersal, we have high population densities in large cities, and city transport systems, plus overcrowding. All of these encourage the spread.

The scientific responses were spectacular in terms of vaccine developments. This should help as this corona-style virus mutates rapidly and will continue to return in the foreseeable future. Experience gained in 2020/2021 may result in better protection for health workers in hospitals and care homes as, initially, they had a high death rate, although in a cohort who were not particularly vulnerable, but succumbed as a result of excessive and repeated exposure to the virus. This was primarily from lack of protective clothing and facilities.

One hopes that governments will learn how to respond to what is a very tricky problem of balancing health resources and lock-down type actions which are totally destructive of people's social and working needs. Current approaches have induced immense national debts, vast numbers of unemployed and collapse of many businesses, plus steep rises in divorces, domestic violence, an explosion in births, depression and mental health issues.

Government information revealed considerable inadequacy in how to respond, and public reactions included large groups who denied there was a pandemic, and/or were too selfish to avoid contracting or spreading the virus if they felt they were not in vulnerable groups. There are also many who are opposed to any vaccinations. Perhaps contributory factors were that TV updates revealed a lack of presentational skill with data that were confusing or misleading, and sometimes appeared to be without foundation. Figure 20.2 highlights one UK TV example where a semi-logarithmic plot implied to most of the public that the incidence rate had saturated. The Chapter 5-style difficulty for the general public was that data were sometimes shown with a logarithmic axis.

Better global planning is essential as we may be totally unable to cope with a pandemic from other corona-style viruses (such as Ebola

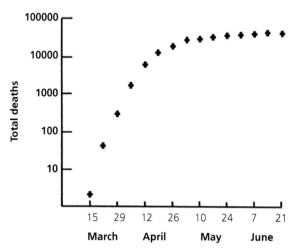

Figure 20.2 UK Covid-19 death totals. This semi-logarithmic plot gives the impression that the rate of increase slowed, whereas from mid-March to May/June 2020 the numbers are compressed 1,000 times.

or SARS). One cannot exclude the possibility of new versions of existing viruses, or further examples of zoonosis (from animals) that could hit us as dramatically, as myxomatosis has destroyed rabbits. Many thousands of such viruses are known to exist. In terms of overpopulation such an event might give us a false sense of progress if there is a major population reduction as a result of a pandemic, but, unless this is matched by other actions to discourage mass reproduction, then it will be just a minor hiccup in the numbers. We must recognize that earlier viruses and bacteria still occasionally resurface. In 2020, bubonic plague was reported. Further, diseases frozen for thousands of years in Arctic soils can re-activate after thawing. This was demonstrated by entire herds of deer dying from anthrax released from frozen ground. Such possibilities are unfortunately not just science fiction, and thus require forward planning and consideration.

World leaders need a clear strategy of how to stop the initial spread of any new pandemic. The original Venetian quarantine worked when travel was slow in the Middle Ages. Modern international transport is immediate and worldwide. Perhaps this implies the need to rapidly close borders at the first sign of the next pandemic. Limiting travel and social gatherings are potentially helpful whilst we are trying to understand the ferocity of a new virus. Crucially, protective facilities for medical staff etc. must be stockpiled, not discussed and ordered a month later with inadequate quality, quantity and distribution. Initial reaction speed is essential as this means limited spread will allow recovery and normal life to resume. The danger of under-reacting, and a lack of any pre-planning and policy, are all too obvious. Planning must also include a response to the fact that the financial consequences of the 2020/21 actions are severe and ongoing. One cannot ignore the major rises in homelessness and depression, plus a vast increase in domestic violence, divorces and suicides. Some estimates of consequent deaths already exceed those from the virus itself. Learning from current mistakes is essential.

My personal feeling of gloom is not that there are new coronatype viruses, but that many governments lacked any understanding of science and the need for forward planning against inevitable natural catastrophes. I hope that in planning responses for the next pandemic, Coronal Mass Ejection or other major debilitating events, the policy makers take care with the information that is needed and consult globally with competent scientists.

The foregoing comments are unfortunately depressing and suffi-ciently serious that they highlight a major imperfection in the way we function, and they require a dramatic change in human activity, a new sense of direction in political and cultural attitudes, and an intense campaign to redefine what should be our unselfish global priorities, to ensure a continuation of both humanity and a habitable planet with a diversity of wildlife. Adding mathematical and science training to po-litical skills should be mandatory, even if it is at just the introductory level of this book.

A hidden problem is to convince not just politicians and the general population, but also industries and multi-national companies, many of whom appear to act as though they are independent nations and are solely interested in commercial profits. The other key step is to induce acceptance of the new world objectives across the globe, even if this is contrary to military and religious factions. To say this is an uphill task is an understatement. At least the advantage of our tribal mentality is that one only needs to convince 10% of the population, and the rest will follow. History records how empires were created by the leadership and aggression of individuals, or nations were totally changed by dictators and demagogues. Purely constructive dictatorships have almost never achieved any long-term global changes, but concerned and informed governments might.

Health services

The UK attempted to introduce a free health service, which is excellent in concept, but clearly failing in many aspects as there is no longer pres-sure on the general population to take care of themselves. The result is an oversubscribed system. Nursing and medical staff are often highly dedicated, but the impression is that the administration is frequently unable to cope with the demands, or has other priorities. I understand it is easier to prescribe treatments than to convince people to change their lifestyles, but since many patients suffer the effects of poor diets, min-imal exercise, smoking, drugs and over-eating etc., it needs a rethink of the system to bring the real effects of self-exacerbated conditions home to the patients. Whether this is financial, or in terms of treatment priority (or both) is unclear.

As one example, local UK hospitals often have a huge burst of week-end 'accident and emergency' patients caused by fighting, drink and drugs. These people are often not just aggressive, but physically attack the staff. A very simple first step for them would be to make an automatic pre-paid charge, before *any* treatment, if these factors seem related to the reason why the patients had come to A and E. I personally would refuse any treatment to patients who attacked staff. The reforms would raise a modest amount of needed income but, more importantly, send out a very strong message.

Lack of exercise and being overweight are equally factors in generating too many patients. Investment in changing these failings should be a priority. Certainly, the UK attitude is often that we do not need to care for ourselves as we have a free medical service (albeit vastly overextended). To underline that our health should primarily be our own responsibility, an element of financial pressure might help. I have experiences related by friends in other countries where the first A and E question is not 'what is wrong?' but 'what is your health insurance policy number?' An element of this would certainly aid the NHS and, more importantly, the attitudes of the public. A very clear feedback from many doctors is that vast numbers of patients have absolutely no interest in improving their lifestyles and assume they are automatically entitled to maximum treatment, even if conditions were self-generated. The public NHS system in the UK is overloaded but probably not understood by senior politicians, who will not recognize the current weaknesses as they rely on private hospitals and consultants.

Finally, I do not understand human psychology as we appear to have lost the will and courage to face up to serious threats. If one looks at nations involved in wars (both past and present) the casualty and death rates, loss of homes and jobs, were/are often vastly higher than from the Covid-19 virus, but the people have had the resilience to continue their lives. In numbers, the UK may have unfortunately experienced some 50,000 virus deaths in the first year (2020), but in terms of wartime bombings, and military battle, ship and air deaths, plus injuries, such number could be matched from maybe just one or two events. Numbers cited for two days of the bombing of Dresden, in 1945, range from around 30,000 of the original population to a quarter of a million, if one includes people who had gone there to take refuge. The two atomic bombs in Japan directly killed 150,000 in Hiroshima, and a further 75,000 in Nagasaki. Overall, around 80 million were killed in the Second World

War, plus vastly more who were injured or traumatized. Current generations seem to lack the fibre to cope with a more minor event than people had managed in the past, although other nations still have this survival instinct in current war zones. It raises an interesting question as to why there are such extreme types of response in peacetime and during a war.

There is a more general oddity that there is immense publicity in the case of the virus, but acceptance of many other causes of avoidable death where the patterns are entrenched. Indeed, there is minimal effective public concern or attempts to reduce them. Just two such are that, worldwide, some 3 million people die each year from excessive alcohol; or in the USA there are ~100 deaths per day (~36,000 per year) from guns (plus many thousands more are injured, including several hundred mass shootings each year). Deaths and destroyed lives from drug addiction are widespread, but there is limited success (and possibly enthusiasm) for actions that tackle this international heinous crime.

Upgrading governments

Fundamental improvements are feasible only if we can upgrade governments and the people or industries that have global power. There is potential to do this. Many countries have a nominally democratic election process, but the actual governments are far less so. Instead there is a continuous split between the different political parties and often a whip system, and/or a local selection committee, which tells the party members how to vote. Failure to do so can cause de-selection from future elections. Clearly this is dictatorship in disguise. I am not making an original comment as precisely the same sentiment was sung by Private Willis in the operetta Iolanthe in 1882 by the scriptwriter W.S. Gilbert.

The elected members should ideally represent an entire constituency, not the tiny fraction of those who voted for them. The UK is probably typical in that this is not what happens. Of the potential UK electorate for a general election, often as many as 20% or more choose not to vote (the 2019 election had 33% who did not vote). With four or five candidates it can then emerge that the overall winner had votes from 30%, or less, of the electorate. Examining the detailed spreadsheet of the 2017 election, which had a low fraction who voted, reveals many examples with elected members who had gained votes from as few as

22% of the electorate. This is not a mandate to ignore the other 70 or 80%.

An inherent serious problem is the confrontational style of the UK Parliament, with two opposing sides separated by a gap (of two sword lengths!) and the resultant childish bickering of two opposing partisan groups. Other countries may have semi-circular seating, but again the grouping goes steadily from far right to far left on the political spectrum and, as in the UK, members are sitting tightly among neighbours of the same political persuasion, and have intense psychological peer pressure not to step out of the party line.

A total solution is difficult, but one could easily arrange that on entering any such chamber, the seating is randomly assigned by simple computer software. Further, ensure that truly secret voting takes place at, and only at, the assigned seat. A random location means one may be sitting next to a person with totally different views. This makes it far more difficult to make extreme denouncements etc. The addition of truly secret electronic voting would further allow one to vote with, against, or abstain, as defined by the issue and personal conscience, or that of their constituents, rather than party pressure. This would be extremely valuable for centrist decisions where there will always be a spread of opinion. Extremists will probably stay together in their voting. A final bonus of the scheme is that those who vote would have to attend the debate, whereas at present a majority often return from other activities, ignorant of the issues raised, and vote according to the party line.

My criticism and thoughts are directed at national level governments, but precisely the same narrow-minded party unity is prevalent in much of local government on issues which are totally local and need common sense, not some party line solution.

The second obvious problem is that many career politicians have never worked outside of politics and have very little first-hand understanding of the problems and needs of the electorate. Again, this is a general problem, but using UK data for higher echelons there is a strong fraternity of similar backgrounds. Over the last 200 years, three quarters of the Prime Ministers had political tutoring at either Oxford or Cambridge universities (around 42 of the 55 at the time of writing). The tutors involved may have had even less experience outside of academia. By contrast, less than 1% of the general population attend these universities. Further, a very high percentage of career politicians went to the same small group of public schools, and in some cases more than 50% of the cabinet has had this identical elitist background. Less obvious is that many senior civil servants, who are advising the ministers, are from precisely the same narrow educational fraternity.

The difficulty in political reform, in any country, is that those in power will be opposed to change. Selling government in terms of democracy has always been fashionable in classical school teaching. It often promotes the values and concept of democracy in terms of ancient civilizations of Greece and Rome. Whilst their systems had input from some of the population, which is clearly preferable to dictatorships, military, inflexible ideological or religious rule, the reality of Greece and Rome were that voting was limited to a small percentage of the adult population who were native citizens. Women were excluded, as were slaves (who were typically 30 to 40% of the total population).

Imperfections are obvious, solutions are not. It is, however, clear that standard political training does not prepare government leaders for matters which involve science, factual data, graphs or tables, or an understanding of the uncertainties with statistics. When this background is needed the response is invariably to find friends, or advisors, from the same social or academic background. The reality is that a small social ruling class often fails to understand both science and the economic problems and attitudes of the majority of the population.

The presidential candidate selection in the USA is even more obviously flawed as, in order to run for the presidency, the operating costs are such that virtually no one who is not a billionaire is likely to succeed. By definition this means the winner has little practical understanding of life for the other 99.99% of the population. In several examples the chosen inner circle has also been in the billionaire category. Dictatorships would also reject these thoughts.

A further action would be to encourage much more early stage cross-party involvement in drafting legislation. This would highlight and help minimize contentious aspects and offer a far wider range of opinions than are currently factored into the politics of many 'democratic' countries. Democracies might also benefit from more access to the data that are used by the leaders to make decisions. As then they would be able to comment if they thought judgements were being made on false or misunderstood facts. Indeed, many ministers hold posts in areas where they are totally ignorant, and one can only hope they have good advice and do not merely rely on TV and Twitter handouts of 'facts'.

Honest and accurate information

Government, understanding, ideals and actions are all based on the information and knowledge that we acquire. False information can only lead to misguided actions, ranging from the personal level to riots, mob rule, dictatorships, revolutions and war. In terms of imperfections, false information is very high on my list of items that need to be addressed. Historically, the evidence is not encouraging as even long before we were literate, the honesty and intent of those with influence defined what the majority believed. It is easy to manipulate people who are ignorant and lack access to several sources. This weakness has always been exploited in every field of control, from religion to politics, warfare and invasions. It extends from global issues down to behaviour within a family or local community. Minor deviations from the truth are generally unchallenged and can then be promoted, but their persistent restatement gives them credence.

A dubious hope might appear to be that with wider access to information we might individually be better informed, and able to separate fact from fiction, or interpretation from bias. In the twenty-first century, we no longer rely on local newspapers, broadcasting and TV but have worldwide access to items available electronically on the internet and other electronic media. In principle we may be able to make our own decisions and not be manipulated, but the flaws in this idealism are several. Firstly, internet access and the sources are censored in many countries. Secondly, we only normally monitor sites that agree with our pre-existing viewpoints. For example, some TV networks and newspapers are heavily politicized, as are internet sources. Rather than them making us open to new ideas we become part of a sub-community that strongly reinforces our views and prejudices (no matter if they are right or wrong). The third factor is that many sites have incorrect 'facts', bias and/or quite deliberately are dishonest and aim to mislead.

Back in Chapter 5 I mentioned the problems and misunderstandings with graphical and other types of data, where the plotting scales are not linear. If this is often a problem for scientists, then it is inevitably very difficult for the general public. In these last chapters I am trying to discuss human errors and bias, and these introduce equally large distortions, even when the data are simply presented. In any topic which has emotive content then facts and interpretations become skewed by our preconceptions. Figure 20.3a offers an approximation of the age distribution of people in the UK (2020). Numbers indicate the total per each ten year group. It is merely a guide, as there are quite obvious regional differences. It is not a smooth curve for reasons as diverse as social, political, medical, emigration and immigration. Nevertheless, it is clear that the population drops rapidly from the mid 60s. Depending on our age we will focus on the pattern where it is most relevant to ourselves. For example, young adults will be less interested in data for the elderly (and *vice versa*). One may visually focus on just 'our' part of the graph. For different audiences one can deliberately skew the view, so a plot with a logarithmic age axis, and/or different strength/size of data indicators. To show the effect this can have, I have taken the data of Figure 20.3a and replotted it on 20.3b with a logarithmic age axis. This compresses data for older people into a smaller region of the graph. Harder to quantify, but almost always true is that we may take a similar limited view subconsciously. A millennial age group who have children may feel that, as young adults, they are the most important section of the community (i.e. a standard view of most people at that phase in life). They will focus on data for their age range (where they imagine big strong graph points) and minimise numbers and value of the elderly (progressively weaker data points). A logarithmic age scale does this for them. The contrast between Figures 20.3a and 20.3b graphically achieves this skewed overview.

By contrast, the more mature, or elderly, may feel they are increasingly important as they have raised and supported families, contributed to the wealth of the nation, and have slowly gained experience and knowledge. Figure 20.3c is a totally different plot which might be imagined by older generations. Rather than a mere linear scale, the emphasis and effect is heightened by a semi-logarithmic plot.

Overall, we can change emphasis both by the way we present information and by the way we choose to view it. Basically, we are highly imperfect and individual in our attempts at information processing. The consequences of non-critical news and data gathering include extremism, conspiracy theories, and often rejection of anything with firmly-based data if it does not fit in with personal prejudice. General

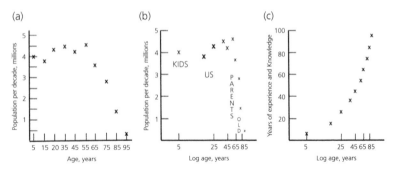

Figure 20.3 Views of age and their relevance to the population distribution. (a) The first figure has a sketch of the UK numbers per decade around 2020. (b) The same data are shown with a logarithmic age axis, and different strength crosses. This may be the view mentally imagined by, say, the millennial generation. (c) The third view, based on maturity, experience and contributions to the nation, only focusses on knowledge and experienced gained throughout life.

ignorance is supported. This is particularly obvious with scientific information as (a) people may not have enough training to understand it, (b) they will not be able to distinguish rubbish and fiction from actual facts, and (c) they will see political divides and assume anything quoted by their party is true and all else is false. Currently there is totally inadequate ability, or willpower, from electronic networking companies to block obviously false or malicious items. The more extreme ones go 'viral' and then, from sheer volume, are believed. Ideas as incongruous as belief in a flat Earth are rapidly increasing despite the evidence, and something so simple as being able to see the curvature of the Earth by looking at the shape of the sea from a high cliff. The early Greek and Islamic estimates from such views determined the size of the globe, and their values were close to the modern data.

Many people are trusting, vulnerable, or may lack the ability to discriminate motivations of politicians and advertising. Indeed, it appears that the general population unquestioningly believe anything they are told or hear that emanates via the TV from their particular party leaders and gurus. They simultaneously reject items from other sources. This pattern is extremely evident in the USA, but realistically it is probably universal. It has certainly been used to drive political power for centuries and has caused conflicts, civil wars and revolutions. Basically, humans are gullible and easily manipulated.

Both mis-information and dis-information are compounded by our desire for sensationalism and the pattern that the more extreme comments and images are rapidly copied, repeated and sent around the world. This spread of false news precisely emulates viral spreads, so the pandemic of false 'facts' is irrevocably in place. Millions will then believe it and never question who has 'shared' the information with them.

Electronics and instant communication networks have contributed significantly to drive this major imperfection of fake news and misinformation, since they rapidly amplify bias, lies and prejudice, and once infected, people rarely retract or modify their views. A corollary of this pattern is that people who believe in one conspiracy theory invariably also believe in many others. Fake information has always existed, the only difference now is that photographs, videos, 'facts' etc. are very easy to falsify, and certainly once they go viral across the world they are embedded in the minds of millions. Negative thoughts on how to exploit these human weaknesses are evident from personal attacks of trolls, newspaper articles which want sensation and sales, to political activities to generate voting decisions. I recognize the problems, am undoubtedly equally guilty of not being able to verify all the information that I quote, and merely rely on instinct and hopefully source material that I trust. I also benefit from a career in science where one expects to be able to have independent verification of statements. It is not perfect, but better than the mechanisms of fake news promotion.

A far less critical example of distortions of facts is obvious in many commercial adverts. For example, with marketing of beauty treatments, or images of how a large car is going to raise our social status etc., we know subconsciously that these are distortions but, since we would like the image they portray, we override our instincts and believe it. The pattern is no different in social and racial prejudice if we think that the difficulties in our lives can be attributed to those who differ from ourselves. Finding a scapegoat removes the responsibility from us. This is the same as assuming the sun and stars shape our destiny, so actions were not our fault. Joining a group of like-minded people means we subconsciously, or actively, increase our unreasonable views. Logic is not considered if we feel better and think that others agree with us. Revolutionaries and dictators have all understood this and exploited such behaviour. Successful ones always have found a suitable minority group (often ethnic or religious) who are blamed for the problems of the nation. Attacking them gives unity to the others and power to the leaders. It is the same as a football team, where a lost match is blamed on the referee, linesmen or poor performance of one member of one's own team, rather than ourselves.

How to drive and change global attitudes

Nothing is perfect and no political action will ever resolve all the problems of the world. Therefore, a consensus with recognition of diversity must be our objective. To make political change often requires pressure and encouragement from outside existing governments, since internally they will tend to remain unchanged as their policies put them in power and change might undermine this. Nevertheless, idealism and motivation has often been apparent from young generations on issues that governments fail to address. Examples from anti-war protests to anti-racism are classic examples where the influence of youth has had impact.

However, to be effective they need to address current and future injustice, as focussing on the past is merely self-indulgence. This is not necessarily recognized in current activities which are directed against *past* actions, such as slavery, exploitation, colonial wars etc. Criticism is well-founded in concept, but irrelevant as we cannot, and should not, attempt to judge the past with the culture of the present. The energy, enthusiasm and motivation must be directed at the current evils of the world, not former ones.

Current examples of destroying statues of people involved with slavery is merely a self-satisfying delusion and a waste of energy and potential emotional pressure. Should one follow through their logic of historic criticisms and say that national institutions, museums, universities, industries etc. that benefited from past slavery should also all be discarded? Slavery is an extreme evil that continues to exist in many guises in the present world and these should be the targets, not historic examples. Remember that virtually all past civilizations were built on slavery, so why ignore Egyptian, Greek, Roman, Chinese, Indian and other examples? Success in reducing current examples might reduce the availability of cheap goods from underpaid workers, people trafficking and the evils of drugs, and then the emotive efforts will be obvious and worthwhile. Genuine changes will impact directly on everyone, including the protestors. Every period in history can be criticized for dishonesty, aggression, war and oppression, so using energy to change current actions that influence the future is a far more valuable use of youthful idealism.

An element of hope

Politics and science often appear to occupy two totally different worlds. Whilst politics has ambitions for particular regions, and clearly has exploited cultural, religious and ethnic differences with objectives of empire, colonialism and domination, by contrast, science tends to have had a more international pattern where personal differences are less important. At the individual level there is still ambition and desire for wealth and fame, but on the broader platform there is often a more equitable interaction than in politics. During my own career, I sense that this global view has slightly improved with far more international objectives and collaborations. These are not just in areas of astronomy or particle physics, where there are immensely costly research facilities that could not be funded and staffed by individual nations. Instead, collaboration and internationalism functions well over a broader spectrum of activities.

Since funding is key to scientific progress, there has been very active collaboration within the European community to have multi-national projects. Participation in them is extremely fruitful, for funding, distributing ideas, new perspectives, interactions and accessing specialized equipment. Further, it offers financial access to people whose work might be blocked by local national politics and cliques. Having participated in, and organized, a variety of such projects, I am deeply sad that these opportunities will be lost to UK scientists with the withdrawal from the European Union. Their benefits were far more than just financial.

I doubt that my own pattern of work and experience is unique, but I can certainly report that I have been able to have very fruitful, friendly and amicable participations in international projects, and collaborative work in a variety of countries. The tangible numbers that support this are that my co-workers have come from more than thirty nations, with men and women of a wide range of religious beliefs, and many different ethnic backgrounds of race and colour. None of these differences have interfered with our working together, or our friendships. I can honestly say this measure of mutual equality is far greater than that discussed and promoted by media and governments. If one can achieve this in science, then it should be possible for the rest of the world to follow. A quantitative measure of successful international working together is my list of some 600 co-authors from across the globe in our publications. This is not some tiny anomaly, but an internationalism matched by many other scientists. If this is feasible in science, then it must also be possible to achieve this where we wish to improve the future of the human population and the quality of life in the world.

My conclusion is that we, the people of this world, are highly imperfect. We must recognize it, try to avoid the pitfalls of easy sensationalism, reject dubious sources and ignore all future information from them. We will never totally succeed, but if we attempt to do this and recognize that all people should be treated fairly, equally and without prejudice, we will have made a positive step that might spread to others. In return they may treat us in a similar acceptable fashion, so there is a very real positive feedback. Failure to do this is not just personal, but will lead to destruction of humanity on a remarkably short time scale.

Additional reading

Current availability of access to thoughts, ideas, information (and lies) from across the world is one benefit of electronic rapid access. Note, however, that not all data are reliable as prejudice, errors, and deliberate distortions are embedded in this morass of information. Separating it is a challenge. During the background reading for this book I have encountered many examples of each type, and I probably erroneously believed, or rejected some. Anything that reeks of conspiracy theories must be treated with extreme suspicion as it typically contains a few true facts and then extrapolations which are unjustified. There is no doubt that conspiracies do, and have, existed. Hindsight applied to actions of many governments will show that the public were frequently misled into wars and/or colonialism for completely dishonest reasons, but many disasters may have arisen from ignorance, rather than a conscious aim to mislead. As an excellent example of our historical errors and failure to learn from the past, I list a book by Simon Jenkins.

Presentation of data needs to be intelligible and obvious without being misleading, even to those with only a modest skill in using graphs, tables and statistics. This is often not the case, so beware. Overall, it implies rapid access can be another example of imperfections, as value judgements are needed in even the most apparently clear articles one can access. In some fields, such as medicine, opinions change, or even reverse on relatively short time scales, so making decisions as to which is correct needs caution. Similarly, industrial and commercial magazines can be very informative but will often highlight their own products.

In terms of books, the following very short list includes items I have read, or used as sources, plus two earlier books by myself. (Both would also make excellent presents for friends!) I have not cited science textbooks as the current aim is to offer thoughts to intelligent readers who might not have a scientific background. The extensive literature of remote sensing will also frequently demonstrate how differences (or imperfections) can be exploited.

Aitken, M.J., *Physics and Archaeology* (Clarendon Press, Oxford, 1974) ISBN 0–19–851922–2.

Carson, Rachel, *Silent Spring* (Penguin Classics, London, 1962; repr. 2020) ISBN 978–0141184944.

Hecht, Jeff, *City of Light, The Story of Fiber Optics* (OUP, New York, 1999) ISBN 978–0195108187.

Jenkins, Simon, *A Short History of Europe from Pericles to Putin* (Penguin Random House, London, 2019) ISBN 978–0241352526.

Lewis, Simon L. and Maslin, Mark A., *The Human Planet, How We Created the Anthropocene*, (Pelican, London, 2018) ISBN 978–02412808801.

MacMillan, Frank, *The Chain Straighteners* (Palgrave Macmillan, London, 1979; repr. 2014) ISBN 978–1349044320.

Klein, Naomi, *This Changes Everything: Capitalism versus the Climate* (Penguin, London, 2015) ISBN 978–0241956182.

Maslin, Mark, *The Cradle of Humanity* (OUP, Oxford, 2017) ISBN 978–0198704522.

Melton, H. Keith, *The Ultimate Spy Book*, (Dorling Kindersley, London, 1996) 978–0751302561.

Quammen, David, *Spillover* (Vintage, London, 2012) ISBN, 978–0099522850.

Rees, Martin, *Our Final Century* (Random House, London, 2003) ISBN 978–0099436867.

Townsend, Peter, *The Darkside of Technology* (OUP, Oxford, 2018) ISBN 978–0198826293.

Townsend, Peter, *The Evolution of Music through Culture and Science* (OUP, Oxford, 2020) ISBN 978–0198848400.

.